Plants of Kananaskis Country

In the Rocky Mountains of Alberta

Plants of Kananaskis Country

In the Rocky Mountains of Alberta

BERYL HALLWORTH & C.C. CHINNAPPA

Botanical Illustrations by SHARON ORSER

Identification Keys by RICHARD DICKINSON

University of Alberta Press

The University of Alberta Press
141 Athabasca Hall
Edmonton, Alberta, Canada T6G 2E8
and
The University of Calgary Press
2500 University Drive N.W.
Calgary, Alberta, Canada T2N 1N4

ISBN 0–88864–297–0

Canadian Cataloguing in Publication Data

Chinnappa, C. C. (Chendanda Chengappa), 1939-
Plants of Kananaskis Country in the Rocky Mountains of Alberta

Includes bibliographical references and index.
Copublished by: University of Calgary Press.
ISBN 0-88864-297-0

1. Botany—Alberta—Kananaskis Country. I. Hallworth, Beryl. II. Title.
QK203.A4C54 1997 581.97123'32 C97-910327-4

Front Cover: (top) Highwood Pass, Kananaskis Country, (bottom) crowberry *(Empetrum nigrum)*.
Back Cover: (top) common horsetail *(Equisetum arvense)*, (middle) sticky purple geranium *(Geranium viscosissimum)*, (bottom) yellow columbine *(Aquilegia flavescens)*.
All cover photos by Cleve Wershler except yellow columbine by H.L. Pidgeon.

Printed and bound in Canada by Quebecor Jasper Printing, Edmonton, Canada.
Filmwork and color separations by Screaming Colour Inc., Edmonton, Canada.
Printed on acid-free paper. ∞

We gratefully acknowledge the support received for our publishing programs from the Department of Canadian Heritage, and the Alberta Foundation for the Arts.

COMMITTED TO THE DEVELOPMENT OF CULTURE AND THE ARTS

Contents

VII FOREWORD

IX ACKNOWLEDGEMENTS

XIII INTRODUCTION TO KANANASKIS COUNTRY
XIII *Location of the Study Area*

XV MAPS OF THE KANANASKIS AREA
XV *Kananaskis Country*
XVI *Bow Valley Provincial Park*
XVII *Spray Lakes/Ribbon Creek Areas*
XVIII *Peter Lougheed Provincial Park*
XIX *East Kananaskis Country*
XX *Highwood/Cataract Areas*

XXI THE GEOLOGY OF KANANASKIS COUNTRY
XXII *Structure*
XXII *Glaciations*
XXIV *The Divisions of Geological Time*

XXV THE CLIMATE OF THE KANANASKIS VALLEY

XXVII THE VEGETATION ZONES OF KANANASKIS COUNTRY
XXVII *Elevation and Topographic Effects*
XXVII *The Foothills Zone*
XXIX *The Montane Zone*
XXIX *The Lower Subalpine Zone*
XXX *The Upper Subalpine Zone*
XXXI *The Alpine Zone*
XXXII *Fire and Weather*
XXXIII *After Fire*
XXXIV *Old Growth*
XXXV *The Impact of Man on Vegetation*

XXXVII **HOW TO USE THIS BOOK TO IDENTIFY A PLANT**

XXXVIII *Use of Scientific Nomenclature*

XXXVIII *Major Groups of Vascular Plants Described in This Book*

XXXIX *Key to the Major Groups*

XL *Life Cycles*

XLV **PLANTS OF SPECIAL INTEREST**

XLV *Insectivorous Plants*

XLVII *Parasitic Plants*

XLVII *Semi-Parasitic Plants*

XLVIII *Saprophytic Plants*

XLVIII *An Endemic Plant*

XLIX **PLANTS AND HABITATS**

1 **PLANT DESCRIPTIONS**

1 *Non-flowering Plants*

23 *Monocots*

73 *Dicots*

277 **APPENDIX 1**
Botanical investigations conducted in Kananaskis Country, prior to the present study

285 **APPENDIX 2**
Species arranged in alphabetical order of scientific family name

311 **GLOSSARY**

341 **REFERENCES**

349 **INDEX**

Foreword

The flora of the Rocky Mountains is spectacular, revealing its dynamic nature throughout the growing season. As the snow cover recedes up the mountain slopes, the wildflowers spread their carpet through the forests, across the subalpine meadows and into the alpine regions. The flora of Kananaskis Country as a whole is particularly rich. Over 900 species of plants have been identified, and the list is still growing as botanical exploration continues.

Plants of Kananaskis Country in the Rocky Mountains of Alberta is a timely contribution to the understanding and enjoyment of the flora of this intriguing area. It provides the visitor, whether botanist, conservationist, naturalist or tourist, with detailed information for more than 400 species. The plant families are described and most are illustrated with a colour photograph of a representative species. Diagnostic keys may be used for identifying an unfamiliar plant to its family and thence to a genus and species. At least one species is described for each family; up to 50 species are included for some of the larger families. In addition to its botanical name (genus, species and authority), each species lists common names and has a description and line drawing showing its characteristic features.

Kananaskis Country was established primarily for nature conservation and recreation. It now contains three provincial parks, with facilities for swimming, fishing, camping, hiking and the general enjoyment of nature. The southernmost park, Peter Lougheed Provincial Park, includes several lakes and a large section of mountain wilderness with snow-covered peaks and glaciers. Maps in the Introduction show the boundaries of Kananaskis Country and the location of the three provincial parks. Background information on geology and climate follows, as well as brief descriptions of the zones of vegetation. Colour photographs illustrate some of the typical plant habitats of the vegetation zones.

Plants of Kananaskis Country includes not just wildflowers but also trees, shrubs, ferns, some grasses and conspicuous sedges and rushes. An illustrated glossary contains diagrams of the structure and arrangement of leaves, inflorescences, flowers, fruits and seeds. This should be very helpful in interpreting the descriptions of the plants and in using the keys.

Beryl Hallworth and C.C. Chinnappa are well qualified to prepare such a field guide. Mrs. Hallworth has a professional background in biology, horticulture and education, with special interests in local history, exploration, archaeology and natural history. Having lived in Calgary for more than 25 years, she has developed an infectious enthusiasm for the conservation of Alberta's native flora and fauna. Dr. Chinnappa is primarily a research biologist and teacher with special training in botany, plant genetics, and biosystematics. He has wide experience in the taxonomy and classification of both vascular and non-vascular plants and is currently Professor and Curator of the Herbarium at the University of Calgary. His research interests include evolution, cytogenetics, population biology, and pollen morphology; his recent studies have dealt with some poorly understood groups of native plants in Alberta. He also has considerable skill as a photographer.

Hallworth and Chinnappa have produced a field guide that combines a number of original features and bridges the gap between a simple picture guide to wildflowers and a purely scientific account of the flora. Above all, this book stresses the need for the conservation of Alberta's native flora.

JAMES H. SOPER
Curator Emeritus & Research Associate
Canadian Museum of Nature
Ottawa

Acknowledgements

We are grateful to Dr. James H. Soper for writing the Foreword to our book. Before he retired, Dr. Soper was the Chief Botanist of the National Museum of Natural Sciences (now the Museum of Nature), Ottawa. He has also reviewed the book and made many useful comments; his help is much appreciated.

In botanical publications, clear illustrations are of paramount importance, and we were fortunately able to enlist the services of botanical artist Sharon Orser, to whom we are indebted. She has prepared drawings of 400 plants found in Kananaskis Country, and it is around these drawings that this book has largely been developed. Sharon deserves a special word of thanks.

The colour photographs are all by C.C. Chinnappa unless otherwise attributed. The following individuals have provided colour photographs: R. Burland (Alberta Environmental Protection), J. Christiansen (Kananaskis Country, Alberta Parks Services), P.K. Anderson, D. Balzer, G. Crawford, P. Eveleigh, J.O. Hrapko, E.J. Kuhn, I.D. Macdonald, H.L. Pidgeon, J. Posey, R.P. Pharis, B.G. Warner, and C. Wershler. The drawings for the glossary were done by B.M. Hallworth and were prepared for publication by P. Allen and A. Brownoff, whose work we are happy to acknowledge.

Drafting botanical keys is an integral part of any botanical publication, and we have been fortunate in enlisting the help of R. Dickinson to construct keys to the families, genera and species. These keys have been presented in a simplified version to help readers with minimal knowledge of botanical terms and are of practical, rather than technical, value. Richard's work is much appreciated, and we would like to express our sincere gratitude.

Several faculty members of the University of Calgary have contributed articles or drawings for the book. In particular, we should like to thank E.A. Johnson for his articles on climate and vegetation, and S.A. Harris for his article on geology. C. L. Curry has provided reference drawings for the life-cycles of the ferns, gymnosperms and angiosperms. We tender our grateful thanks to all these colleagues.

B. Smith has compiled Appendix 1 from the many check-lists of plants found in Kananaskis Country; this has involved a great deal of checking and re-checking. She has also provided technical assistance with S. Orser's drawings. Her work is much appreciated.

The grass family, Poaceae (Gramineae), and the sedge family, Cyperaceae, are well represented in the flora of Kananaskis Country: there are 89 species of grasses and 96 species of sedges. We have selected 32 grasses and 10 sedges common in Kananaskis Country and relatively easy to determine, to represent these groups. The species descriptions were written by A.H. Weerstra; we are indebted to her for her work in this section.

Maps are essential for any publication dealing with the flora of a particular area, and we are grateful to the Director of Kananaskis Country, Alberta Environmental Protection, for permission to use the area maps, some colour photographs of the various habitats found there and a few photographs of certain plant species. The map of vegetation zones was researched by E.A. Johnson.

We have occasionally used modifications of some line drawings from the following textbooks: *Native Trees of Canada*, by R.C. Hosie (for conifer outlines); *Vascular Plants of the Pacific Northwest*, by Hitchcock et al.; *Fundamentals of Plant Systematics*, by A.E. Radford; and *The Taxonomy of Flowering Plants*, by C.L. Porter. We are grateful to the publishers of these books for permission to use the drawings.

All scientific names are taken from the second edition of Moss's *Flora of Alberta*, by John G. Packer. All the common plant-names have been taken from the *Alberta Vegetation Species List*, edited by David Ealey.

Accurate, up-to-date check-lists are essential in preparing a botanical guide, and we are indebted to several naturalists who have helped us to make, as far as possible, a complete inventory of the plants found in Kananaskis Country—over 900 species. We should like to thank the following: D. Allen, C.D. Bird, N. Kondla, I. MacDonald, O. Droppo, J. Williams, N. Emery, S.A. Harris, D. Jacques, B. Danielson, C. Wallis and C. Wershler. We owe a special debt of gratitude to J. Corbin, former Curator of the Herbarium at the Kananaskis Centre for Environmental Research (now the Kananaskis Field Stations), who has been very supportive. As well as providing check-lists, he proofread the book and made many helpful comments. Several other naturalists who collected plants in the area are mentioned in Appendix One. We are grateful to R.T. Ogilvie, C. Wallis and J. Williams for their help in providing us with references for this section.

We are grateful to D. Allen, not only for his check-lists, but for his meticulous proofreading of the original manuscript and his helpful comments. We are also indebted to J. Christiansen, the Manager, Interpretation and Education, of Kananaskis Country, not only for his careful proofreading and useful comments, but for his unstinting support for the last five years. We also thank L. Roll, J. Ramamoorthy and T.S. Bakshi for their help in preparation of the manuscript.

We should like to thank R.W. Davies, former Department Head of Biological Sciences at the University of Calgary, and A.P. Russell, present Department Head, for their support and for allowing us to use the secretarial facilities of the Department. Eileen Muench, who has done the typing for this book, has proved an excellent typist, with the necessary qualities of patience and cheerfulness. We are truly grateful for her work. L. Marin, J. Redlin and S. Stauffer have also given some help with the typing, and we should like to thank them also.

We gratefully acknowledge the financial support provided by the University of Calgary Publication Subvention Grant, the Kananaskis Field Stations and the Alberta Sport Recreation, Parks and Wildlife Foundation.

To all those people who have helped, in many ways, to produce this book, we tender our sincere thanks. The authors take full responsibility for any errors or omissions and will be happy to remedy them if occasion offers.

As John Gerard wrote in the Preface to his well-known *Herball* in 1597, "*Accept this at my hands, loving countriemen, as a token of my goodwill, and trusting that the best and well-minded will not rashly condemn me, although some things have passed worthie reprehension.*" Although these words were written 400 years ago, the sentiments expressed are still appropriate.

Introduction to Kananaskis Country

Kananaskis Country was established by the Alberta Government in 1977 under the Premiership of Peter Lougheed. Kananaskis Provincial Park was designated at the same time and was renamed Peter Lougheed Provincial Park in 1986, to honour the Premier, who devoted much of his time to ensure that Kananaskis Country was successful. The capital development was paid for by the Alberta Heritage Fund.

LOCATION OF THE STUDY AREA

Kananaskis Country covers 4160 km^2 *(see maps, pages XV-XX)*. The western boundary is well defined: it follows the eastern boundary of Banff National Park, and further south, the interprovincial boundary between Alberta and British Columbia. It extends as far west as 115°30'. The eastern boundary has an irregular outline through the foothills of the Rocky Mountains. It lies west of Bragg Creek, Millarville, Turner Valley and Longview, and extends as far east as 114°14'W. The northern boundary is also irregular, lying south of the Bow River and extending as far north as latitude 51°07'N. The town of Canmore lies 3 km to the north; Bow Valley Provincial Park is included in this northern section. The southern boundary is irregular too, partly because it has been extended to include the area around Pasque Mountain and the Plateau Mountain Natural Area. There are no nearby towns or hamlets to use as reference points: the town of Coleman lies 60 km south. The boundary extends as far south as 50°05'N. The township-range limits include townships 11 to 25 and ranges 3 to 11, west of the 5th Meridian.

Kananaskis Country was created to preserve a beautiful area of the Rockies: the rugged western mountains with extensive glaciers; the Kananaskis, Elbow, Sheep and Highwood rivers with their many tributaries; the beautiful lakes; and the foothills scenery. The area is ecologically diverse, supporting significant populations of wildlife and an extensive flora. Wildlife species include moose, deer, elk, mountain goat, bighorn sheep and both black and grizzly bears. There is also a high concentration of cougars in some places. The Sheep River Wildlife Sanctuary lies in the southeastern part of Kananaskis Country. The bird life is extensive and varied, typical of the mountains and foothills. A major migration of golden eagles takes place along the Front Ranges of the Rocky Mountains in spring and fall. The Fisher Range and Mount Lorette lie on the route. In spring 1993, over 4000 golden eagles were recorded. The plant life is equally diverse and features alpine species, grassland plants, marsh plants and forest species as well as aquatic plants in the lakes and rivers. The southernmost extension of black spruce in Alberta is found in the Bragg Creek area.

There is a wide range of recreational services offered in Kananaskis Country. For information concerning Kananaskis Country's facilities and recreational activities contact:

KANANASKIS COUNTRY
Suite 201, 800 Railway Avenue
Canmore, Alberta
T1W 1P1

Phone: (403) 678-5508
E-mail: rchamney@env.gov.ab.ca
or call the Barrier Lake Information Centre at (403) 673-3985.

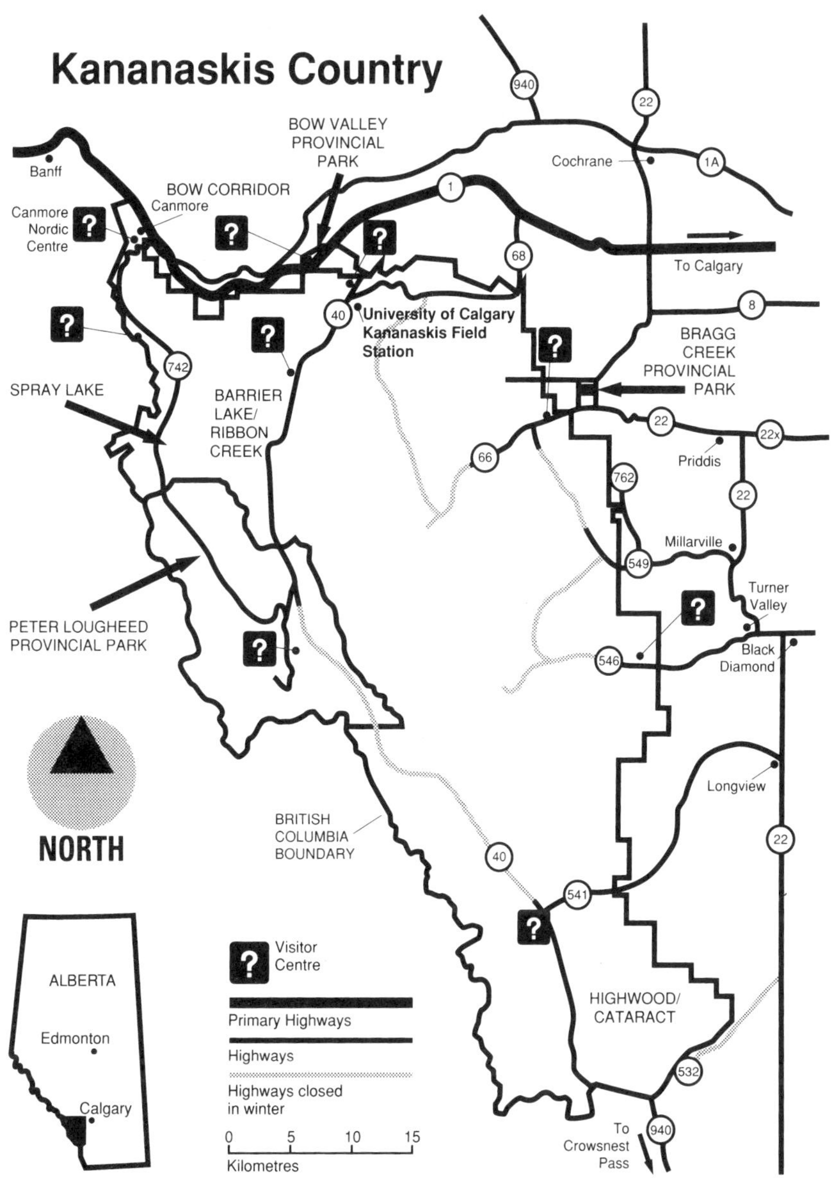
Kananaskis Country
BOW VALLEY PROVINCIAL PARK
Banff
BOW CORRIDOR
Canmore
Canmore Nordic Centre
Cochrane
To Calgary
University of Calgary Kananaskis Field Station
BRAGG CREEK PROVINCIAL PARK
SPRAY LAKE
BARRIER LAKE/ RIBBON CREEK
Priddis
Millarville
Turner Valley
Black Diamond
PETER LOUGHEED PROVINCIAL PARK
Longview
NORTH
BRITISH COLUMBIA BOUNDARY
ALBERTA
Edmonton
Calgary
Visitor Centre
Primary Highways
Highways
Highways closed in winter
0 5 10 15
Kilometres
HIGHWOOD/ CATARACT
To Crowsnest Pass
940
22
1A
1
68
8
40
742
66
762
549
546
541
532
22x

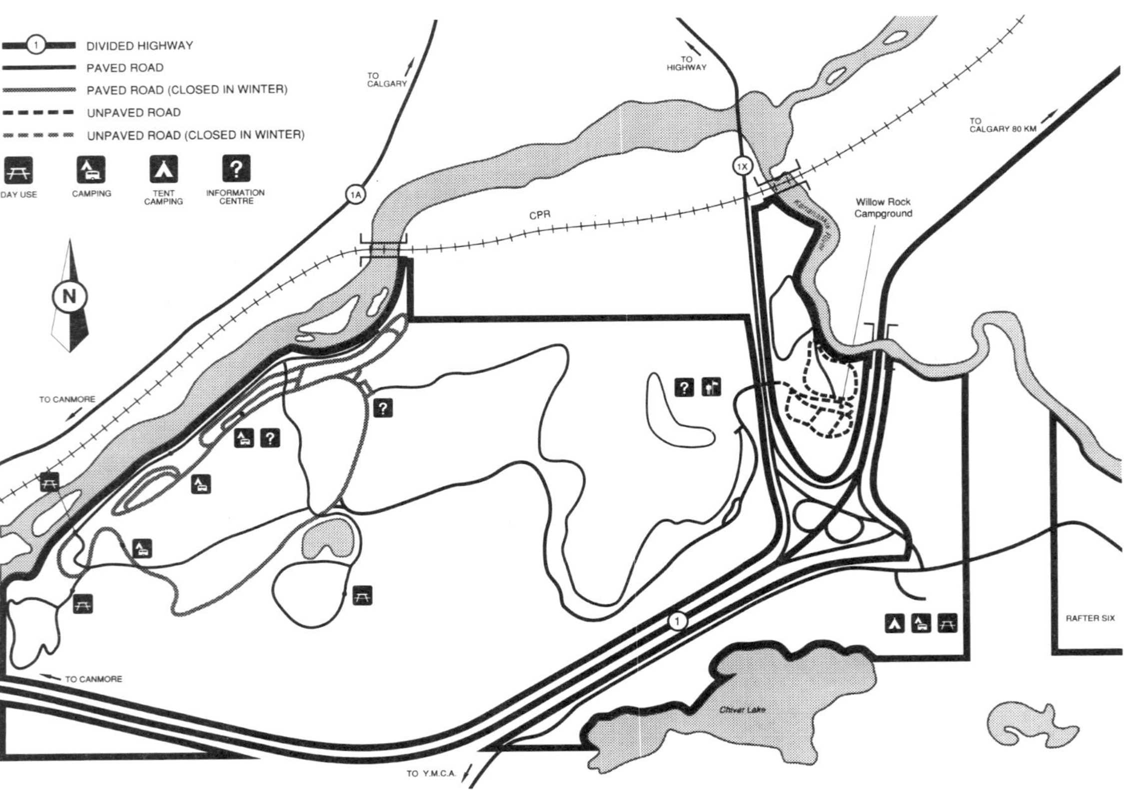
BOW VALLEY PROVINCIAL PARK
DIVIDED HIGHWAY
PAVED ROAD
PAVED ROAD (CLOSED IN WINTER)
UNPAVED ROAD
UNPAVED ROAD (CLOSED IN WINTER)
DAY USE
CAMPING
TENT CAMPING
INFORMATION CENTRE
N
TO CALGARY
1A
TO HIGHWAY
1X
TO CALGARY 80 KM
CPR
Willow Rock Campground
TO CANMORE
TO CANMORE
1
RAFTER SIX
TO Y.M.C.A.

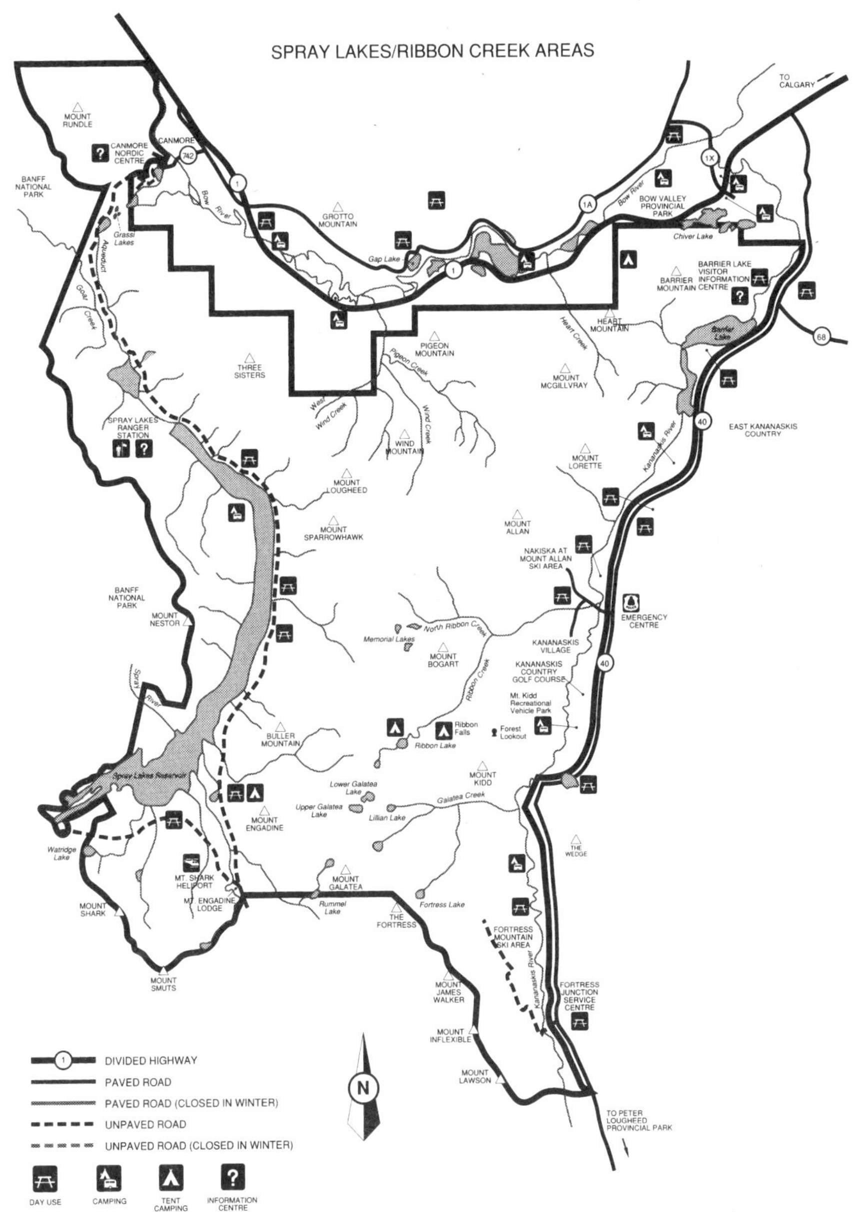
SPRAY LAKES/RIBBON CREEK AREAS
MOUNT RUNDLE
CANMORE NORDIC CENTRE
CANMORE
BANFF NATIONAL PARK
Grassi Lakes
Aqueduct
Goat Creek
Bow River
GROTTO MOUNTAIN
Gap Lake
THREE SISTERS
PIGEON MOUNTAIN
Pigeon Creek
West Wind Creek
Wind Creek
WIND MOUNTAIN
SPRAY LAKES RANGER STATION
MOUNT LOUGHEED
MOUNT SPARROWHAWK
TO CALGARY
BOW VALLEY PROVINCIAL PARK
Chiver Lake
BARRIER MOUNTAIN
BARRIER LAKE VISITOR INFORMATION CENTRE
Barrier Lake
HEART MOUNTAIN
Heart Creek
MOUNT MCGILLVRAY
EAST KANANASKIS COUNTRY
Kananaskis River
MOUNT LORETTE
MOUNT ALLAN
NAKISKA AT MOUNT ALLAN SKI AREA
EMERGENCY CENTRE
KANANASKIS VILLAGE
KANANASKIS COUNTRY GOLF COURSE
Mt. Kidd Recreational Vehicle Park
Forest Lookout
BANFF NATIONAL PARK
MOUNT NESTOR
Spray River
Memorial Lakes
North Ribbon Creek
MOUNT BOGART
Ribbon Creek
Ribbon Falls
Ribbon Lake
BULLER MOUNTAIN
Spray Lakes Reservoir
MOUNT KIDD
Lower Galatea Lake
Upper Galatea Lake
Lillian Lake
Galatea Creek
MOUNT ENGADINE
THE WEDGE
Watridge Lake
MT. SHARK HELIPORT
MT. ENGADINE LODGE
MOUNT GALATEA
Rummel Lake
THE FORTRESS
Fortress Lake
MOUNT SHARK
MOUNT SMUTS
FORTRESS MOUNTAIN SKI AREA
MOUNT JAMES WALKER
FORTRESS JUNCTION SERVICE CENTRE
MOUNT INFLEXIBLE
MOUNT LAWSON
TO PETER LOUGHEED PROVINCIAL PARK
1
1A
1X
742
40
68
N
DIVIDED HIGHWAY
PAVED ROAD
PAVED ROAD (CLOSED IN WINTER)
UNPAVED ROAD
UNPAVED ROAD (CLOSED IN WINTER)
DAY USE
CAMPING
TENT CAMPING
INFORMATION CENTRE

PETER LOUGHEED PROVINCIAL PARK

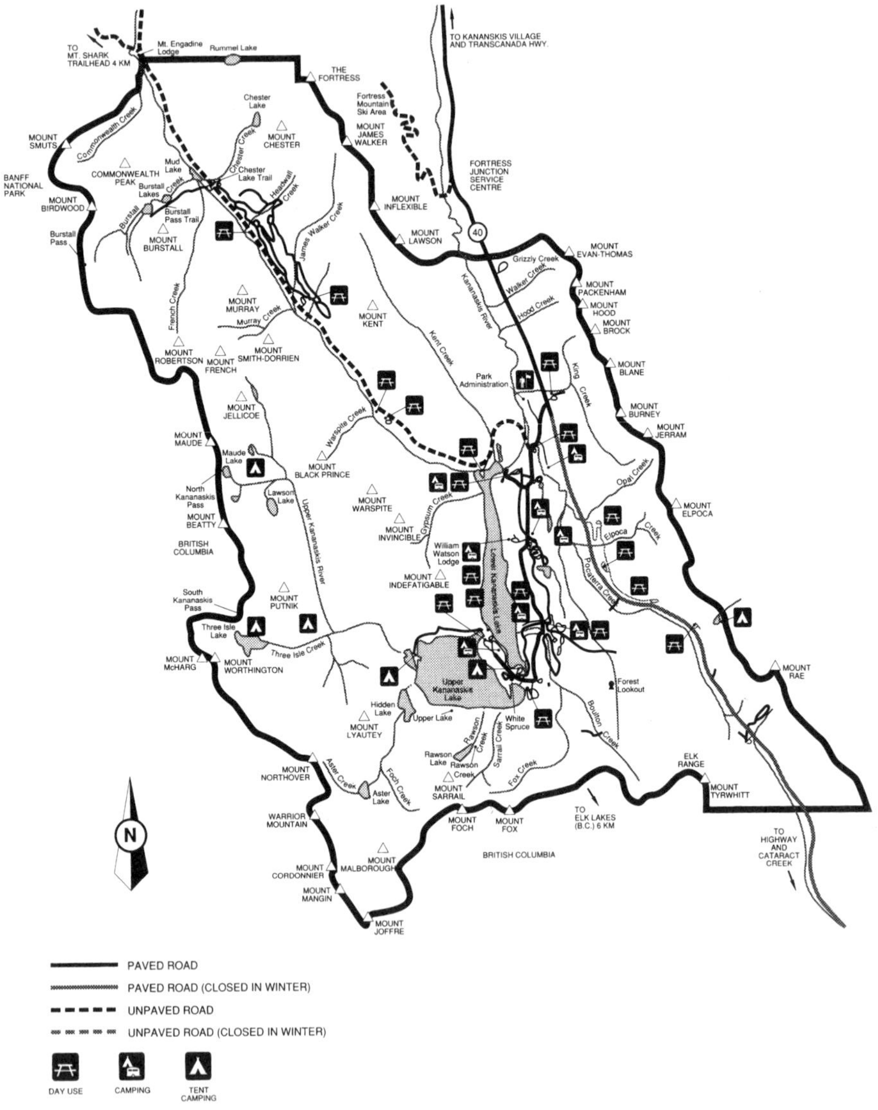

TO MT. SHARK TRAILHEAD 4 KM
Mt. Engadine Lodge
Rummel Lake
TO KANANSKIS VILLAGE AND TRANSCANADA HWY.
THE FORTRESS
Fortress Mountain Ski Area
Chester Lake
MOUNT CHESTER
MOUNT JAMES WALKER
MOUNT SMUTS
Commonwealth Creek
COMMONWEALTH PEAK
Mud Lake
Chester Creek
Chester Lake Trail
Headwall Creek
FORTRESS JUNCTION SERVICE CENTRE
BANFF NATIONAL PARK
Burstall Lakes
Burstall Creek
Burstall Pass Trail
MOUNT BIRDWOOD
MOUNT INFLEXIBLE
Burstall Pass
MOUNT BURSTALL
James Walker Creek
MOUNT LAWSON
40
Grizzly Creek
MOUNT EVAN-THOMAS
Kananaskis River
Walker Creek
MOUNT PACKENHAM
French Creek
MOUNT MURRAY
Murray Creek
MOUNT KENT
Hood Creek
MOUNT HOOD
MOUNT BROCK
Kent Creek
MOUNT ROBERTSON
MOUNT FRENCH
MOUNT SMITH-DORRIEN
King Creek
MOUNT BLANE
Park Administration
MOUNT JELLICOE
MOUNT BURNEY
Warspite Creek
MOUNT JERRAM
MOUNT MAUDE
Maude Lake
MOUNT BLACK PRINCE
Opal Creek
North Kananaskis Pass
Lawson Lake
Upper Kananaskis River
MOUNT WARSPITE
Gypsum Creek
MOUNT ELPOCA
MOUNT BEATTY
MOUNT INVINCIBLE
Elpoca Creek
BRITISH COLUMBIA
William Watson Lodge
Lower Kananaskis Lake
MOUNT INDEFATIGABLE
Pocaterra Creek
South Kananaskis Pass
MOUNT PUTNIK
Three Isle Lake
Three Isle Creek
MOUNT McHARG
MOUNT WORTHINGTON
MOUNT RAE
Upper Kananaskis Lake
Forest Lookout
Hidden Lake
Upper Lake
MOUNT LYAUTEY
White Spruce
Rawson Creek
Sarrail Creek
Boulton Creek
Rawson Lake
MOUNT NORTHOVER
Aster Creek
Foch Creek
Fox Creek
MOUNT SARRAIL
ELK RANGE
MOUNT TYRWHITT
Aster Lake
WARRIOR MOUNTAIN
MOUNT FOCH
MOUNT FOX
TO ELK LAKES (B.C.) 6 KM
N
TO HIGHWAY AND CATARACT CREEK
MOUNT MALBOROUGH
BRITISH COLUMBIA
MOUNT CORDONNIER
MOUNT MANGIN
MOUNT JOFFRE

PAVED ROAD
PAVED ROAD (CLOSED IN WINTER)
UNPAVED ROAD
UNPAVED ROAD (CLOSED IN WINTER)

DAY USE

CAMPING

TENT CAMPING

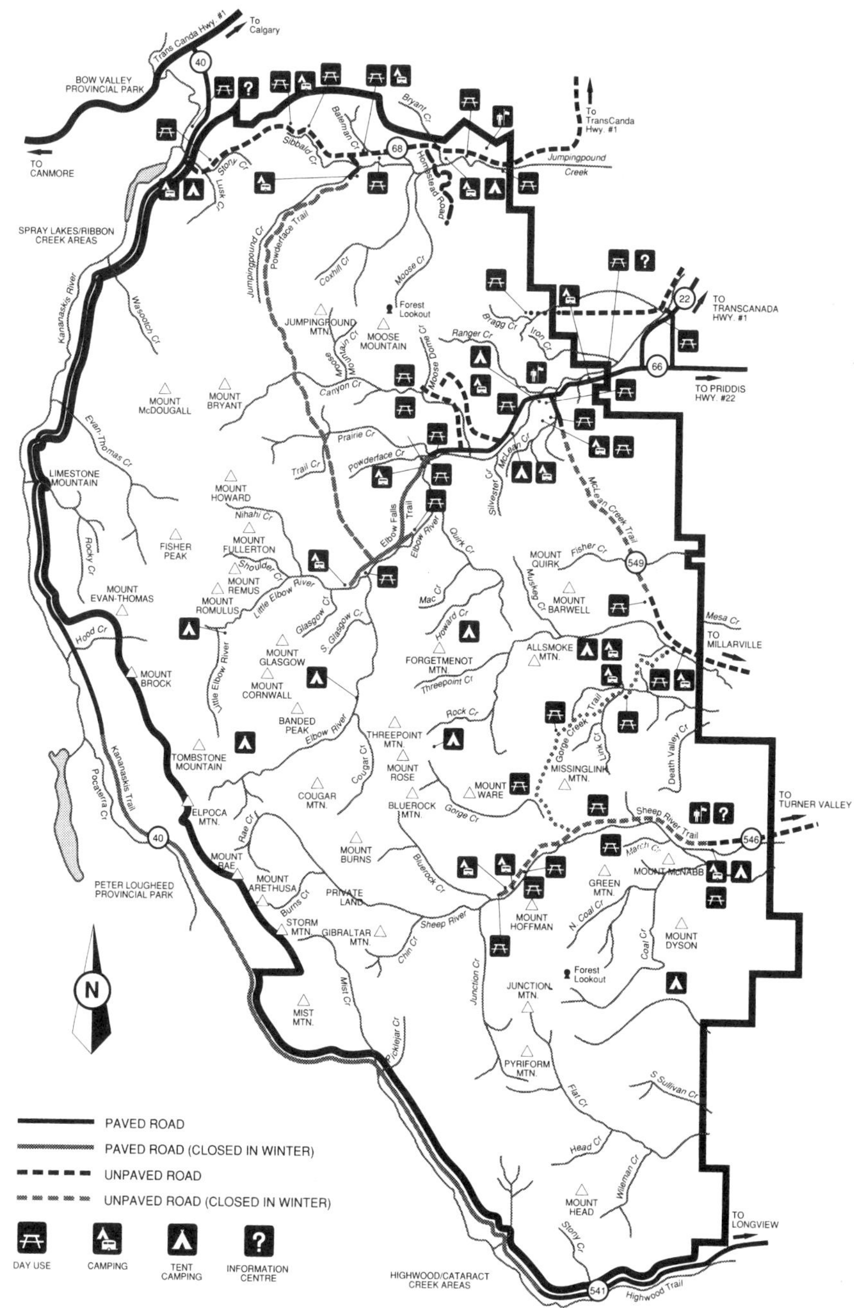
EAST KANANASKIS COUNTRY
Trans Canda Hwy. #1
To Calgary
BOW VALLEY PROVINCIAL PARK
TO CANMORE
To TransCanda Hwy. #1
Jumpingpound Creek
SPRAY LAKES/RIBBON CREEK AREAS
TO TRANSCANADA HWY. #1
TO PRIDDIS HWY. #22
TO MILLARVILLE
TO TURNER VALLEY
TO LONGVIEW
PETER LOUGHEED PROVINCIAL PARK
HIGHWOOD/CATARACT CREEK AREAS
Highwood Trail
Kananaskis Trail
Kananaskis River
Powderface Trail
Homestead Road
Elbow Falls Trail
McLean Creek Trail
Gorge Creek Trail
Sheep River Trail
Forest Lookout
MOUNT McDOUGALL
MOUNT BRYANT
JUMPINGPOUND MTN.
MOOSE MOUNTAIN
LIMESTONE MOUNTAIN
MOUNT HOWARD
FISHER PEAK
MOUNT FULLERTON
MOUNT REMUS
MOUNT ROMULUS
MOUNT EVAN-THOMAS
MOUNT QUIRK
MOUNT BARWELL
ALLSMOKE MTN.
FORGETMENOT MTN.
MOUNT GLASGOW
MOUNT CORNWALL
MOUNT BROCK
BANDED PEAK
THREEPOINT MTN.
MOUNT ROSE
TOMBSTONE MOUNTAIN
MISSINGLINK MTN.
MOUNT WARE
COUGAR MTN.
BLUEROCK MTN.
ELPOCA MTN.
MOUNT BURNS
MOUNT RAE
MOUNT ARETHUSA
PRIVATE LAND
GREEN MTN.
MOUNT McNABB
MOUNT HOFFMAN
STORM MTN.
GIBRALTAR MTN.
MOUNT DYSON
JUNCTION MTN.
MIST MTN.
PYRIFORM MTN.
MOUNT HEAD
PAVED ROAD
PAVED ROAD (CLOSED IN WINTER)
UNPAVED ROAD
UNPAVED ROAD (CLOSED IN WINTER)
DAY USE
CAMPING
TENT CAMPING
INFORMATION CENTRE

HIGHWOOD/CATARACT

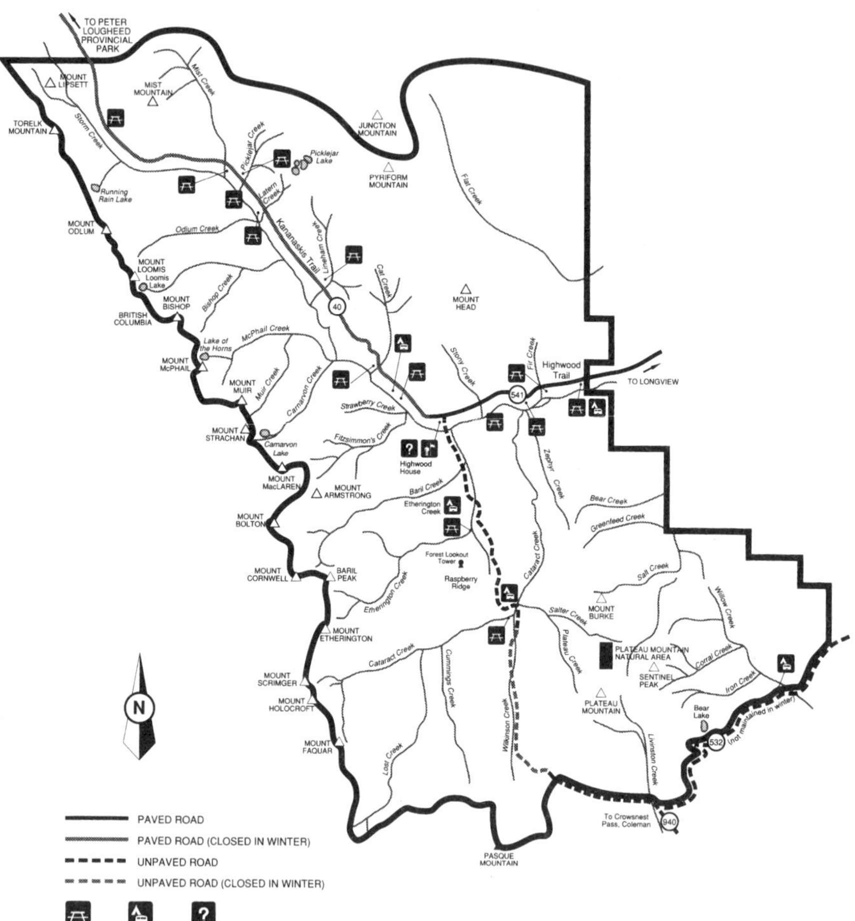

The Geology of Kananaskis Country

The Kananaskis area was one of repeated subsidence and deposition through much of the Paleozoic and Mesozoic eras. In general, the area now occupied by the mountains underwent considerably more subsidence than the adjacent foothills and prairies. During the Devonian and Mississippian periods, the sediments consisted mainly of carbonates, whereas the Mesozoic period saw the deposition of clastic sediments derived from the land area to the east.

The mountains first appeared in Lower Cretaceous times as a result of uplift to the west, providing an influx of sediments from that direction. The zones of deposition of these sediments in shallow epicontinental seas moved slowly eastwards until the Paleocene period, when the seas disappeared completely. By then the main structures we see today in Peter Lougheed Provincial Park had been formed, but erosion gradually removed many of the original mountain tops. Streams flowed northeastward, and the remains of the outwash gravels on the Tertiary (Miocene) valley floors are now preserved on the tops of the Hand Hills, Wintering Hills and Cypress Hills (Storer, 1978). These coarse, well-rounded gravel deposits would require a fast-flowing, braided stream to transport the sediments from the mountains. Subsequent erosion reduced the landscape to the present state, with glaciation occurring during the last phases of erosion.

The bedrock formations found in this area are listed in Gordy, Frey and Ollerenshaw (1975). Details of the thicknesses and distributions of these formations will be found in McCrossan et al. (1964).

STRUCTURE

Kananaskis Country straddles three of the four major structural provinces in this part of Alberta. To the east of Kananaskis Country is the interior plains region, usually referred to as the prairies. This region is characterized by flat-lying or gently folded sediments overlying a Precambrian basement.

To the west are the foothills, which consist of extensive, west-dipping thrust sheets of rock. At the surface, these thrusts take the form of faults in imbricated (overlapping and broken), soft Mesozoic rocks. At depth, however, these faults change to thrusts in Paleozoic carbonate rocks which do not appear at the surface. The rocks strike northwestwards, and the more resistant beds are eroded into the bedrock ridges which are the main characteristic of this structural province.

The outer margin of the true mountains is characterized by Paleozoic carbonate rocks thrust over the Mesozoic sediments (e.g., at Pigeon Mountain). These are the Front Ranges. The softer Mesozoic strata (e.g., the Kootenay Formation) is more easily eroded and has produced the pronounced strike valleys separating the individual ranges. It is along these valleys that extensive coal deposits may be found. The carbonate rocks provide important habitats for calciphilous plants, such as flame-coloured lousewort (*Pedicularis flammea*), which require high calcium-content soils.

Near the continental divide in the northwest part of Peter Lougheed Provincial Park, the imbricated thrusts are developed in Lower Paleozoic rocks (carbonates and clastics, which are fragments of pre-existent rocks). The area occupied by these thrusts is quite small, however.

GLACIATIONS

Most of Kananaskis Country has been covered by mountain glaciers at some time during the latter part of the Pleistocene period. The ice originated both within Peter Lougheed Provincial Park and to the north and west, but some of the crests of the outer (eastern) peaks, such as Plateau Mountain, may have remained unglaciated for at least 100,000 years. This information is important to botanists since these peaks would have acted as *refugia* for plants during glaciations, and it offers an explanation of the origin of the diverse alpine flora present. Species such as golden fleabane (*Erigeron aureus*) that occur only in southwestern Alberta presumably evolved on these mountains. When the climate was cold, but no major glaciers blocked their way, alpine plants were able to enter the region from as far away as California and the Arctic (Harris, 1990).

The last (Late Wisconsin) glaciation was extensive, but the ice had commenced a retreat from the area in the outer foothills by about 11,500 years before the present (B.P.) (MacDonald et al., 1987), and in the Elk Valley by about 12,000 years B.P. (Harrison, 1976). However, in the case of the Elk Valley, proglacial lakes persisted well into the Holocene period, and sediment dams also caused substantial ponding in the Crowsnest Pass until about 5120 years B.P. (Miller, 1991). Incomplete deglaciation and local readvances of ice on the foothills and prairies produced proglacial lakes, such as glacial Lake Calgary in the area north and east of Kananaskis Country, that may have persisted into the Holocene (Harris, 1985).

Subsequently, a warm period (the Altithermal) occurred between about 7000 years B.P. and 4000 years B.P., followed by cooler, wetter conditions, as shown by pollen evidence and the nature of macrofossils in lake sediments in Peter Lougheed Provincial Park. There is also evidence for minor fluctuations of the ice fronts, the largest being the Cavell advance which ended near the end of the last century (Luckman & Osborn, 1979).

THE DIVISIONS OF GEOLOGICAL TIME

CENOZOIC	Quaternary	Holocene	
			0.011 ma
		Pleistocene	
			1.67 ma
	Tertiary	Pilocene Miocene Oligocene Eocene Paleocene	
			66 ma
MESOZOIC	Cretaceous Jurassic Triassic		
			245 ma
PALEOZOIC	Permian Devonian Silurian Ordovician Cambrian	Pennsylvanian Mississippian	} Carboniferous
			570 ma
PROTEROZOIC			
			2500 ma
ARCHAEAN			

ma = millions of years

The Climate of the Kananaskis Valley

The climate of the Kananaskis valley is transitional between the continentality of the plains to the east and the wetter, milder climate on the west side of the Continental Divide. The Kananaskis valley has continental conditions at its mouth and increasingly cordilleran conditions moving (south) up the valley.

Consequently, the climate of the Kananaskis watershed has two aspects. The Front Range (north end) of the valley has a transitional plains-cordilleran climate with cold winters, warm summers, and well-defined summer precipitation maximum and winter minimum. The Main Range (south end) of the valley has a more cordilleran climate of cold winters, cool summers, winter and summer peaks in precipitation and poorly defined minima in February to March and September to October. Both ranges have elevational effects on temperature and precipitation, but the Main Range, because of its higher mountains and more westerly position, intercepts the Pacific air masses first. Thus, the Main Range has increased precipitation and creates a rain shadow in the adjacent Front Ranges. Glaciers (icefields) are common in the Main Range, but non-existent in the Front Range in the Kananaskis. The Smith-Dorrien valley and Kananaskis lakes receive runoff from the glaciers and greater precipitation because the North and South Kananaskis Passes allow the moister Pacific air mass to enter more directly.

The average temperature in January is -10°C and in July is 14°C at Kananaskis Field Stations, a more continental location. Pocaterra, a more cordilleran station, has an average temperature in January of -14°C and in July of 12°C. Temperature patterns within the valley show *adiabatic* cooling with elevation ($-0.0098°C/m^{-1}$) except in the valley bottoms, which are subject to cold air drainage from higher elevations at night. This cold air drainage can result in as much as a 7°C decrease in temperature. Absolute humidity of the valley during these cold air drainage events is that of the upper elevation source areas. A temperature inversion on windless nights can be found about 100 m above the valley floor. The isothermal layer is about 150 m thick. Slopes with closed forest show little differences in temperature or humidity with changes in slope or aspect. However, open, south-facing slopes are warmer, while open, north-facing slopes are cooler and moister.

Precipitation in the Kananaskis valley reflects the transitional position between the plains and cordillera. The continental climate of the Front Range has a single summer maximum and winter minimum of precipitation, while the cordilleran climate of the Main Range has a well-defined summer maximum and a less well-defined winter maximum. The Kananaskis has two peaks of precipitation at lower elevations: one in June, the other in August. (At higher elevations, the early peak occurs in May.) The lowest amount of precipitation occurs in October or November. Kananaskis Field Stations has peak precipitation in June of 98.9 mm and in August of 67.5 mm, a minimum in November of 28.4 mm, and an annual average precipitation of 618.1 mm. Pocaterra has peak precipitation in May of 68.8 mm, August of 69.1 mm and December of 59.2 mm. The October minimum is 22.9 mm and the annual average precipitation is 550.9 mm. In general, the amount of precipitation increases by about 20 mm for each 100 m increase in elevation in the valley.

Chinook winds recur on average 25 times a year, primarily during the winter. Because of the SW-NE alignment of the main valley and the passes in the south, the drying effects of the chinook winds are most pronounced on the west-facing slopes at approximately 1700 m elevation.

The Vegetation Zones of Kananaskis Country

ELEVATION AND TOPOGRAPHIC EFFECTS

The vegetation of the Kananaskis valley is primarily determined by elevation, topographic position, substrate and fire. These factors influence both the composition and age of the forests. Vegetation has been divided elevationally into five zones: foothills, montane, lower subalpine, upper subalpine and alpine *(see map, page XXVIII)*. In each vegetation zone, topography and substrate affect vegetation composition along gradients of soil moisture and nutrients, but fire determines the forest's age.

THE FOOTHILLS ZONE

The foothills zone is an important region, covering about ten percent of the study area. It is characterized by forests of lodgepole pine (*Pinus contorta* var. *latifolia*), trembling aspen (*Populus tremuloides*) and white spruce (*Picea glauca*). It is a zone of transition, because it divides the conifer-dominated vegetation of the subalpine and boreal uplands from the aspen-dominated vegetation of the aspen grovelands; it also supports plants normally found in the rough fescue grasslands. Shrubs found in this zone include wolf willow (*Elaeagnus commutata*), junipers (*Juniperus* spp.), bearberry (*Arctostaphylos uva-ursi*), glandular bogbirch (*Betula glandulosa*) and willows (*Salix* spp.). Herbaceous plants found here include cow parsnip (*Heracleum lanatum*), strawberry (*Fragaria* sp.), yellow mountain avens (*Dryas drummondii*) and common fireweed (*Epilobium angustifolium*). Along the streamsides sweet cicely (*Osmorhiza* spp.), currants (*Ribes* spp.), gooseberries (*Ribes* spp.), meadow rue (*Thalictrum* spp.),

VEGETATION ZONES

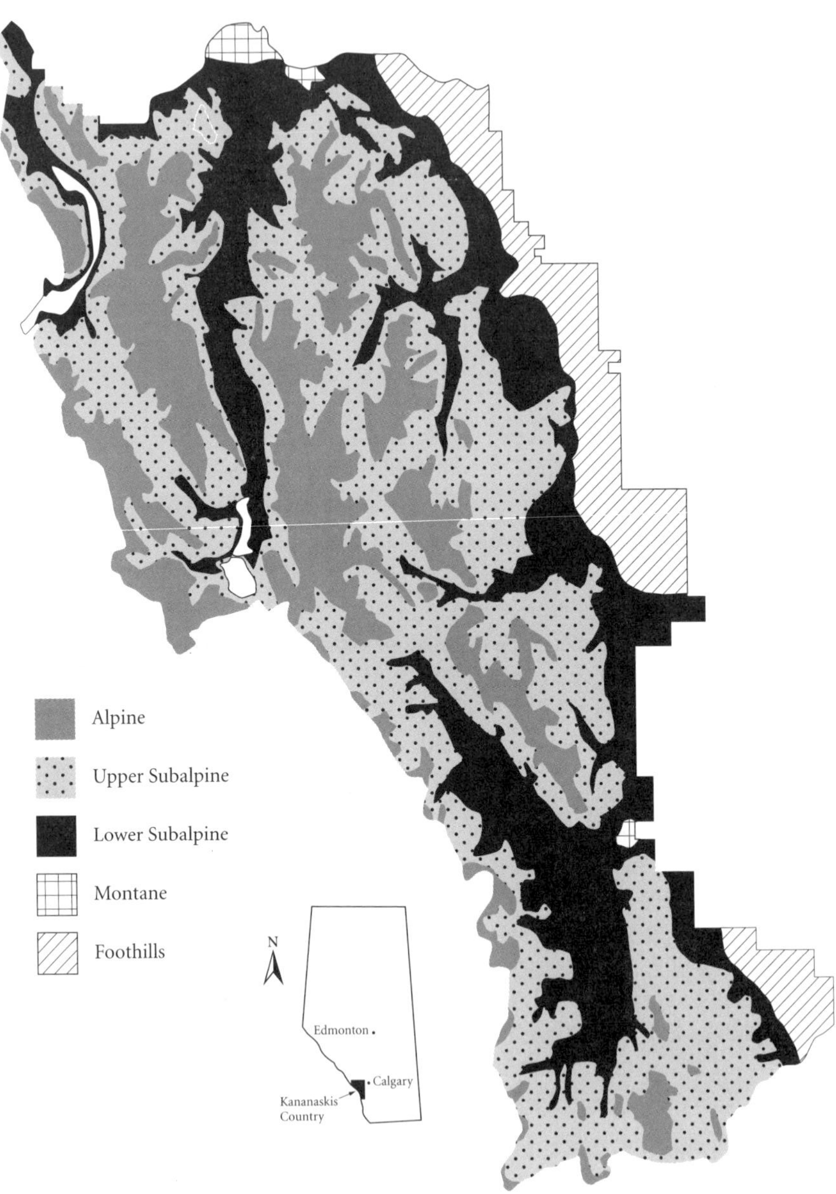

common pink pyrola (*Pyrola asarifolia*), cotton-grasses (*Eriophorum* spp.) and horsetails (*Equisetum* spp.) are found.

THE MONTANE ZONE

The montane zone extends into the mouth of the Kananaskis valley from the Bow valley and consists of forests of Douglas fir (*Pseudotsuga menziesii*), white spruce, lodgepole pine and trembling aspen. Because these forests have been grazed by cattle and extensively logged, they have a large proportion of aspen, but Douglas fir is the characteristic tree of this zone. Above Barrier Lake, Douglas fir is much less common and is confined to particular sites, for example, on scree around the Upper Kananaskis Lake and on steep canyon sides in King's Canyon.

The shrubs found in this zone include juniper, bearberry, prickly rose (*Rosa acicularis*) and birch-leaved spiraea (*Spiraea betulifolia*). The herbaceous plants include northern bedstraw (*Galium boreale*), prairie groundsel (*Senecio canus*), northern twinflower (*Linnaea borealis*), hairy wild rye (*Elymus innovatus*), June grass (*Koeleria macrantha*) and sweet grass (*Hierochloe odorata*). Feather-mosses are found in woodlands on north-facing slopes.

THE LOWER SUBALPINE ZONE

The lower subalpine zone occurs from about 1200 m to about 1800 m. Very dry sites, such as exposed bedrock, scree and gravel streambeds, are covered with Engelmann spruce (*Picea engelmannii*), lodgepole pine and limber pine (*Pinus flexilis*). On such sites the trees are short and the forest is open, with sunlight reaching the ground between the trees. Very wet sites adjacent to the Kananaskis River, along persistent streams or in areas of loamy soil are dominated by Engelmann spruce. Because these wet sites always have adequate moisture, the trees grow rapidly and at 50 to 60 years of age can be as large as 0.5 m in diameter. These wet forests are easily recognized by a ground-cover of horsetails, abundant lichens on the trees and deep shade cast by the spruce. Most of the forested areas in the lower subalpine zone are neither very dry or very wet, but are moderate in soil moisture and covered by lodgepole pine and Engelmann spruce. The proportion of pine and spruce depends on several factors, specifically fire (see After Fire section). The only other significant forest type in the lower subalpine zone is aspen forest. Mature aspen stands are characteristic of fine-textured silty soils often associated with the sediments of old glacial lake-bottoms and stream deltas. The best example of these forests is seen at the south end of White Man's Flats, at Ribbon Creek.

The shrubs found in the dense coniferous forests of the lower subalpine zone include Labrador tea (*Ledum groenlandicum*), bracted honeysuckle (*Lonicera involucrata*) and Canada buffaloberry (*Shepherdia canadensis*). Feather-mosses and horsetails flourish in the deep shade, the mosses forming deep carpets. Herbaceous plants found in this zone include bronze-bells (*Stenanthium occidentale*) and bunchberry (*Cornus canadensis*).

THE UPPER SUBALPINE ZONE

The upper subalpine zone extends from about 1800 m to about 2300 m or treeline. At approximately 1800 to 1900 m, lodgepole pine becomes scarce in the forest, leaving Engelmann spruce and subalpine fir (*Abies lasiocarpa*). Fir increases in abundance starting at approximately 1700 m. Alpine larch (*Larix lyallii*) is also present in the upper subalpine but only for several hundred metres below treeline. The best time to see the distribution of alpine larch is in the fall, when this deciduous conifer's needles turn bright yellow.

Dry sites in the upper subalpine zone are occupied by limber pine and sometimes by white-bark pine (*Pinus albicaulis*). Very wet sites feature subalpine meadows. Moderately moist sites are populated by Engelmann spruce and subalpine fir forests, whose canopies open up nearer to the treeline.

In the upper part of this zone, where the subalpine merges with the alpine zone, the *Krummholz* region of stunted, bushy, "wind-flagged" trees is found. *Krummholz* is the German word for "crooked wood," and this description is appropriate, because trees such as Engelmann spruce and subalpine fir can survive only in a stunted condition under such harsh conditions. Where conditions are better, at lower elevations, "tree-islands," small groups of trees, are found.

On *Krummholz* slopes and tree-islands, arctic willow (*Salix arctica*) is found with 4-angled cassiope (*Cassiope tetragona*), white mountain avens (*Dryas octopetala*) and alpine bistort (*Polygonum viviparum*). Cream mountain heather (*Phyllodoce glanduliflora*) and red mountain heather (*Phyllodoce empetriformis*) are also found, with Sitka valerian (*Valeriana sitchensis*), grouseberry (*Vaccinium scoparium*) and alpine speedwell (*Veronica alpina*). On south-facing avalanche slopes, hairy wild rye, rough fescue (*Festuca scabrella*), awnless brome (*Bromus inermis*), bluegrasses (*Poa* spp.), sulphur hedysarum (*Hedysarum sulphurescens*), harebell (*Campanula rotundifolia*), yarrow (*Achillea millefolium*) and wild strawberry (*Fragaria virginiana*) grow. On north-facing avalanche slopes, where it is much colder, net-leaved dwarf willow (*Salix reticulata* spp. *nivalis*), arctic willow, alpine milk vetch (*Astragalus alpinus*), northern goldenrod (*Solidago multiradiata*) and alpine bistort are found. The glandular bog birch, near timberline, provides important winter feed for moose.

THE ALPINE ZONE

The alpine zone is characterized by long, very cold winters and short, cool summers. It is above the treeline and snow accumulation is high, except on wind-swept slopes and ridges. Many herbaceous species have adapted to these harsh conditions, and in the brief summer, there is plenty of colour in the landscape. In the moist alpine meadows, there are paintbrushes (*Castilleja* spp.), buttercups (*Ranunculus eschscholtzii*), alpine veronica (*Veronica alpina*), chalice flower (*Anemone occidentalis*) and red mountain heather. The alpine meadow at the Highwood Pass is a good place to find these species.

In moist parts of the alpine zone, dense hummocks of sedges (*Carex* spp.) are found; in better-drained areas, carpets of white mountain avens can be seen.

At higher elevations, where there is a high wind-chill factor, drought conditions can develop. Three types of plant adaptations have emerged:

a. **"cushion" plants,** like moss campion (*Silene acaulis*) and cut-leaved fleabane (*Erigeron compositus*);
b. **rosette plants:** dandelions (*Taraxacum* spp.), dwarf saw-wort (*Saussurea densa*), sweet-flowered androsace (*Androsace chamaejasme*), and golden fleabane (*Erigeron aureus*);
c. **mat plants:** white mountain avens, net-veined dwarf willow, common selaginella (*Selaginella densa*) and prickly saxifrage (*Saxifraga bronchialis*).

On the very highest peaks, lichens on rock boulders are the only form of plant-life. Lichens are plants that consist of two components: one, a green or blue-green alga, the other, a fungus. Both live together in a harmoniously balanced life-form called *symbiosis.*

FIRE AND WEATHER

Weather, through its effect on lightning occurrence and fuel moisture, is the main cause of natural fires in the Kananaskis valley. Before the mid-1700s, the Kananaskis valley burned on average once every 60 years. After the mid-1700s, the Kananaskis valley burned once every 90 years. (This does not mean that every location burned every 60 or 90 years, but that an area *equal* to the size of Kananaskis valley burned every 60 or 90 years. That is, some areas may have burned more than once, while others not at all.) The fire frequency changed in the mid-1700s due to a cooler, moister climate, which decreased the chance of fire occurrence and spread and also resulted in the re-advance of the Haig, Mangin and Petain glaciers. Before the mid-1700s, the climate had been warmer and drier, allowing a greater chance of fire occurrence and spread.

In most years fires, if they occur at all, are small and do not spread. At infrequent intervals, however, weather conditions lead to extremely dry conditions and lightning strikes, creating the potential for large, fast-moving crown fires. It is these large fires which determine the fire-frequency. The last large fire was the Galatea Creek Fire, which occurred in August 1936, burning more than 8400 ha. Fires of similar size also occurred in 1920, 1904, 1890, 1881, 1870, 1858, 1848, 1828, 1803 and 1791.

The Galatea Creek Fire appears to have been typical of large fires in the Canadian Rockies. It started by lightning on August 3, 1936. It was preceded by a summer of abnormally hot, dry weather and was the result of a high-pressure system which had been stationary over Alberta, diverting the normal sequence of high (dry-warm) and low (wet-cool) pressure systems north and south. The fire spread slowly as a crown fire, from August 3rd to the 9th. As is characteristic of these fires, the largest area burnt was during a short period of time, in this case during the afternoon of August 9. This period of rapid fire-spread and extreme crowning was due to high winds during the transition period between the movement eastward of the stationary high pressure system and its replacement by a low pressure system. Such transition between highs and lows is always characterized by high, variable winds due to the changes in pressure. As a result of the wind and the extremely dry condition of the fuels, the fire moved so quickly and gave off so much heat that differences in vegetation types, age or elevation had little effect in determining the direction and spread of the fire. Some areas survived the fire by chance alone or because they were upwind of the fire (upper Ribbon Creek Valley, for example).

AFTER FIRE

After a fire, tree recruitment occurs for five to eight years. After this time, few seeds germinate and almost no seedlings will survive to reach the canopy. Clearly, then, the seeds that arrive and germinate first will have the best chance of becoming part of the forest canopy and surviving to provide seeds for recruitment after the next fire. The proportion of lodgepole pine and Engelmann spruce to germinate after fire is determined by the nearby seed sources, the amount of mineral soil exposed and the specific temperature and moisture conditions that occurred during seedling establishment. Lodgepole pine cones open only after a fire, so if pine is present in the pre-fire forest, its seedlings will certainly be present in the post-fire forest. On the other hand, Engelmann spruce and subalpine fir seed must disperse from surviving individuals at the edge of the burn or from unburned remnants of the forest within the burn. Each of these three species have different but overlapping requirements of exposed mineral soil for successful germination and establishment. Consequently, the amount of forest floor litter and duff ashed by the fire will, to some extent, determine the forest composition. Lodgepole pine germinates and establishes best on completely exposed mineral soil, while subalpine fir germinates best on moss-covered surfaces (although it can germinate on exposed mineral soil). Engelmann spruce is intermediate in soil surface requirements, but is more similar to lodgepole pine than subalpine fir.

Because of the narrow period in which establishment can occur, all trees will be approximately the same age. They will be of different sizes, however, because each species has different height growth-rates, shade tolerance and mortality rates. Lodgepole pine has the most rapid height growth, while Engelmann spruce and subalpine fir have slower rates of height growth. It is not surprising that lodgepole pine, which grows rapidly and hence tends to remain in the canopy, is also the least shade-tolerant and suffers heavy mortality when shaded. On the other hand, Engelmann spruce has a slower growth-rate and cannot keep up with lodgepole pine in the canopy. Spruce is, however, more shade-tolerant and consequently suffers less mortality from shading. Subalpine fir is similar to Engelmann spruce in both growth rate and shade tolerance.

A mature 60-year old forest will often consist of a canopy of pine and an understory of spruce with some fir, all of the trees approximately the same age. Lodgepole pine reaches its mature height at approximately 100 years of age, but Engelmann spruce and subalpine fir will continue to grow for another 50 to 100 years. This allows them to reach the canopy eventually. Engelmann spruce and subalpine fir do not shade out and replace the lodgepole pine when they reach the canopy; they all continue to coexist.

The lack of any effective recruitment below the spruce-fir layer is somewhat surprising. There will be a few seedings, but most of these have very high mortality rates and generally have a very small chance of reaching the canopy. As the trees in the canopy age, there is little in the understorey to replace them. This clearly indicates that this forest depends on frequent fires within the lifespan of the canopy to replace itself.

OLD GROWTH

Given the high frequency of fire, is there any possibility of finding old growth forests? Old forests (greater than 400 years old) occur in two situations: either in areas that have survived fire by chance, or in areas that tend not to burn because of the specifics of their location.

The first area is made up of scattered clumps of surviving individuals that are not associated with specific habitats. Because of the low chance of surviving fire, they make up less than one percent of the forested area in the Kananaskis Valley. For example, there are Engelmann spruce trees greater than 400 years of age which have survived at least two fires that destroyed the forest around them, found adjacent to the Fortress Road. They stand out markedly in the 1936 burn, seen just before the Fortress Ski Lodge.

Really old trees, that is, trees greater than 400 years old, are confined to specific sites that tend not to burn. These locations are not conducive to fire spread because they have little vegetative cover and fuel accumulation. These sites also tend to be very low in nutrient and moisture content. Most of these sites are found near treeline, on bedrock, on scree, on dry gravel streambeds (for example, Wasootch or Porcupine Creek) or on landslides (for example, adjacent to Upper Kananaskis Lake).

Old trees tend to be primarily Douglas fir or limber pine, and they can be easily identified by their short stature, lack of needles and dead terminal leaders.

THE IMPACT OF MAN ON VEGETATION

The impact of native peoples on the vegetation is inconclusive. Before the smallpox epidemic in 1781, the Front Range and foothills of the Rocky Mountains were occupied by the Kootenay people. This tribe spent its summers on the Columbia and Kootenay Rivers, west of the Rocky Mountains, fishing on large lakes. In the fall and winter they hunted buffalo, which had moved into the foothills for the winter. The Kananaskis valley was occupied only seasonally and was used primarily as a corridor between their summer and winter camps, either because the valley did not attract many buffalo or have a climate suitable for their limited agriculture. After 1781, the Kootenays withdrew to the west side of the Rockies, by which time their numbers had been reduced by approximately three-fourths. The Front Range appears to have been largely unoccupied by natives until approximately the 1840s, when the Stoney (Assiniboine) peoples migrated into the area from the north. The Stoneys were primarily hunters and gatherers, dependent on buffalo, elk, deer and bighorn sheep for meat, skins and horn, and on herbs and shrubs for berries, seeds and other plant materials.

Neither the Kootenays nor the Stoneys appear to have intruded greatly on their environment, based on the lack of physical and cultural evidence for significant effects on the vegetation. The only exception to this *may* be that certain meadows, for example White Man's Flats at Ribbon Creek, were kept free of trees by burning. This burning would have occurred in the spring when the forest still had snow in it. These grassy areas would have attracted elk and deer in early spring and summer, and later provided forage for horses.

The first visit of Europeans to the Kananaskis valley was in 1787, when David Thompson explored the area. Until 1883, however, the valley was mainly used as a corridor between the plains and the interior valleys of British Columbia through the Kananaskis passes, just as native peoples had used it for generations before.

In 1883, the Eau Claire and Bow Lumber Company received four leases in the Kananaskis Valley. The lumber company was primarily interested in spruce for saw timber and pine as pole lumber. Loggers used axes and crosscut saws to fell the trees and cut them into proper lengths. Then skidders, using horses, dragged logs directly to rollways on the river or to skidways, where the logs could be loaded on sleds and taken to the river. The logs were transported to the sawmill in Calgary in the spring when the river thawed, with log drives taking approximately two months.

The way logging was conducted left minimal impact on the Kananaskis valley. The areas were logged only during the winter, using horses to drag the logs to the river. Permanent roads were not necessary for the sleds which carried the timber. Steep slopes were usually not logged, and only more level areas close to the river were cut. Two timber surveys were conducted up the river, one in 1883 and another in 1972. A comparison of the composition and size of trees in these two surveys found that the forest had not suffered any significant change in that time. Since logging was conducted only in areas close to the river, the Kananaskis valley sustained minimal change due to logging.

Coal developments in the 1940s and 1950s and the construction of two hydro-electric dams, one on the Upper Kananaskis Lake in 1932 and another creating Barrier Lake in 1945, had localized impact and did not extensively affect the vegetation in the valley.

Until the construction of Highway 40 (the Kananaskis Trail), the Kananaskis Valley was largely inaccessible without difficulty. There were no farms or ranches in the valley because these had been forbidden by the Dominion Lands Act of 1879, which reserved the Front Range as a source of timber and water for settlers on the plains. The deep snow, lack of forage and poor agricultural potential also discouraged permanent settlement.

How to Use This Book to Identify a Plant

1. Read through the Glossary, pp. 311–339, paying special attention to the illustrated pages. The labelled illustrations of leaf shapes, flower sections, inflorescences and fruits will help explain the botanical terms used in the plant descriptions.
2. Using the key to the major groups (p. XXXIX), decide which group a particular plant belongs to.
3. Use the key to the families in that group to determine which family has the characteristics of that plant.
4. Check the description of the family, then read through the key to the genera belonging to that family. Some families have only one genus, but large families, like Asteraceae (Compositae), have many, and care is needed. Use the key provided to find the genus to which a particular plant belongs.
5. Check the description of the genus and ensure that the characteristics described match those of the plant. Then use the key to the species of that genus; this is often the most difficult part. Some genera have many species, and a good lens will be useful here.
6. Having decided what species the plant belongs to, check the line-drawing of that species. Nearly all the species described in the text are illustrated. Read the description of the plant and check it to see if the description matches the specimen. Many families, such as Brassicaceae (Cruciferae) and Apiaceae (Umbelliferae), require mature fruits as well as flowers, stems and leaves to determine the species.
7. Line drawings are marked with scale bars:
   (|) indicates one centimetre;
   (I) indicates one millimetre.

NOTE: *The keys may include several genera that are not described or illustrated in the text, but that are found in Appendix 2.*

USE OF SCIENTIFIC NOMENCLATURE

In botanical publications, the scientific names of the plants are given and another name, that of the author (or authority), is added. This refers to the botanist who named the plant. For example, the tufted daisy is known as *Erigeron caespitosus* Nuttall, because the plant was named by Thomas Nuttall. He decided that the plant was a new species belonging to the genus *Erigeron*, and as it was tufted, he gave it the species name *caespitosus* (Latin for "tufted"). Authors' names are usually abbreviated, so textbooks give the name as *Erigeron caespitosus* Nutt. Sometimes two authors' names are given; this means that two botanists have collaborated to document the traits of the new species and have named it jointly. For example, if the initials T. and G. appear after a plant name, it means that the botanists John Torrey and Asa Gray have worked together to name it. The initial L. appears after a very large number of plant names. The L. stands for Linnaeus, the father of plant taxonomy.

MAJOR GROUPS OF VASCULAR PLANTS DESCRIBED IN THIS BOOK

Division	*Class*	*English Name*	*Group Name*
PTERIDOPHYTA	**Lycopsida**	Club-mosses	Pteridophytes
	Sphenopsida	Horsetails	(Ferns & Fern Allies)
	Pteropsida	Ferns	
SPERMATOPHYTA	**Gymnospermae** (Coniferopsida)	Conifers	Gymnosperms (Conifers)
	Angiospermae		Flowering Plants (Dicots & Monocots)
	Subclass		
	Dicotyledoneae (Magnoliopsida) (2 seed-leaves)	Dicotyledons	
	Monocotyledoneae (Liliopsida) (1 seed-leaf)	Monocotyledons	

KEY TO THE MAJOR GROUPS

1. Plants reproducing by spores (no seeds), mostly herbaceous. . . . . . . Ferns & Fern Allies (see Non-flowering Plants, p. 1)
   Plants reproducing by seeds; trees, shrubs and herbs. 2
2. Flowers absent; seeds in dry or berry-like cones; leaves resinous, usually needle- or scale-like . . . . Conifers (see Non-flowering Plants, p. 1)
   Flowers present; seeds inside fruits; leaves not resinous. 3
3. Leaves with parallel venation; flower parts in multiples of 3 . . . . Monocots
   Leaves with net venation; flower parts in multiples of 4 or 5 . . . . Dicots

LIFE CYCLE OF A CLUB-MOSS (*LYCOPODIUM*)

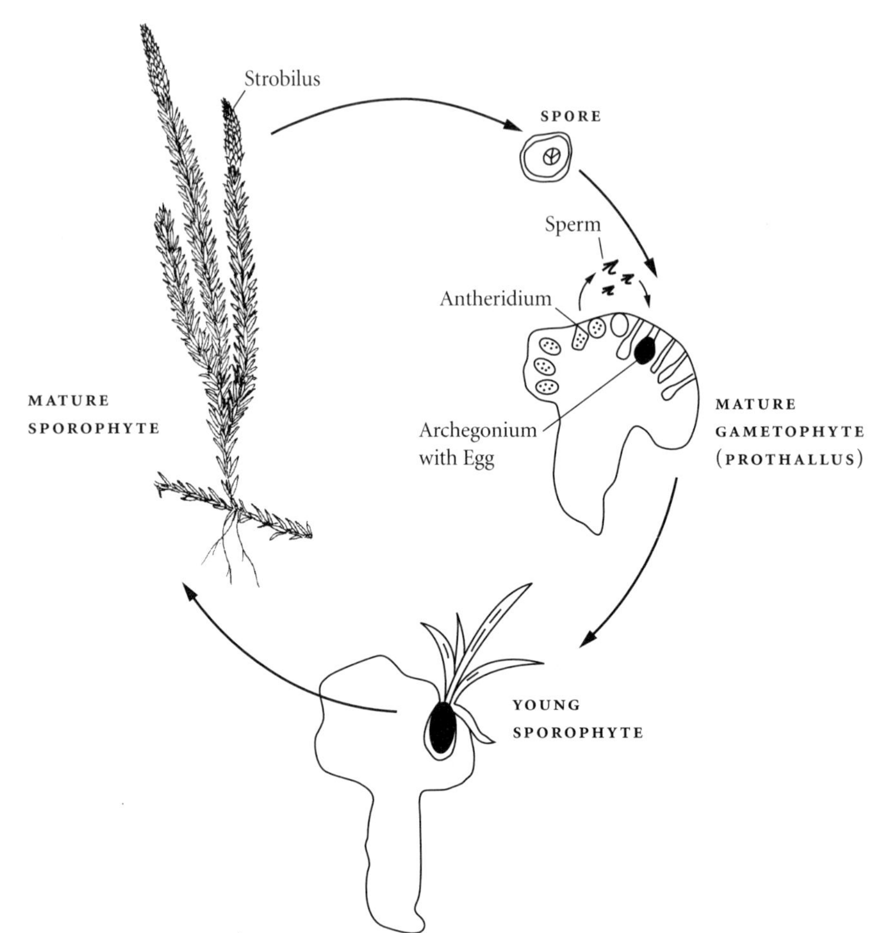

LIFE CYCLE OF A HORSETAIL *(EQUISETUM)*

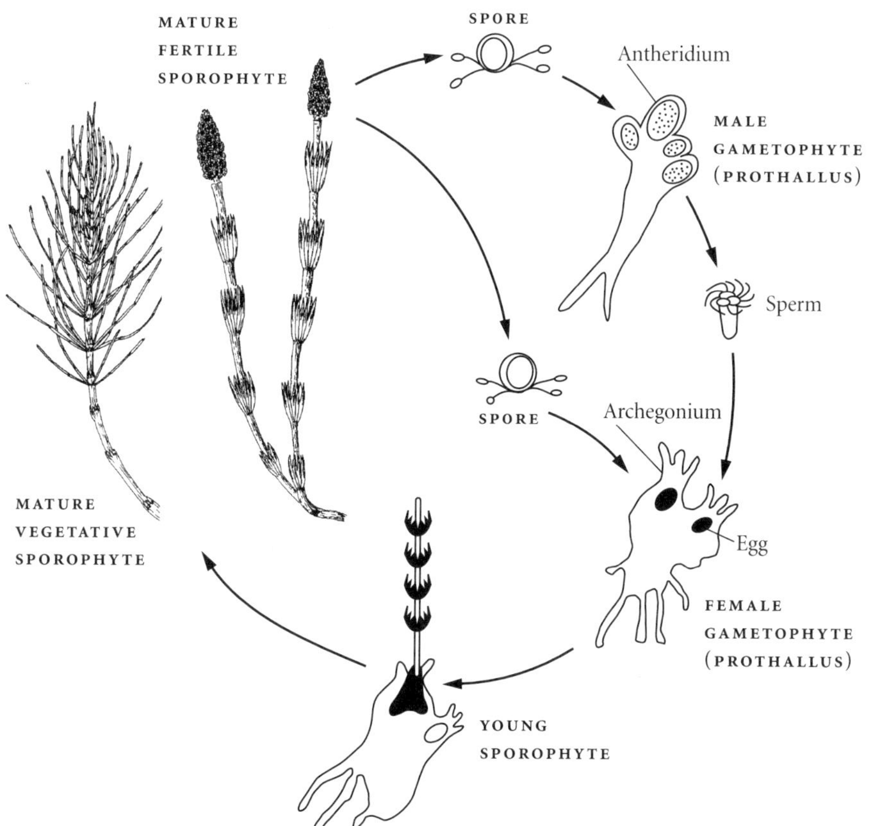

LIFE CYCLE OF A FERN

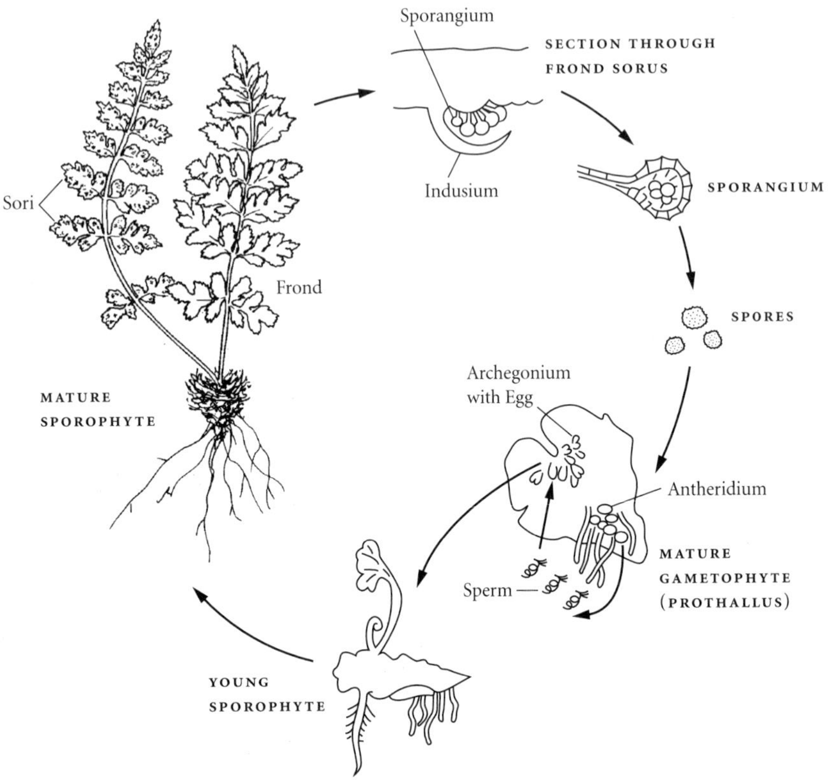

LIFE CYCLE OF A CONIFER

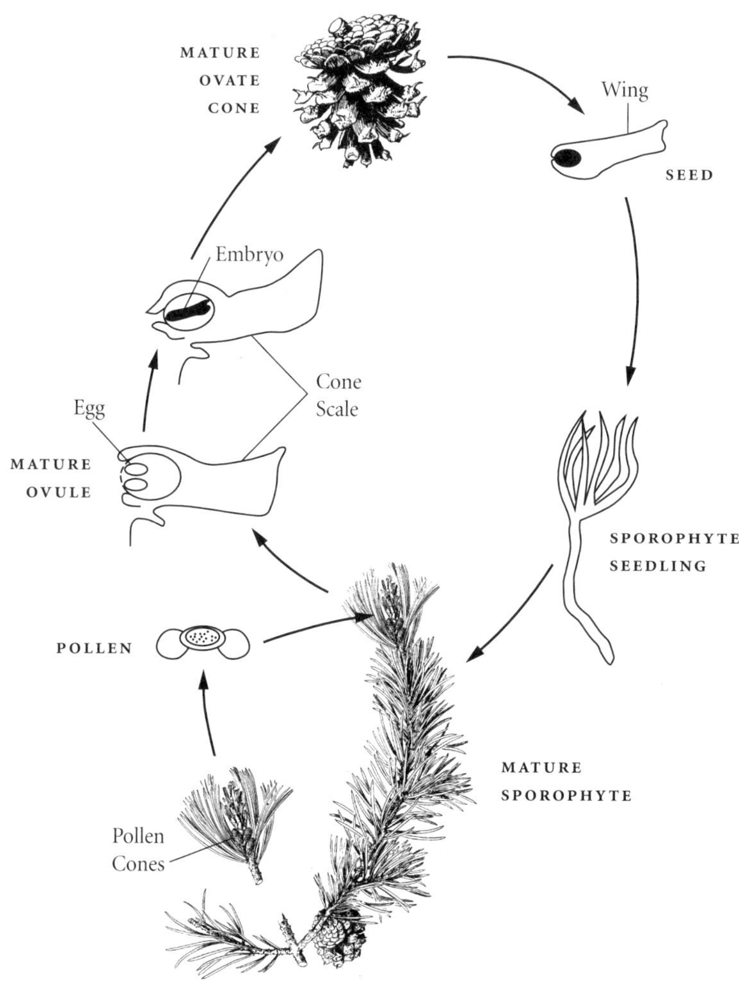

LIFE CYCLE OF A FLOWERING PLANT

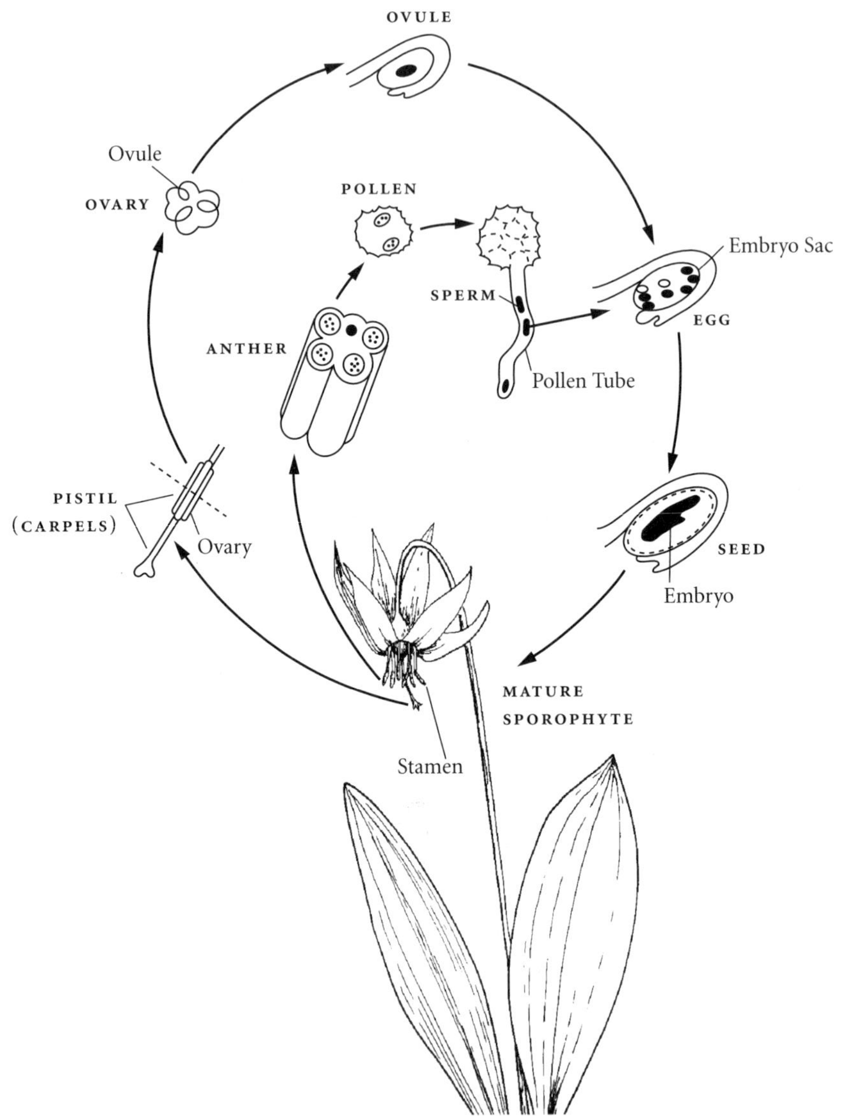

Plants of Special Interest

INSECTIVOROUS PLANTS

Insectivorous plants entrap and digest insects to supplement their nutrition. These plants grow in habitats such as acid swamps, calcareous springs and bogs that are often deficient in essential nutrients such as nitrogen. Nitrogen is obtained from the protein in the bodies of the prey; the passage of this element into the plant tissues has been confirmed by experiments with radioisotope tracers.

Only the sundews, butterworts and bladderworts will be considered here, as these are the only insectivorous plants found in Kananaskis Country.

Long-leaved or oblong-leaved sundew (*Drosera anglica* Huds.) is a striking plant. The leaves form a prostrate rosette, and the leaf-blades are covered with reddish, sticky glands or "tentacles" which entrap insects. The white flowers are not very conspicuous, but the reddish leaf-blades with the tentacles, each with a "dew drop" of sticky exudate, are quite eye-catching. Long-leaved sundew is described and illustrated on pages 130–131.

Butterworts (*Pinguicula vulgaris* L.) have evolved a different method of trapping insects. The pointed, broadly lanceolate leaves form a rosette; they have no leaf-stalks, and their surfaces are covered with microscopic glands which produce a sticky exudate. The edges of the leaves are inrolled. Small insects land on the leaves, mire down in the glandular secretions and are digested. Butterwort is illustrated and described on page 232.

The leaves of common bladderwort (*Utricularia vulgaris* L.) bear complicated insect-traps, completely different from the sticky traps of sundew and butterwort. Bladderwort is an aquatic plant with feathery, finely dissected leaves, and many of these underwater leaflets have been modified into bladders or utricles (see page 233). The bladder has a "door" (a layer of mucilage) with sensitive hairs attached to it. Water is drawn from the inside of the bladder, creating a negative pressure. (The glands probably absorb the water; see Stage 1 below.) When an insect touches the sensitive hairs, the door opens and the negative pressure causes an inrush of water which carries the insect in with it. A small structure called a *velum* attaches the door to the bladder wall and probably controls the opening and closing of the door; it takes 1/460 of a second to open. After a period of 15 to 30 minutes, the "trap" is reset. The trapped insect is digested after several days and its nutrients absorbed. The glands are believed to produce the digestive juices.

STAGE 1

Suction-force develops inside bladder. Prey touches sensitive hairs.

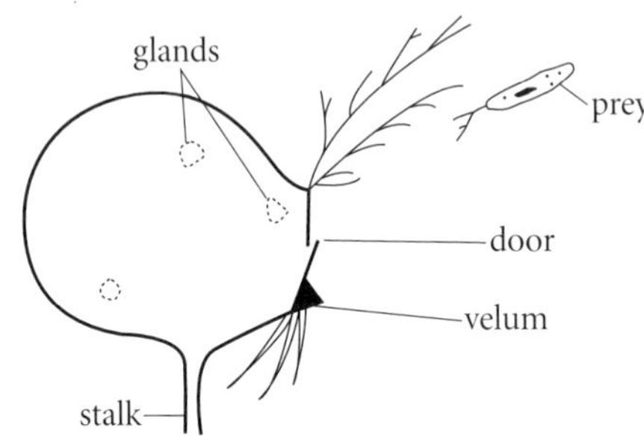

STAGE 2

The trap is sprung. Door opens. Water rushes in, carrying the prey.

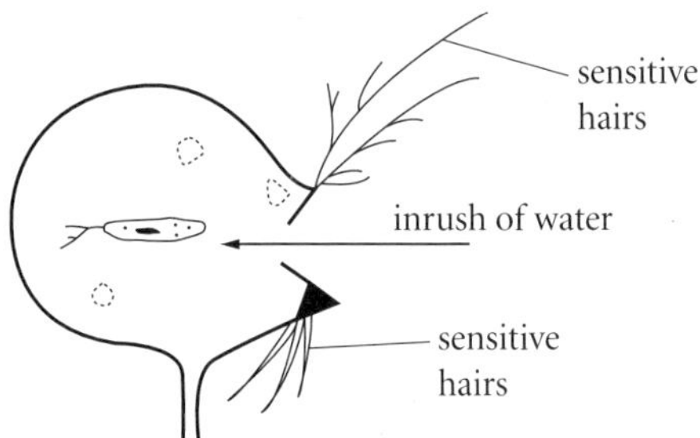

STAGE 3

Door closes. Trap is reset. Digestion of prey takes place.

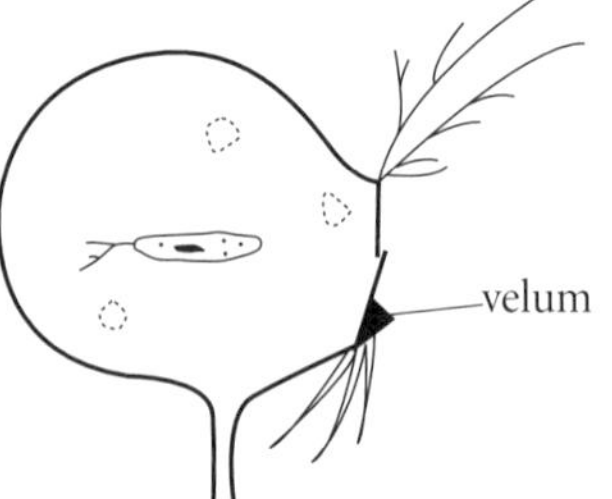

PARASITIC PLANTS

These plants obtain all their nourishment from another living organism; they possess no chlorophyll and cannot make their own food. Two parasitic plants grow in Kananaskis Country: clustered broomrape (*Orobanche fasciculata*) and one-flowered broomrape (*Orobanche uniflora*). Both belong to the family Orobanchaceae. These plants are fleshy and their leaves are reduced to scale-leaves, as they have lost their photosynthetic function. The plants attach to the roots of various host-plants by means of suckers (*haustoria*), by which they obtain water and nutrients. Both species are described and illustrated on pages 230–231.

Pine dwarf mistletoe (*Arceuthobium americanum*) is not a complete parasite like the broomrapes. The stem is yellowish green and contains a certain amount of chlorophyll with which to carry on photosynthesis, but the leaves are reduced to tiny scales. The plant is parasitic upon the stems of lodgepole pine (*Pinus contorta*). When it infects a tree, it induces the growth of large, unruly masses of branches, called "witches' brooms." Single sticky seeds are violently expelled when the fruit is ripe and become imbedded in the branches, causing a fresh infestation. This plant is described and illustrated on page 90.

SEMI-PARASITIC PLANTS

The paintbrushes, *Penstemon* spp, are semi-parasites. They look like normal plants, with green leaves, but they attach to the roots of adjacent plants by means of suckers and are probably never completely independent. Paintbrushes belong to the family Scrophulariaceae and are described and illustrated on pages 220–222. Other semi-parasites belonging to this family are *Orthocarpus*, *Pedicularis* and *Rhinanthus*, described and illustrated on pages 223–225 and 228.

Pale comandra (*Comandra umbellata* Nutt.) is another semi-parasite. Its rootstocks bear roots with sucker-like organs that attach to the roots of neighbouring plants and obtain nourishment from them. Related *Geocaulon lividum* is also a semi-parasite. Both plants belong to the sandalwood family, Santalaceae, and are described and illustrated on pages 89–90.

SAPROPHYTIC PLANTS

These plants lack chlorophyll and obtain their nourishment from non-living organic matter with the aid of mycorrhizal fungi. Indian pipe (*Monotropa uniflora* L.) is an example of a saprophytic plant. The plant belongs to the Indian pipe family, Monotropaceae, and is described and illustrated on page 189.

The coral-root orchids (*Corallorhiza* spp.) are also saprophytic. They obtain food and water from decaying humus with the aid of mycorrhizal fungi. All orchids have a relationship with such fungi, but the coral-root orchids have become dependent on them (although *C. trifida* also contains some chlorophyll). *Corallorhiza* spp. are described and illustrated on pages 66–67.

AN ENDEMIC PLANT

Kananaskis whitlow grass (*Draba kananaskis* G.E. Mulligan) is an endemic plant, that is, a rare plant that is restricted geographically. The species was recognized in 1970 and was described as a new species by Gerald Mulligan in the *Canadian Journal of Botany* 48: 1897-1898. The *type* location was described as "Near Snow Ridge Ski Resort, 21 miles South of Highway 1 (the Trans-Canada) on the Kananaskis-Coleman Rd, on the Eastern slope of the Kananaskis Range, about 50°48'N, 115°12'W, altitude 7250'." Chromosome number *n*=32, G.A. & D.G. Mulligan, 3477 August 8, 1969. Holotype DAO. (Snow Ridge Ski Resort is now Fortress Mountain Ski Resort and the Kananaskis-Coleman Rd is now Highway 40.)

Kananaskis whitlow grass is a member of the mustard family, Brassicaceae (formerly Cruciferae). Brassicaceae is a large family, with some 3000 species, and *Draba* is a large genus with about 300 species; many of these are found in alpine "rock gardens."

There are 23 species of *Draba* in Alberta, the only province in Canada in which *D. kananaskis* is known to occur. The combination of characteristics that separates this species from other *Draba* species in Kananaskis Country are its yellow flowers, 1 or 2 stem leaves, loosely tufted basal rosette, and nearly sessile cruciform hairs on the surface of the basal leaves. It is similar to *D. longipes*, but that species has white petals. *D. kananaskis* has also been recorded from the upper Evan-Thomas Valley in Kananaskis Country by Peter Lee, 1980 and was found in 1985 by Achuff and Corns at a site near Maligne Lake in Jasper National Park, 250 km northwest of the Kananaskis Range.

Plants & Habitats

Representative Species and Landscapes

1.

2.

3.

4.

5.

6.

1. CLUB-MOSS FAMILY (Lycopodiaceae) STIFF CLUB-MOSS (*Lycopodium annotinum*) *p. 3*

2. LITTLE CLUB-MOSS FAMILY (Selaginellaceae) LITTLE CLUB-MOSS (*Selaginella densa*) *p. 4*

3. HORSETAIL FAMILY (Equisetaceae) WOODLAND HORSETAIL *Equisetum sylvaticum* *p. 5*

4. ADDER'S-TONGUE FAMILY (Ophioglossaceae) VIRGINIA GRAPEFERN *Botrychium virginianum* *p. 7*

5. FERN FAMILY (Polypodiaceae) FRAGILE BLADDER FERN (*Cystopteris fragilis*) *p. 7*

6. CYPRESS FAMILY (Cupressaceae) GROUND JUNIPER (*Juniperus communis*) *p. 12*

7. **PINE FAMILY**
(Pinaceae)
WHITE SPRUCE
(*Picea glauca*)
p. 14

8. **CATTAIL FAMILY**
(Typhaceae)
COMMON CATTAIL
(*Typha latifolia*)
p. 26

9. **BUR-REED FAMILY**
(Sparganiaceae)
NARROW-LEAVED BUR-REED
(*Sparganium angustifolium*)
p. 27

10. **PONDWEED FAMILY**
(Potamogetonaceae)
CLASPING-LEAF PONDWEED
(*Potamogeton richardsonii*)
p. 28

11. **ARROW-GRASS FAMILY**
(Juncaginaceae)
SEASIDE ARROW-GRASS
(*Triglochin maritima*)
p. 29

12. **WATER-PLANTAIN FAMILY**
(Alismataceae)
ARROWHEAD
(*Sagittaria cuneata*)
p. 30

7.

8.

9.

10.

11.

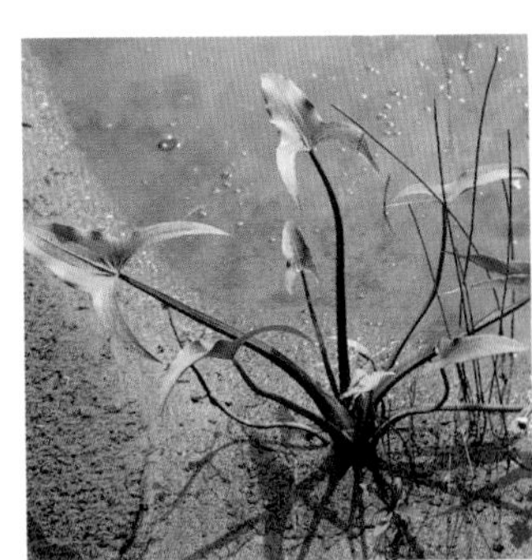

12.

PHOTO CREDITS
All photos by C.C. Chinnappa except:
P.K. Anderson: H6
D. Balzer: 28
G. Crawford: 69
P. Eveleigh: 29
J.O. Hrapko: 46, 62
Kananaskis Country, Alberta Parks Protection: H1, H3, H4, H5
E.J. Kuhn: 68
I.D. Macdonald: 60
M. O'Shea: 9, 10, 50
H.L. Pidgeon: 26, 47
J. Posey: 4
R.P. Pharis: 31
B.G. Warner: 15, 25
C. Wershler: 12, 44, 52

13.

14.

15.

16.

17.

18.

13. **GRASS FAMILY** (Poaceae) **WATER FOXTAIL** (*Alopecurus aequalis*) *p. 32*

14. **SEDGE FAMILY** (Cyperaceae) **AWNED SEDGE** *Carex atherodes* *p. 47*

15. **DUCKWEED FAMILY** (Lemnaceae) **COMMON DUCKWEED** (*Lemna minor*) *p. 52*

16. **RUSH FAMILY** (Juncaceae) **WIRE RUSH** (*Juncus balticus*) *p. 53*

17. **LILY FAMILY** (Liliaceae) **GLACIER LILY** (*Erythronium grandiflorum*) *p. 55*

18. **IRIS FAMILY** (Iridaceae) **COMMON BLUE-EYED GRASS** (*Sisyrinchium montanum*) *p. 63*

19. ORCHID FAMILY (Orchidaceae) YELLOW LADY'S-SLIPPER (*Cypripedium calceolus*) *p. 64*

20. WILLOW FAMILY (Salicaceae) BEBB'S WILLOW (*Salix bebbiana*) *p. 81*

21. BIRCH FAMILY (Betulaceae) WATER BIRCH (*Betula occidentalis*) *p. 85*

22. NETTLE FAMILY (Urticaceae) COMMON NETTLE (*Urtica dioica*) *p. 87*

23. SANDALWOOD FAMILY (Santalaceae) BASTARD TOADFLAX (*Comandra umbellata*) *p. 88*

24. MISTLETOE FAMILY (Loranthaceae) DWARF MISTLETOE (*Arceuthobium americanum*) *p. 90*

19.

20.

21.

22.

23.

24.

25.

26.

27.

28.

29.

30.

25. **BUCK-WHEAT FAMILY**
(Polygonaceae)
WATER SMARTWEED
Polygonum amphibium
p. 91

26. **GOOSEFOOT FAMILY**
(Chenopodiaceae)
STRAWBERRY BLITE
Chenopodium capitatum
p. 97

27. **AMARANTH FAMILY**
(Amaranthaceae)
RED-ROOT PIGWEED
(*Amaranthus retroflexus*)
p. 99

28. **PURSLANE FAMILY**
(Portulacaceae)
ALPINE SPRING BEAUTY
(*Claytonia megarhiza*)
p. 100

29. **PINK FAMILY**
(Caryophyllaceae)
MOSS CAMPION
(*Silene acaulis*)
p. 101

30. **BUTTERCUP FAMILY**
(Ranunculaceae)
TALL BUTTERCUP
(*Ranunculus acris*)
p. 108

31. **POPPY FAMILY** (Papaveraceae) **ALPINE POPPY** (*Papaver kluanensis*) *p. 118*

32. **FUMITORY FAMILY** (Fumariaceae) **GOLDEN CORYDALIS** (*Corydalis aurea*) *p. 119*

33. **MUSTARD FAMILY** (Brassicaceae) **GOLDEN DRABA** (*Draba aurea*) *p. 121*

34. **STONECROP FAMILY** (Crassulaceae) **ROSEROOT** (*Tolmachevia integrifolia*) *p. 131*

35. **SAXIFRAGE FAMILY** (Saxifragaceae) **TUFTED SAXIFRAGE** (*Saxifraga caespitosa*) *p. 132*

36. **GRASS-OF-PARNASSUS FAMILY** (Parnassiaceae) **NORTHERN GRASS-OF-PARNASSUS** (*Parnassia palustris*) *p. 138*

31.

32.

33.

34.

35.

36.

37.

38.

39.

40.

41.

42.

37. CURRENT/ GOOSEBERRY FAMILY (Grossulariaceae) GOLDEN CURRANT (*Ribes aureum*) *p. 139*

38. ROSE FAMILY (Rosaceae) WHITE DRYAD (*Dryas octopetala*) *p. 141*

39. PEA FAMILY (Fabaceae) SHOWY LOCOWEED (*Oxytropis splendens*) *p. 154*

40. GERANIUM FAMILY (Geraniaceae) STICKY PURPLE GERANIUM (*Geranium viscosissimum*) *p. 164*

41. FLAX FAMILY (Linaceae) WILD BLUE FLAX (*Linum lewisii*) *p. 165*

42. MILKWORT FAMILY (Polygalaceae) SENECA SNAKEROOT (*Polygala senega*) *p. 165*

43. **SPURGE FAMILY** (Euphorbiaceae) **LEAFY SPURGE** (*Euphorbia esula*) *p. 166*

44. **CROWBERRY FAMILY** (Empetraceae) **CROWBERRY** (*Empetrum nigrum*) *p. 168*

45. **MAPLE FAMILY** (Aceraceae) **MOUNTAIN MAPLE** (*Acer glabrum*) *p. 169*

46. **MALLOW FAMILY** (Malvaceae) **SCARLET MALLOW** (*Sphaeralcea coccinea*) *p. 169*

47. **VIOLET FAMILY** (Violaceae) **BOG VIOLET** (*Viola nephrophylla*) *p. 170*

48. **OLEASTER FAMILY** (Elaeagnaceae) **WOLF WILLOW** (*Elaeagnus commutata*) *p. 172*

43.

44.

45.

46.

47.

48.

49.

50.

51.

52.

53.

54.

49. EVENING PRIMROSE FAMILY (Onagraceae) GREAT WILLOWHERB (*Epilobium angustifolium*) *p. 173*

50. WATER-MILFOIL FAMILY (Haloragaceae) SPIKED WATER-MILFOIL (*Myriophyllum exalbescens*) *p. 177*

51. MARE'S-TAIL FAMILY (Hippuridaceae) COMMON MARE'S-TAIL (*Hippuris vulgaris*) *p. 177*

52. GINSENG FAMILY (Araliaceae) WILD SARSAPARILLA (*Aralia nudicaulis*) *p. 178*

53. CARROT FAMILY (Apiaceae) HEART-LEAVED ALEXANDERS (*Zizia aptera*) *p. 179*

54. DOGWOOD FAMILY (Cornaceae) BUNCHBERRY (*Cornus canadensis*) *p. 184*

55. **WINTER-GREEN FAMILY** (Pyrolaceae) **ONE-SIDED WINTERGREEN** (*Orthilia secunda*) *p. 185*

56. **INDIAN-PIPE FAMILY** (Monotropaceae) **INDIAN PIPE** (*Monotropa uniflora*) *p. 188*

57. **HEATH FAMILY** (Ericaceae) **COMMON BEARBERRY** (*Arctostaphylos uva-ursi*) *p. 189*

58. **PRIMROSE FAMILY** (Primulaceae) **MOUNTAIN SHOOTING STAR** (*Dodecatheon conjugens*) *p. 197*

59. **GENTIAN FAMILY** (Gentianaceae) **FOUR-PARTED GENTIAN** (*Gentianella propinqua*) *p. 200*

60. **BUCKBEAN FAMILY** (Menyanthaceae) **BUCKBEAN** (*Menyanthes trifoliata*) *p. 202*

55.

56.

57.

58.

59.

60.

61.

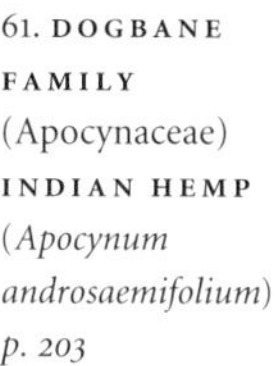

62.

63.

64.

65.

66.

61. **DOGBANE FAMILY** (Apocynaceae) **INDIAN HEMP** (*Apocynum androsaemifolium*) *p. 203*

62. **MILKWEED FAMILY** (Asclepiadaceae) **SHOWY MILKWEED** (*Asclepias speciosa*) *p. 204*

63. **PHLOX FAMILY** (Polemoniaceae) **SHOWY JACOB'S-LADDER** (*Polemonium pulcherrimum*) *p. 205*

64. **WATERLEAF FAMILY** (Hydrophyllaceae) **MOUNTAIN PHACELIA** (*Phacelia sericea*) *p. 209*

65. **BORAGE FAMILY** (Boraginaceae) **TALL MERTENSIA** (*Mertensia paniculata*) *p. 211*

66. **MINT FAMILY** (Lamiaceae) **WILD BERGAMOT** (*Monarda fistulosa var. menthaefolia*) *p. 216*

67. **FIGWORT FAMILY**
(Scrophulariaceae)
YELLOW BEARDTONGUE
(*Penstemon eriantherus*)
p. 219

67.

68. **BROOM-RAPE FAMILY**
(Orobanchaceae)
ONE-FLOWERED CANCER-ROOT
(*Orobanche uniflora*)
p. 230

68.

69. **BLADDER-WORT FAMILY**
(Lentibulariaceae)
COMMON BUTTERWORT
(*Pinguicula vulgaris*)
p. 232

69.

70. **PLANTAIN FAMILY**
(Plantaginaceae)
COMMON PLANTAIN
(*Plantago major*)
p. 233

70.

71. **MADDER FAMILY**
(Rubiaceae)
NORTHERN BEDSTRAW
(*Galium boreale*)
p. 234

71.

72. **HONEY-SUCKLE FAMILY**
(Caprifoliaceae)
TWINFLOWER
(*Linnaea borealis*)
p. 236

72.

73.

74.

75.

76.

77.

73. **VALERIAN FAMILY** (Valerianaceae) **NORTHERN VALERIAN** *(Valeriana sitchensis)* *p. 241*

74. **TEASEL FAMILY** (Dipsacaceae) **BLUE BUTTONS** *(Knautia arvensis)* *p. 242*

75. **BLUEBELL FAMILY** (Campanulaceae) **HAREBELL** *(Campanula rotundifolia)* *p. 243*

76. **COMPOSITE FAMILY** (Asteraceae) **PARRY'S TOWNSENDIA** (*Townsendia parryi*) *p. 245*

77. *Draba kananaskis* A rare endemic plant growing on the scree slopes of Fortress Mountain *p. XLVIII, 126*

H1. Grassland north of Bragg Creek

H1.

H2. Sibbald Creek, wetland habitat (bladderwort in bloom)

H2.

H3. Sibbald Flats, Moose Mountain

H3.

H4.

H4. Highwood River Valley, lower subalpine habitat

H5.

H5. Hogarth Lake and Mount Chester

H6.

H6. Marmot Creek Basin, timberline (spruce and larch)

H7. View from Hailstone Butte towards the foothills and Chain Lakes

H7.

H8. Hailstone Butte, alpine habitat

H8.

H9. Plateau Mountain, alpine tundra habitat, fellfield

H9.

Non-flowering Plants

FERNS & FERN ALLIES

CONIFERS

Ferns & Fern Allies

Ferns and fern allies include the club-mosses, horsetails and ferns. They are herbaceous plants that reproduce by spores, which develop inside sporangia.

KEY TO FERNS AND FERN ALLIES

1. Leaves small, usually scale-like 2
   Leaves broad, never scale-like.......................... 4
2. Stems jointed and hollow .... Equisetaceae
   Stems not jointed 3
3. Stems less than 6 mm wide; plants less than 3 cm tall ....
   .... Selaginellaceae
   Stems more than 7 mm wide; plants taller than 4 cm........
   .... Lycopodiaceae
4. Sporangia (spore sacs) produced on leaf underside.........
   .... Polypodiaceae
   Sporangia produced in terminal clusters on a specialized stalk
   .... Ophioglossaceae

Lycopodiaceae CLUB-MOSS FAMILY

Club-mosses are low, evergreen, moss-like plants, often with prostrate stems trailing over the ground or spreading by rhizomes, with adventitious roots. Branched shoots usually arise from the prostrate stems, and the branches are usually dichotomous (forked). The small leaves are usually linear or lanceolate and are overlapping. The reproductive organs (*sporangia*) are produced in club-shaped spikes (*strobili*), hence the common name "club-moss." The sporangia burst, releasing minute, yellow spores which develop into green *prothalli* (gametophytes) with sexual organs. When the female eggs are fertilized, new plants are formed: sporophytes. *(See diagram of the Life Cycle, page XL.)*

Leaves in whorls of 4; appressed .... *L. annotinum*
Leaves in whorls of 6 or more; spreading .... *L. alpinum*

Lycopodium alpinum L.

ALPINE CLUB-MOSS

The ascending branches arise from underground stems (rhizomes) and are repeatedly forked; the leaves of the fertile branchlets are appressed and 6-ranked. The fertile branchlets bear 1 to 3 strobili (cones), club-like in shape; the spores are yellow and powdery. This club-moss is found on alpine slopes and in alpine meadows.

Lycopodium annotinum L.

STIFF CLUB-MOSS

This species can be distinguished from the alpine club-moss because the branches are not repeatedly forked. They have 2 or 3 side-branches only. The leaves are quite different because they are not appressed—they stand out at an angle. The habitats of the two plants are also different: the stiff club-moss is found in damp woods.

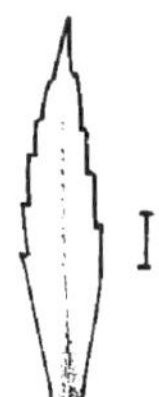

Selaginellaceae LITTLE CLUB-MOSS FAMILY

The plants belonging to Selaginellaceae are similar to those of Lycopodiaceae, but the leafy shoots in the latter are mostly 7 to 15 mm wide, while those of Selaginellaceae are less than 6 mm wide. Selaginellaceae plants are heterosporous (with 2 kinds of spores, megaspores and microspores, produced inside megasporangia and microsporangia), while those of Lycopodiaceae are homosporous, with only 1 kind of spore; this distinction is very important. The sporangia are found in the strobili (cones), which arise from the branches. The sporangia-bearing leaves in the cone are called *sporophylls.* The megaspores produce prothalli that bear female organs, and the microspores produce prothalli that bear male organs. When the eggs are fertilized, new plants are formed. *(See diagram of the Life Cycle, page XL.)*

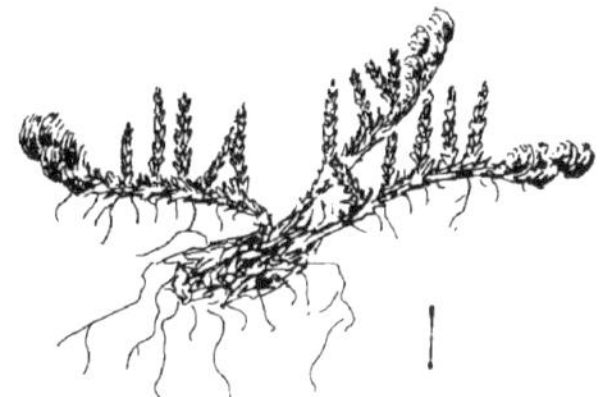

Selaginella densa Rydb.

LITTLE CLUB-MOSS, PRAIRIE SELAGINELLA

The little club-moss gets its common name because it looks like a moss, and the strobili (cones) are like elongated clubs. It is a small perennial plant, mat-forming, and the tiny leaves each end in a white bristle. The sporophylls of the cone are also tipped with a bristle. The little club-moss is a plant of the dry prairies and open sandhills, but it can also be found in alpine areas. NOTES: *In Kananaskis Country, little club-moss occurs near the summit of Plateau Mountain and elsewhere.*

Equisetaceae HORSETAIL FAMILY

The horsetails are rush-like plants, found in damp places, arising from rhizomes or tubers. The aerial stems are jointed, with scale-like leaves at the nodes; these are whorled and fused laterally into a sheath. When branches occur, they are in whorls from the nodes—a distinctive feature. The terminal cones (strobili) bear the sporangia, with green spores; the spores germinate into prothalli, bearing the sexual organs.

1. Stems of 1 type; mostly unbranched 2
   Stems of 2 types; fertile stems brown, appearing in early spring; vegetative stems green, with numerous branches. .... *E. arvense*
2. Stems tufted and often zigzagged, less than 16 cm tall, less than 2 mm thick. . ... *E. scirpoides*
   Stems straight, generally taller than 16 cm, more than 2 mm thick. .... *E. hyemale*

Equisetum arvense L.

COMMON HORSETAIL, FIELD HORSETAIL

Common horsetail is a plant with creeping rhizomes bearing fertile stems in early spring; these are erect, bearing terminal cones, enclosing sporangia with spores. The sterile shoots are green, slender, ridged and branched—the numerous branches are in dense whorls, and the sheaths have 12 sharp teeth. Common horsetail is found in numerous habitats: banks, roadsides and meadows.

NOTES: *This plant often grows as a weed in cultivated fields.* *A related species, woodland horsetail* (E. sylvaticum), *is easily distinguished from common horsetail. The branches of* E. sylvaticum *branch repeatedly, and the stems are rough to the touch.*

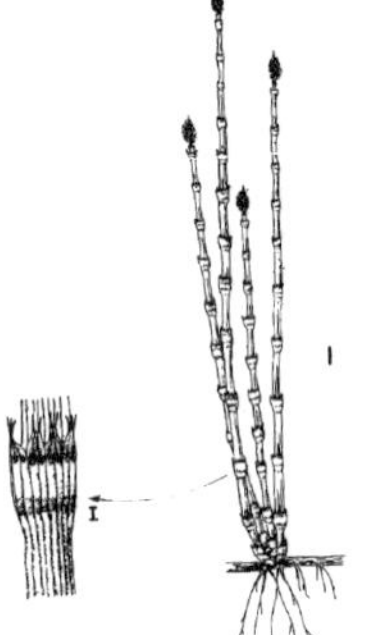

Equisetum hyemale L.

COMMON SCOURING RUSH

Scouring rush has a slender rhizome, blackish, with stems mostly unbranched, so it differs from the common horsetail. The stems are rough, with 18 to 40 ridges with siliceous cross-bands, and the sheaths have 2 distinct black bands (common horsetail has only 1).

NOTES: *Scouring rush is not a true rush.*

Equisetum scirpoides Michx.

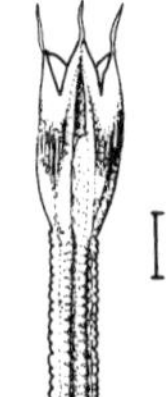

DWARF SCOURING RUSH

This species can be distinguished from the other horsetails because its stems are very slender and the strobili and sheaths are smaller. The stems are often curved, quite unlike the straight, stiff stems of *E. hyemale*, and they are clustered.

Ophioglossaceae ADDER'S-TONGUE FAMILY

These plants arise from a cluster of fleshy roots, and they consist of a frond (a single sterile blade) and a sporophyll (a fertile blade bearing spores). The frond is usually dissected or compound, and the sporophyll bears clusters of globose sporangia containing yellow spores.

Botrychium lunaria (L.) Sw.

MOONWORT

Moonwort is a perennial, growing from a short rootstock, which gives rise to a simple, erect stalk. This stalk divides into 2 parts, the green sterile blade (leaf), and the fertile sporophyll (a spore-bearing leaf). The sterile leaf has several pinnate lobes, characteristically fan-shaped (flabellate) and often overlapping. The fertile sporophyll bears clusters of large, globose sporangia arranged in double rows. Moonwort is found in open woods and in meadows.

NOTES: *The specific name* lunaria *comes from the Latin word* luna, *"a moon." This probably refers to the broad, fan-shaped lobes of the sterile leaf.* *A related species, Virginia grapefern* (B. virginianum), *has large, highly dissected fronds and is a much larger plant (15-80 cm tall).*

Polypodiaceae FERN FAMILY

These are the true ferns. The plants arise from rhizomes (underground stems) which produce leaf-like fronds, usually bearing the sporangia on their undersurface. The fronds are coiled when in bud, giving rise to the name "fiddleheads." The spores germinate and produce minute, green *prothalli* which bear sexual organs. When the female *oogonia* are fertilized, a new fern plant is produced, and the cycle starts afresh. Some ferns, such as *Cryptogramma*, produce their sporangia in structures that do not look leaf-like, but have fertile pinnules.

NOTE: *Botanical terms specific to ferns can be found in the Glossary, pages 311 to 339.*

LIFE HISTORY OF A FERN

Gametophyte Stage: When the spores germinate, each spore produces a minute green organism called a *prothallus*. The prothallus bears male and female organs (*antheridia* and *archegonia*) and is called the gametophyte, because it produces the gametes.

Sporophyte Stage: When the female gamete (egg) is fertilized by the male gamete, the resulting zygote develops into the mature fern, bearing spores in the sporangia. *(See diagram of the Life Cycle, page XLII.)*

1. Sporangia (spore-sacs) located along leaf-margin .... *Pellaea*
   Sporangia not located along leaf-margin 2
2. Fronds once-pinnate 3
   Fronds 2 to 3-pinnate 5
3. Fronds evergreen .... *Polystichum*
   Fronds dying in autumn 4
4. Sporangia 4 to 8 per pinna; frond stalk (stipe) reddish brown .... *Asplenium*
   Sporangia many per pinna; frond stalk (stipe) yellow, green or brown .... *Woodsia*
5. Ferns growing in moist, wooded areas.......................... *Athyrium*
   Ferns growing on rocky ledges, crevices or slopes 6
6. Veins reaching edge of pinna .... *Cystopteris*
   Veins not reaching edge of pinna .... *Woodsia*

Asplenium viride Huds.

GREEN SPLEENWORT

Green spleenwort is a dainty, attractive little fern, with tufted fronds up to 15 cm long rising from a rhizome. These fronds have 10 to 20 pairs of pinnae, with sori on the undersurfaces. There are 2 to 4 pairs of sori on a pinna, each covered with a delicate indusium. Green spleenwort grows in rock crevices, especially on calcareous rocks.

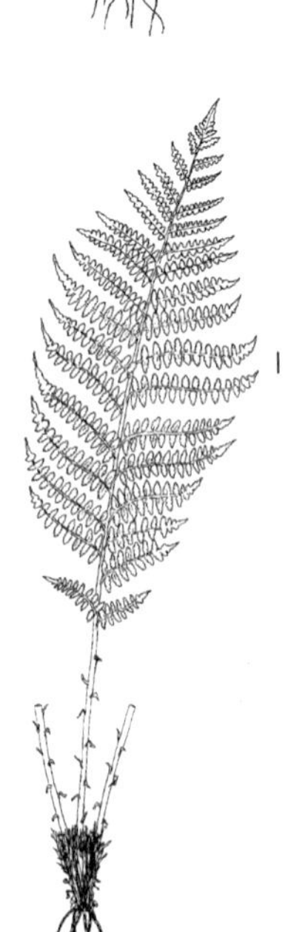

Athyrium filix-femina (L.) Roth.

LADY FERN

Lady fern grows from thick rhizomes covered with scales. The fronds that arise from these rhizomes are large—sometimes up to 2 m; they each bear 20 to 30 pairs of pinnae. The sori are found on the undersurface of the fronds; each sorus consists of a group of sporangia, containing spores, and is covered by an indusium. Most ferns are found in moist places, and the lady fern is no exception: it is found in damp woods and thickets.

Cystopteris fragilis (L.) Bernh.

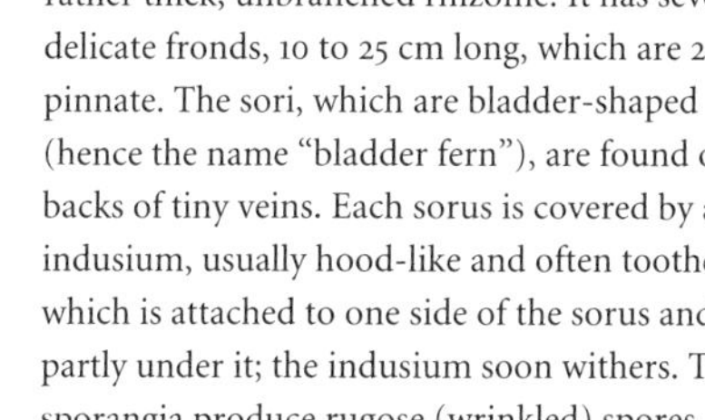

FRAGILE BLADDER FERN

Bladder fern is a dainty plant that grows from a rather thick, unbranched rhizome. It has several delicate fronds, 10 to 25 cm long, which are 2 to 3-pinnate. The sori, which are bladder-shaped (hence the name "bladder fern"), are found on the backs of tiny veins. Each sorus is covered by an indusium, usually hood-like and often toothed, which is attached to one side of the sorus and partly under it; the indusium soon withers. The sporangia produce rugose (wrinkled) spores. Bladder fern is found on moist ledges and slopes, up to 2500 m.

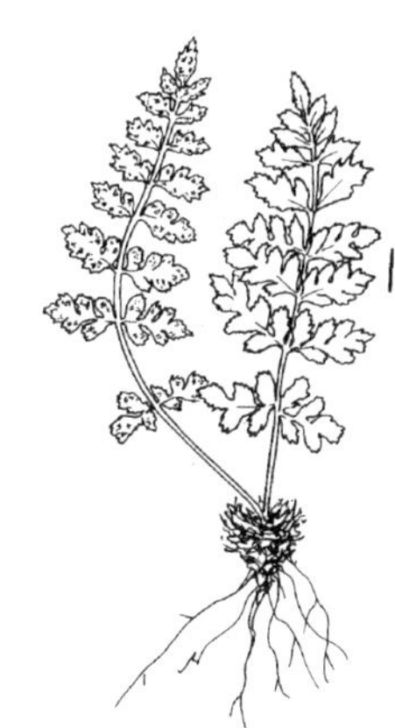

Pellaea Link — CLIFF-BRAKE

Short scaly rhizomes give rise to fronds which are of 2 kinds: those with sporangia in sori (fertile), and those with no sori (sterile). The sori are marginal; the revolute (downward rolled) pinna margin forms a common indusium.

Fronds of 1 type; stipes brown . . . . *P. glabella*
Fronds of 2 types; stipes purplish black . . . . *P. atropurpurea*

Pellaea atropurpurea (L.) Link

PURPLE CLIFF-BRAKE

This fern has a short rhizome, covered with scales, and the fronds are pinnate, stiffly erect. The sori are found on the margins of the pinnae. The stipes (leaf stalks) are slender, 5 to 20 cm long, and are purplish black, hence the species name *atropurpurea.* The plants are found on dry limestone rocks.

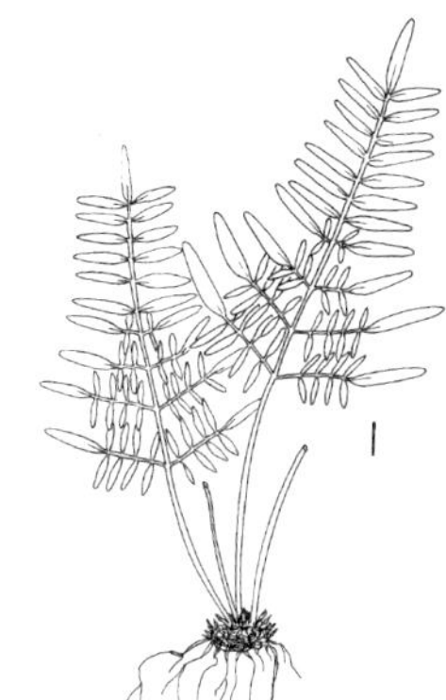

Pellaea glabella Mett. ex Kuhn

SMOOTH CLIFF-BRAKE

P. glabella can be distinguished from *P. atropurpurea* because although both species have fertile fronds with sporangia in sori, the fronds of *P. glabella* are shorter than those of the other species; also, the stipes are brownish, not purplish black. The habitats of both species are similar: both found on dry limestone rocks.

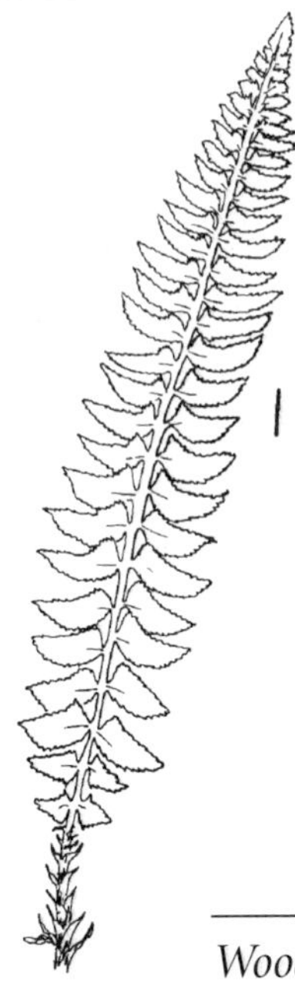

Polystichum lonchitis (L.) Roth

NORTHERN HOLLY FERN

Northern holly fern has a thick, more or less erect rhizome which produces tufted pinnate fronds 1 to 4 dm long; these fronds have short stipes. The pinnae are alternate, 25 to 40 on each side of the rachis. The sori are usually in 2 rows on the upper pinnae, each with the peltate indusium attached to the centre of the sorus. Northern holly fern is found on rocky slopes.

Woodsia R.Br. — WOODSIA

Woodsias are low-growing ferns, with short, ascending rootstocks that give rise to densely tufted fronds, with 1 to 2-pinnate blades. The sori are rounded, each with an indusium attached by its base under the sporangia, open at the top, and often fringed or lobed-edged.

Pinnae have midribs with white hairs . . . . *W. scopulina*
Pinnae have midribs without white hairs . . . . *W. oregana*

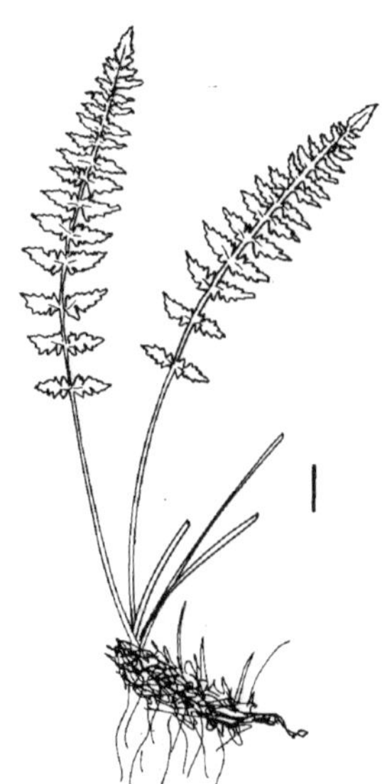

Woodsia oregana D.C. Eat.

OREGON WOODSIA

Oregon woodsia is a low-growing fern with tufted fronds, usually 5.5 cm long and sometimes glandular (this trait distinguishes it from *W. scopulina*). The fronds are pinnate, with the sori on the undersurfaces of the pinnae. These sori are each associated with an indusium. The indusium is attached by its base under the sporangium; it is disc-like and open at the top, with a fringed edge. Oregon woodsia is found on rocky cliffs and ledges.

Woodsia scopulina D.C. Eat.

MOUNTAIN WOODSIA

Mountain woodsia is a low fern with densely tufted fronds, usually 15 cm long. These fronds are pinnate, with the sori on the undersurfaces at the tips of the tiny veins. The veins have long, whitish, jointed hairs which are not found in Oregon woodsia. The sori each have an indusium attached by its base under the sporangia and open at the top, with a fringed edge. Mountain woodsia is found on rocks and in cliff crevices.

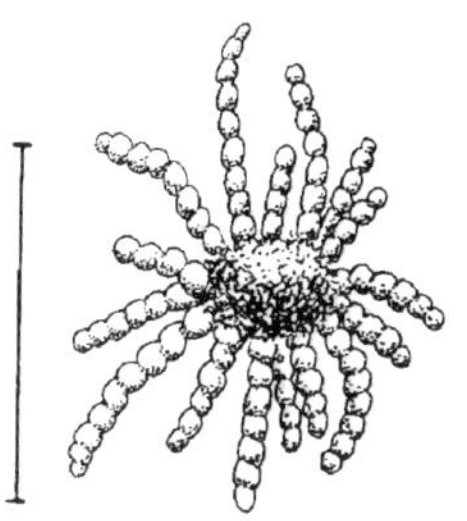

Conifers

KEY TO THE CONIFERS

1. Leaves scale-like, overlapping and pressed against the stem . . . . *Cupressaceae*
   Leaves needle-like. 2
2. Cones small, berry-like, blue. . . . . *Cupressaceae*
   Cones dry and woody . . . . *Pinaceae*

Cupressaceae CYPRESS FAMILY

Members of the cypress family are evergreen trees and shrubs. Their leaves are either needle-like or scale-like and grow either opposite or in whorls. The branches bear cones of 2 kinds: male (staminate) and female (carpellary). The male cones are small and are usually terminal on the branches. They have 2 to 5 pollen sacs, called *microsporangia*, borne on the scales; they produce pollen. The scales of the cone are *peltate*, that is, they are flat and attached to a stalk in from the margin. The female cones, or seed cones, are small and either dry or fleshy. They bear ovules in *megasporangia*. In junipers, when the ovules are fertilized by the pollen, these "fleshy cones" look like berries.

Seeds in a bluish green, berry-like cone; shrubs and small trees . . . . *Juniperus*
Seeds in a dry cone; large trees . . . . *Thuja*

Juniperus L. JUNIPER

Junipers grow as shrubs or small trees. The leaves may be needle-like or scale-like; they may be opposite or in whorls of 3. The pollen-cones and seed-cones may be borne either on the same plant or on separate plants. On the female cone, the scales coalesce to form a resinous, waxy, blue "berry" which has 1 to 6 seeds.

1. Leaves sharp-pointed, whorls of 3 . . . . *J. communis*
   Leaves not sharp-pointed, scale-like, opposite 2
2. Prostrate shrub, branches growing flat on the ground . . . . *J. horizontalis*
   Erect shrub or small tree. . . . . *J. scopulorum*

Juniperus communis L.

GROUND JUNIPER, COMMON JUNIPER

This shrub is often prostrate and forms large clumps, although it sometimes forms a semi-erect shrub 1 m high, with shredding bark. The leaves are sharp-pointed needles with a whitish bloom on the upper surface. Male and female cones are usually on separate plants. The female cones are pale blue and fleshy; they look like berries when ripe and contain 1 to 3 seeds. The male cones are small and look like catkins. Common juniper is found in open woods and dry, rocky, gravelly soil. It is widespread in Alberta.

Juniperus horizontalis Moench

CREEPING JUNIPER

This shrub can be distinguished from ground juniper because the leaves are not needle-like: they are small, oval, pointed and imbricate (overlapping). It is a prostrate shrub and forms large, low mats. The fleshy female cones ("berries") are dark blue with 2 to 6 seeds. Creeping juniper is found on dry, sandy, rocky slopes and on flood plains. It is widespread in Alberta.

NOTES: *In Kananaskis Country, creeping juniper often grows with* J. scopulorum.

Juniperus scopulorum Sarg.

ROCKY MOUNTAIN JUNIPER

Rocky Mountain juniper is a shrub or small tree with erect branches. In Kananaskis Country it grows to a height of 3 to 3.6 m. The branches bear scale-like leaves with sub-acute tips. This juniper can be distinguished from ground juniper because the leaves are not awl-shaped in whorls of 3, and from creeping juniper because the latter species is a prostrate shrub with scale leaves that are strongly apiculate (ending in a sharp point). The "berry" (actually a fleshy cone) of the Rocky Mountain juniper is bright blue with a bloom. This shrub is found in open, rocky areas.

Pinaceae PINE FAMILY

Members of the pine family are either trees or shrubs. They are usually evergreen (the larches are an exception), and they bear resin-ducts. The leaves are needle-shaped and may be spirally arranged, single or in clusters. Pines are typical conifers (i.e., they bear cones). There are two types of cones (strobili): male (staminate) and female (carpellary). The male cones are small and have two pollen sacs, called *microsporangia*, on each cone-scale; these produce pollen. The female cones, or seed-cones, are larger and bear megasporangia with ovules that turn into seeds when fertilized by the pollen. There are 2 seeds, usually winged, on the upper surface of each scale. The scales of both male and female cones are called *sporophylls.*

1. Needles in groups of 2 or more 2
   Needles borne singly on the stem 3
2. Needles in groups of 2 to 5 . . . . *Pinus*
   Needles in groups of 10 or more . . . . *Larix*
3. Needles 4-sided . . . . *Picea*
   Needles flat 4
4. Bark rough and furrowed; cones drooping, with numerous 3-lobed bracts. . . . . *Pseudotsuga*
   Bark smooth, resin blisters present; cones erect, without 3-lobed bracts . . . . *Abies*

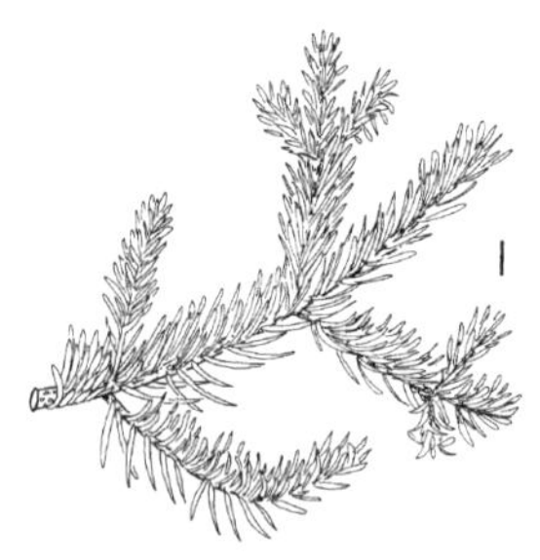

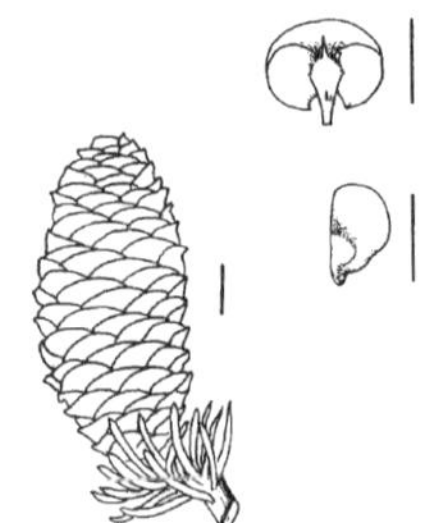

Abies lasiocarpa (L.) Mill.

SUBALPINE FIR

Subalpine fir is a tree that can reach a height of 25 m, but near treeline it often grows as a prostrate shrub. The young bark is grey and smooth, with resin blisters. The leaves are linear and often appear 2-ranked; they are curved upwards with notched or blunt tips. The male and female cones grow near the top of the tree at the tips of branches. The male cones are small and hang from the underside of branches, but the larger female cones are characteristically erect. They are cylindrical, dark purplish brown, and the scales are deciduous; the central axis remains after the scales have fallen. The trees are found in moist woods.

NOTES: *The Blackfoot burned the needles of the subalpine fir as incense at their ceremonies. They used the resin as an antiseptic for wounds.*

Larix Adans. **LARCH**

Larches are trees with rough bark and branches of 2 kinds: long branches of the current year, with a few leaves, and spur-like branches, producing 10 to 40 leaves. The male cones are small, nearly round, and the seed-cones are often reddish. Most conifers are evergreen but the larches are an exception: the leaves turn golden in the autumn and drop off. In Kananaskis Country, the larches produce a lovely show of colour at this season.

Needles 3-sided, in groups of 15 to 30; subalpine elevations . . . . *L. occidentalis*
Needles 4-sided, in groups of 30 to 40; lower elevations. . . . . *L. lyallii*

Larix lyallii Parl.

SUBALPINE LARCH, LYALL'S LARCH

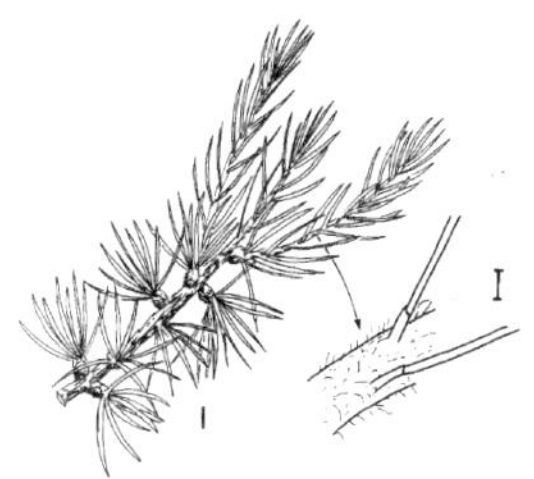

Lyall's larch is a small tree. It can grow to 15 m but is usually only 10 m tall. The needle-like leaves grow in clusters of 30 to 40 on dwarf twigs and are bluish green and 4-sided. The female cones have wavy scales and slender, long-awned bracts; the seeds are winged. The male cones are smaller and yellow. Subalpine larch, as its name implies, grows in the subalpine zone in the mountains; it is usually found on rocky, gravelly soil.

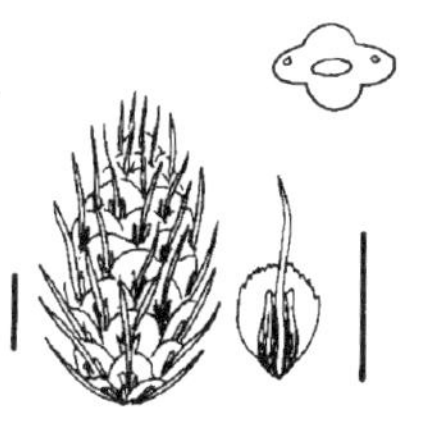

NOTES: *The species name* lyallii *honours David Lyall (1817-1895), a Scottish botanist and surgeon who collected many specimens in North America, particularly while serving with the North American Boundary Commission.*

Larix occidentalis Nutt.

WESTERN LARCH, WESTERN TAMARACK

Western larch can be distinguished from subalpine larch because it is tall (up to 40 m high) and its bark is deeply furrowed. The leaves are 3-angled and the twigs are smooth, while the subalpine larch has 4-angled leaves and its twigs are long-hairy. The cones are also different: those of subalpine larch bear bracts that exceed the cone scales, while those of the western larch do not. Western larch is found on moist mountain slopes and is rare in Alberta.

NOTES: *Western larch is one of the most important timber-producing trees in western Canada.*

Picea A. Dietr. — SPRUCE

Spruces are large evergreens with rough, flaky bark. The leaves are needle-like, 4-sided and spirally arranged; they grow from persistent peg-like bases called *sterigmata.* The seed-cones are often terminal and drooping; they mature the first season.

1. Cones less than 2.5 cm long, remaining on stems for several years; trees up to 10 m tall . . . . *P. mariana*
   Cones more than 2.5 cm long, falling off by the winter; trees over 25 m tall 2
2. Needles less than 1.5 cm long; cone-scales with stiff edges . . . . *P. glauca*
   Needles more than 1.5 cm long; cone-scales with papery edges. . . . . *P. engelmannii*

Picea engelmannii Parry ex Engelm.

ENGELMANN SPRUCE

Engelmann spruce is a tall tree. It grows up to 30 m high and has a spire-like crown and drooping lower branches. At treeline it is often a small shrub. The needle-like leaves are blue-green and 4-angled; they are curved with flattened tips. The female cone has ragged scales, tapered at both ends; they are not woody. The seeds are winged. Male cones are small and yellow. Engelmann spruce is found in subalpine forest and interbreeds with white spruce in areas where their ranges overlap.

NOTES: *The specific name* engelmannii *honours George Engelmann, a 19th century physician and botanist who was born in Germany and became an authority on conifers.*

Picea glauca (Moench) Voss

WHITE SPRUCE

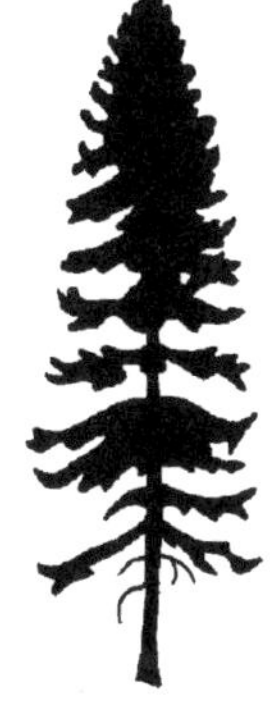

White spruce is a tall tree; it can reach 40 m if growing conditions are favourable. The 4-sided leaves are similar to those of Engelmann spruce; they are sharp-pointed, are spirally arranged and often have a white bloom—hence the specific name *glauca.* Male and female cones are found on the same tree. The female cones are deciduous, with reddish brown cone-scales, and the seeds have 2 thin wings. The male cones are pale red and grow at the ends of branches. White spruce grows on north-facing slopes in the northern forests and at low to middle elevations in the mountains. It is widespread throughout Alberta and Canada.

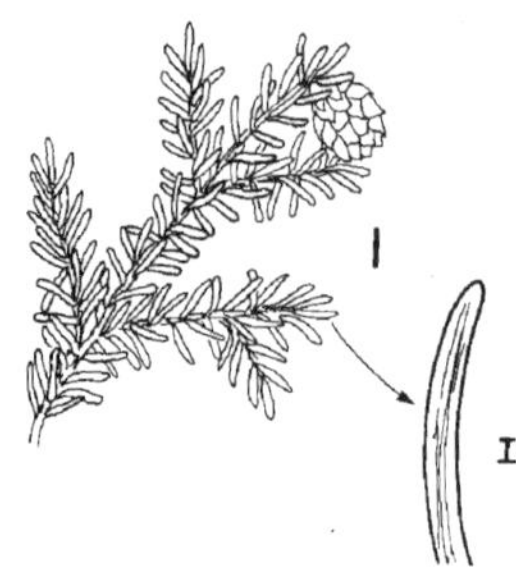

Picea mariana (P. Miller) B.S.P.

BLACK SPRUCE, BOG SPRUCE

Black spruce, often known as bog spruce or swamp spruce, is a small, slow-growing tree, up to 10 m tall, with grey-brown bark. It often has a club-shaped crown, which helps to distinguish it from white spruce (which has a narrow, steeple-shaped crown). Black spruce also has rust-coloured hairs on the bark and twigs; these are absent in white spruce. The blue-green needles are up to 1.5 cm long and 4-sided, arranged all around the twig. Hanging cones, 2 to 3 cm long, persist on the tree for 20 to 30 years. The seeds are winged and nearly black. Black spruce is associated with muskegs and *Sphagnum* bogs. It is a boreal forest tree, abundant and widespread, but it can sometimes be found much further south and has been reported from Kananaskis Country.

NOTES: *The cones, like those of the lodgepole pine* (Pinus contorta), *can withstand great heat. Heat softens the resin holding the cone-scales together, so the scales open and the seeds escape.* ~ *Black spruce can also reproduce asexually by "layering": the lower branches become covered with moss or leaf-litter and develop roots. Eventually, a new tree is produced.* ~ *When Jacques Cartier and his fellow explorers arrived at the future site of Montreal in the 1500s, "spruce tea," made from the bark and leaves of the black spruce, is said to have cured their scurvy.* ~ *The native peoples of eastern Canada boiled twigs, new shoots and cones of the black spruce with maple syrup or honey to make "spruce beer," a good source of vitamins and minerals. They used the resin as chewing gum.* ~ *The wood of this tree is used in the pulp and paper industry: the long fibres add strength to paper.* ~ *The generic name* Picea *comes from the Latin word* pix, *meaning "resin" or "pitch."* ~ *The specific name* mariana *was given by Philip Miller, an English botanist, in 1731. It means "of Maryland." Miller believed the forests of Maryland were representative of North American forests. Unfortunately, this particular species does* not *grow in Maryland.*

Pinus L. — PINE

Pine trees are evergreens with scaly or fissured bark. The leaves are needle-like, usually longer than 3 cm, in fascicles (bundles) of 5 or 2. The fascicles are enclosed at the base in membraneous sheaths. Pollen-cones in clusters grow at the base of expanding twigs. The seed-cones are small and greenish; they mature in 2 years, when they become very woody and conspicuous.

1. Needles in groups of 2 . . . . *P. contorta*
   Needles in groups of 5 2
2. Bark smooth, greyish white; cones 3 to 7 cm long. . . . . *P. albicaulis*
   Bark becoming rough with age; cones 8 to 20 cm long . . . . *P. flexilis*

Pinus albicaulis Engelm.

WHITE-BARK PINE, SCRUB PINE

White-bark pine is a small tree that grows up to 10 m. If growing in exposed situations on rocky ridges, it forms a stunted, bushy shrub. The lower branches are often prostrate. The twisted trunk has whitish bark, hence the common name. The needle-like leaves grow in bundles of 5 on hairy twigs. The female cones are purplish brown with thick scales 3 to 7 cm long. The scales disintegrate to release the seeds, which are large, brown and wingless. White-bark pine is occasionally found in the timberline belt of the Rocky Mountains, at high subalpine elevations.

NOTES: *The specific name* albicaulis *comes from 2 Latin words:* albus *(white) and* caulis *(stem).*

Pinus contorta Loudon

LODGEPOLE PINE

Lodgepole pine can reach 20 to 30 m in height and often grows in dense stands. The twigs are orange to dark brown and bear needles in bunches of 2, which makes it easy to distinguish from limber or white-bark pine. The female cones of this pine are 2 to 5 cm long; they are tan-coloured, woody and often curved backward on the branch. Cones remain on the tree when mature. The thick cone-scales bear small prickles. The scales are tightly closed until the heat from a forest fire softens the resin that holds the scales together and the winged seeds are released. Lodgepole pine is common in the Rocky Mountains; it is found at the lower elevations of the subalpine zone.

NOTES: *Lodgepole pine wood was used for fires by the Blackfoot and Blood tribes because it is resinous and burns easily. The lodgepole pine was also for tipi poles because of its tall, straight trunk. They used the inner bark for food. The resin was used to waterproof their baskets.*

Pinus flexilis James

LIMBER PINE

Limber pine is a small, slow-growing tree, up to 12 m high, and is often twisted; it may be solitary or grow in small stands. The needles are clustered at the ends of the branches and are 3 to 8.5 cm long, in clusters of 5. White-bark pine also bears needles in clusters of 5, but can be distinguished from limber pine because the cones differ in size: those of white-bark pine has are only 3.7 cm long, while those of limber pine are 8 to 20 cm long. Limber pine grows on rocky slopes and ridges in montane and subalpine zones. In exposed situations it presents a characteristic gnarled appearance.

NOTES: *The common name "limber pine" comes from the supple young branches: the specific name* flexilis *means "flexible."*

Pseudotsuga menziesii (Mirb.) Franco

DOUGLAS FIR

The needle-like leaves of Douglas fir are dark green and flat; they are spirally arranged but appear 2-ranked because of a twist at the base. The female cones are pendent, 4 to 10 cm long and light reddish brown; they bear distinctive 3-pronged bracts between the scales. The male cones are 5 to 10 mm long. Douglas fir is found in the montane zone and subalpine zone; it is essentially a mountain species but occasionally extends into the foothills along the river valleys.

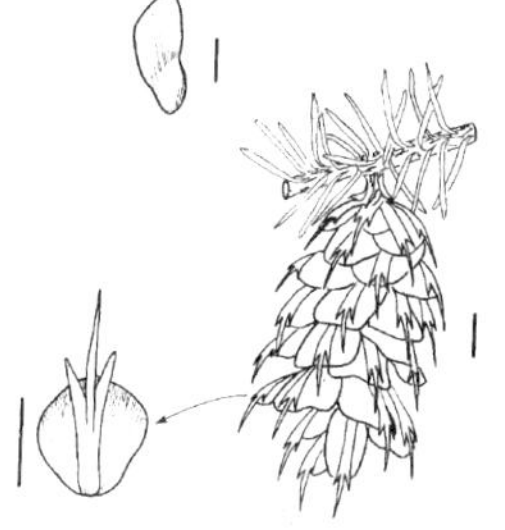

NOTES: *Douglas fir is named for a Scottish botanist, David Douglas, who was sent to North America by the Horticultural Society of London (later the Royal Horticultural Society). Douglas made many botanical explorations in North America along the Pacific Coast.* ~ *The generic name* Pseudotsuga *is derived from the Greek word* pseudos, *which means "false," and* tsuga, *the Japanese name for hemlock. But as the trees are quite different in appearance, the name seems a misnomer.* ~ *The specific name* menziesii *honours Archibald Menzies, a Scottish physician and naturalist, who sailed with Captain Vancouver. He discovered the Douglas fir.* ~ *Although called a "fir," this species does not belong to the firs* (Abies). ~ *Douglas fir is the largest tree in Canada, but in Alberta it does not attain the heights it achieves in British Columbia, where it can reach 70 m. It has thick, deeply furrowed bark and is an important timber tree.*

Flowering Plants

~ MONOCOTS

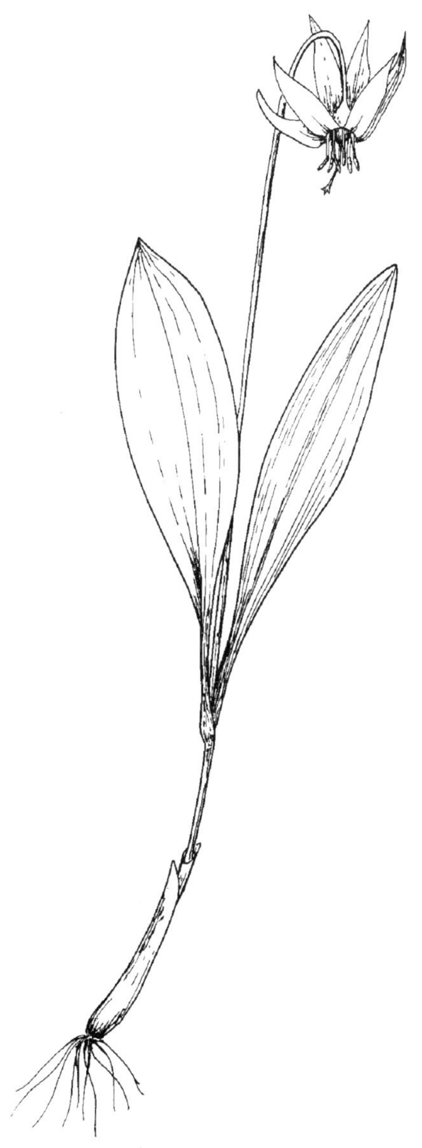

Monocots

Monocots are distinguished by a number of features. In monocots, the embryos of the seeds have only one cotyledon (seed-leaf). The plant leaves usually have parallel veins. The vascular bundles of the stems are irregularly arranged, and the cambium is absent. The flower-parts are arranged in threes or sixes, never in fives. Monocots are usually herbs; they are rarely shrubby.

KEY TO THE MONOCOTS

1. Flowers showy, some floral parts coloured 2
   Flowers not showy, floral parts usually green or brown..... 5
2. Ovary superior (or partly inferior in some species) 3
   Ovary inferior..................................... 4
3. Plants of marshy habitats; sepals 3, green; petals 3, white; leaves arrowhead-shaped or broad, strongly nerved .... Alismataceae
   Plants of marsh habitat, sepals petaloid, leaves not arrow head-shaped................................. Liliaceae
4. Sepals and petals essentially the same shape, size and colour .... Iridaceae
   Sepals and petals of different shapes, sizes and colours .... Orchidaceae
5. Plants aquatic (free-floating or submersed) 6
   Plants not aquatic (sometimes rooting in water, but with the stems emerging).................................. 8
6. Plants free-floating, plant-body flat, rounded and small (less than 6 mm across)......................... Lemnaceae
   Plants submersed, differentiated into stems and leaves 7
7. Stems jointed; flowers perfect (i.e., with both sexes) .... Potamogetonaceae
   Stems not jointed; flowers imperfect (i.e., with a single sex, male or female) .... Sparganiaceae
8. Leaves basal....................................... 9
   Leaves alternate or opposite 10
9. Flowers in dense cylindrical clusters .... Typhaceae

	Flowers borne singly along the flowering stem	Juncaginaceae
10.	Flowers in globe-shaped clusters	Sparganiaceae
	Flowers not in globe-shaped clusters	11
11.	Stems mostly triangular; leaves 3-ranked	Cyperaceae
	Stems round; leaves 2-ranked	12
12.	Stems hollow; nodes swollen	Poaceae
	Stems solid; nodes not swollen	Juncaceae

Typhaceae CATTAIL FAMILY

These are perennial, herbaceous plants found in marshes or aquatic habitats. They grow 1 to 2 m tall, with linear, sheathing leaves. Their tiny flowers grow in dense, cylindrical, apical spikes.

Typha CATTAIL

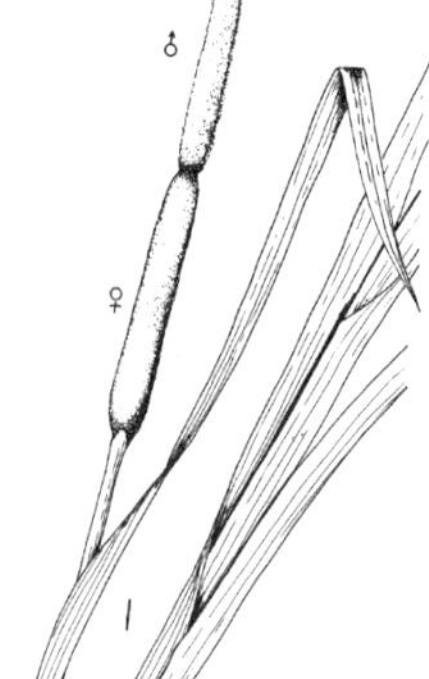

Typha latifolia L.

COMMON CATTAIL

Cattails are found in marshes and shallow water. They have large, creeping rhizomes. The rhizomes produce large colonies of plants with long, strap-like leaves. The leaves sheathe the stem, which is 1 to 2 m tall. The flowers are tiny and are borne in a characteristic dense, cylindrical spike at the top of the stem: the brown carpellary (female) flowers below and the yellow staminate (male) above. When the male flowers have shed their pollen, they fall off, leaving a naked stipe. The female flowers produce masses of minute, fluffy achenes, which are dispersed by the wind. Cattails are found in the eastern areas of Kananaskis Country but not in the mountains.

NOTES: *Several parts of the cattail were used for food by native peoples. The pollen (rich in protein) was used for flour, the young inflorescence was cooked as a vegetable, and the leaves were used to make mats. The rhizomes, containing inulin (a starch-like compound), were also used for food, and native women used the "fluff" of the fruiting-spikes for diapers and padding.*

Sparganiaceae BUR-REED FAMILY

These are aquatic or marsh plants that grow from rhizomes. The leafy stems are either erect or floating. The grass-like leaves are floating and 2-ranked; they are sheathed at the base. The flowers grow in dense, round heads, which bear either male or female flowers, and are found on the upper part of the stem.

Sparganium L. BUR-REED

Bur-reeds are perennial swamp and aquatic plants with erect or floating stems. The leaves are grass-like and usually floating. Staminate and carpellary flowers grow in separate, dense, round heads (the staminate heads above the carpellary ones) on the leafy stem. Male flowers have 3 to 5 stamens and brownish scales; female flowers have 3 to 6 sepals and an ovary. Ovaries turn into beaked achenes. The female head becomes bur-like.

Plants mostly floating; male flower-heads solitary. . . . . *S. minimum*
Plants mostly erect; male flower-heads 2 to 5 . . . . *S. angustifolium*

Sparganium angustifolium Michx.

NARROW-LEAVED BUR-REED

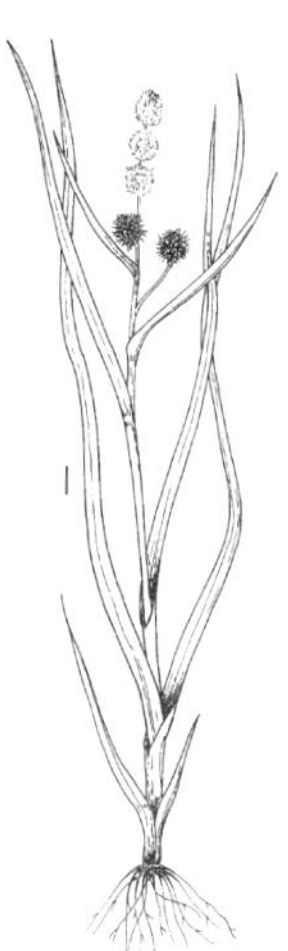

Bur-reed is a perennial plant, which may grow erect or float in deep water. The stems are 2 to 10 dm long and bear grass-like leaves, 2 to 15 mm wide, sheathing at the base. The tiny flowers are borne in dense heads; there are 2 to 5 carpellary (female) heads and 2 to 5 staminate (male) heads. The fruits are tiny achenes with beaks, hence the common name "bur-reed": the fruiting head looks like a round burr. The plant is found in ponds, streams and shallow lakes.

NOTES: *The species name* angustifolium *means "narrow-leaved."*

Sparganium minimum (Hartm.) Fries

SLENDER BUR-REED

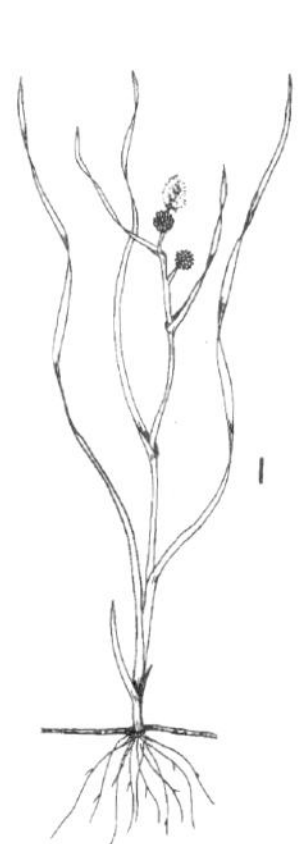

Slender bur-reed is similar to *S. angustifolium* but it has slender stems, only 1 to 8 dm long, and its leaves are smaller and narrower, only 1 to 7 mm wide. The achene beaks are only 1 to 1.5 mm long, while those of *S. angustifolium* are 1.5 to 4.0 mm long. The slender bur-reed is found in shallow water.

NOTES: *Both species of bur-reed are found in the stream running through Sibbald Flats in Kananaskis Country and in other places.*

Potamogetonaceae PONDWEED FAMILY

Pondweeds are water plants with jointed stems, often bearing adventitious roots; the leaves are either floating or submerged. The flowers are arranged in a spike-like cluster raised above the water-level; the greenish flowers are in whorls. Many species carry on vegetative reproduction by tubers or detached winter-buds. Pondweeds grow in ponds and sluggish, shallow streams. The different species can often be distinguished by leaf shape.

Potamogeton L. PONDWEED

Pondweeds are perennial, rooted aquatic plants, floating or submerged. The leaves are variable, from lanceolate to filiform, usually with sheathing base. The flowers are borne in whorls on unbranched erect stalks and are emergent; they are small, greenish and bisexual. Staminate flowers have 4 stamens; carpellary flowers have 4 ovaries. The fruit is a single-seeded achene. Reproduction is also carried on by winter buds and root-tubers. The root-tubers are eaten by waterfowl and muskrats.

1. Leaves less than 2 mm wide . . . . *P. filiformis*
   Leaves more than 5 mm wide 2
2. Leaves with 3 prominent veins; floating leaves absent . . . . *P. richardsonii*
   Leaves with 7 prominent veins; floating leaves present . . . . *P. alpinus*

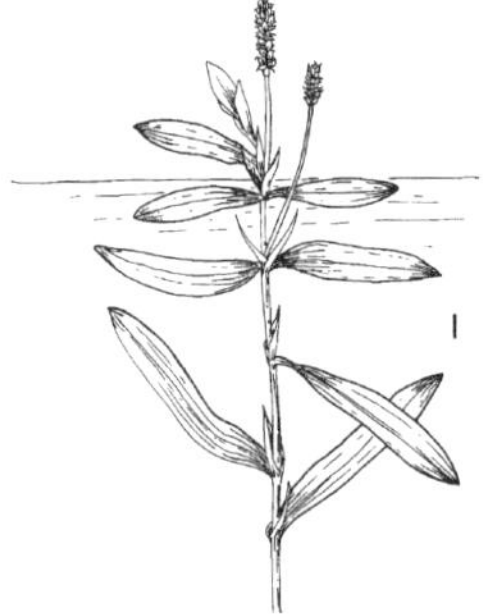

Potamogeton alpinus Balbis

ALPINE PONDWEED

This pondweed has a reddish colour. The leaves are lance-shaped, tapering to a slender base. The flower-spikes, with long stalks, arise from the leaf-axils. The flowers are tiny and greenish; they grow in whorls. Each flower has 4 stamens and 4 ovaries. The tiny fruits are dry and single-seeded. This pondweed is found in ponds and shallow streams.

Potamogeton filiformis Pers.

THREAD-LEAVED PONDWEED

This species is easily distinguished from *P. alpinus*, because it is a smaller, delicate, submerged plant, with very long, narrow, dainty leaves; the plant is many-branched. The specific name *filiformis* means “thread-like” and refers to the leaves. The flower spikes are erect, on long stalks above the water-level. The plant is found in shallow ponds and sluggish streams.

Potamogeton richardsonii (Benn.) Rydb.

CLASPING-LEAF PONDWEED

Clasping-leaf pondweed looks quite different from the other 2 species, because the leaf-bases, as the common name implies, clasp the stem. The leaves are large and broad at the base, taper to a point and have wavy edges. The tiny flowers are arranged in dense spikes, above the submerged leaves, and the flowers turn into minute fruits, 3 mm long, with short beaks. The plant is found in ponds.

Juncaginaceae ARROW-GRASS FAMILY

These plants are perennial and are found in brackish marshes. The leaves arise from the base (a rhizome) and are linear and rather fleshy. The small, greenish flowers are arranged in a spike-like raceme at the top of a tall, main flower-stalk or peduncle.

Triglochin L. ARROW-GRASS

Arrow-grasses are glabrous, perennial marsh plants with linear basal leaves, sheathing and rather fleshy. Flowers are arranged in an elongated spike-like raceme; each greenish flower has 2 whorls of perianth-leaves (3 in each), 6 stamens, 3 to 6 carpels with plumose stigmas; the fruits are 3 to 6 follicles.

Stigmas 6; caudex with persistent leaf bases; plants stout, usually over 50 cm tall . . . . *T. maritima*

Stigmas 3; caudex without persistent leaf bases; plants slender, up to 20 cm tall . . . . *T. palustris*

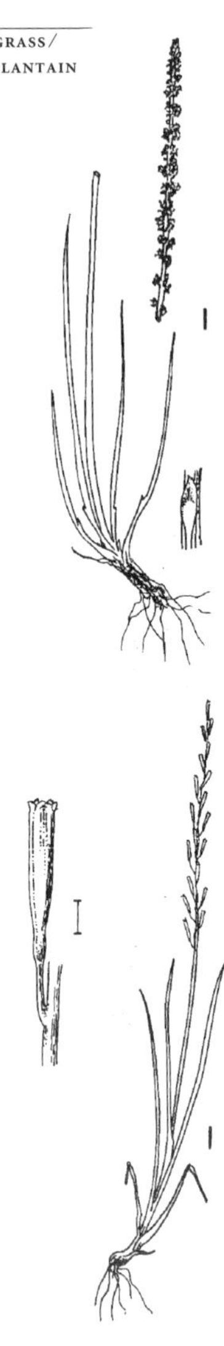

Triglochin maritima L.

SEASIDE ARROW-GRASS

Arrow-grass is a grass-like perennial with several basal leaves up to 5 dm long, arising from stout rhizomes. The main flowering-stalk bears a spike-like raceme 1 to 4 dm long. The tiny flowers are greenish; each flower has 6 perianth-leaves in 2 whorls of 3, representing the sepals and petals. There are 6 stamens attached to the bases of the perianth leaves. In the centre of the flower are 6 carpels, attached to an axis, with 6 plumose stigmas; these develop into follicles (like legumes, but opening down only 1 side), with persistent stigmas. Arrow-grass is found in boggy areas, beaver ponds and brackish marshes.

Triglochin palustris L.

SLENDER ARROW-GRASS

Slender arrow-grass has a short rhizome with very slender *stolons* growing from it. The basal leaves are linear and very narrow. The small, greenish flowers are arranged in a spike-like raceme at the top of a long, erect flower-stalk; each flower has 6 perianth-segments in 2 whorls (these represent the petals and sepals), 6 stamens and 3 carpels attached to an axis with plumose stigmas. The fruits are follicles. Slender arrow-grass is found in bogs, lakeshores and in marshy places.

NOTES: *Slender arrow-grass can be distinguished from the other species because it has 3 carpels and 3 stigmas instead of 6 carpels and 6 stigmas. Also, the fruit has a winged axis.*

Alismataceae WATER-PLANTAIN FAMILY

Members of this family are aquatic perennials and are also found on marshy ground. They are tufted plants with fibrous roots, basal leaves with long stalks, and erect stems. The leaves may be erect or floating; in arrowhead they may be submerged.

The flowers are arranged in whorls on the main flower-stalk and are usually in clusters. Each flower has 3 green sepals and 3 white petals; there are 6 to 25 stamens and numerous carpels, which develop into achenes. The flowers may be bisexual or unisexual.

Sagittaria cuneata Sheld.

ARROWHEAD, WAPATO

Arrowhead is an aquatic plant, 2 to 4 dm tall. Its arrow-shaped leaves have long leaf stalks *(petioles)* which raise them above water-level. The submerged leaves are lanceolate. Both types of leaves arise from slender rhizomes, which produce tubers. The large, white flowers are conspicuous. They often grow in whorls of three along the main flower-stalk; the lower flowers are usually female (carpellary or pistillate), while the upper flowers are male (staminate). Each flower has 3 small sepals, 3 white petals up to 12 mm long, 15 to 25 stamens, with a dense cluster of carpels in the centre. The fruits are winged achenes. Arrowhead plants are found in shallow water in ponds, or in streams.

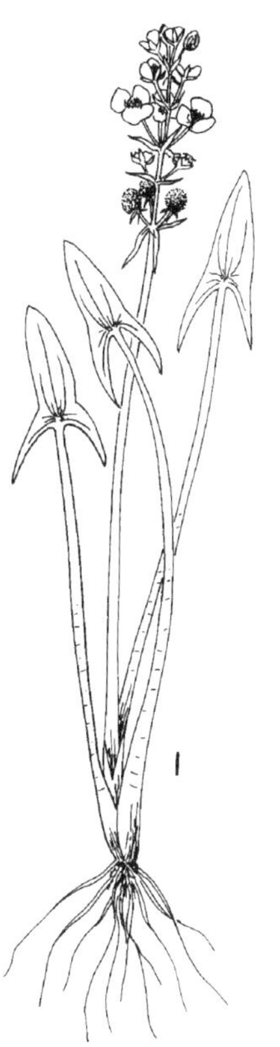

NOTES: *Arrowhead is a descriptive name for this plant, because the leaves are shaped like arrowheads and are characteristic. The Latin word* sagittus *means "an arrow," hence the generic name* Sagittaria. *The specific name* cuneata *means "wedge-shaped."* Wapato *is the native name for arrowhead.* ~ *Native peoples found the tubers on the rhizomes a good source of food because they are rich in starch. The tubers were loosened in shallow water using sticks, but in deeper water, the women clung to the sides of their canoes, and loosened them with their feet—a wet and dreary task!* ~ *On their journey through the Rockies to the Pacific in 1805/6, explorers Lewis and Clark found that these tubers were the main vegetable food of the natives on the Lower Columbia River; during one winter they themselves lived largely on the tubers, which are often called "duck potatoes."* ~ *Muskrat and beaver are very fond of these tubers.*

Poaceae (Gramineae) GRASS FAMILY

This is one of the most important plant families because many of our food crops are grasses (wheat, barley, oats, rye, maize, millet, sugar). The family has a worldwide distribution. Grasses may be annual or perennial and have hollow stems, except at the nodes; the leaves are in 2 ranks, sheathing at the base. The leaves are usually linear, they always show parallel veining, and the sheath grows around the stem like a split tube. There is usually a tiny ligule where the sheath and the leaf-blade meet. The tiny flowers (florets) are arranged in spikelets, often clustered together in racemes, panicles, or spikes. Two empty bracts on the spikelet, below the florets, are called "glumes." Each floret consists of 2 bracts, called the *lemma* and *palea*, enclosing 3 stamens, and an ovary with 2 styles and plumose stigmas. The fruit, the "grain," is called a *caryopsis.*

NOTE: *The following key is highly simplified so that all users will be able to identify grasses. Some liberties have been taken during its simplification.*

1. Panicle open 2
   Panicle compact 14
2. Plants of wet habitats 3
   Plants of dry habitats 6
3. Leaf-tips boat-shaped . . . . *Poa*
   Leaf-tips not boat-shaped 4
4. Spikelets purple. . . . . *Vahlodea*
   Spikelets not purple 5
5. Spikelets 1-flowered. . . . . *Agrostis*
   Spikelets 2 or more-flowered. . . . . *Glyceria*
6. Leaf tips boat-shaped. . . . . *Poa*
   Leaf tips not boat-shaped 7
7. Spikelets 1-flowered 8
   Spikelets 2 or more-flowered 9
8. Leaves 30 to 70 cm long; ligule pointed . . . . *Oryzopsis*
   Leaves shorter; ligule not pointed . . . . *Agrostis*
9. Spikelets purple, 3 to 5-flowered; sheaths closed. . . . . *Schizachne*
   Spikelets not purple 10
10. Spikelets many-flowered. . . . . *Bromus*
    Spikelets 2 to 8-flowered. 11
11. Spikelets 2-flowered . . . . *Deschampsia*
    Spikelets 3 or more-flowered 12
12. Lemmas without an awn . . . . *Hierochloe*
    Lemmas awned 13
13. Ligule reduced to a tuft of hairs . . . . *Danthonia*
    Ligule not as above. . . . . *Festuca*
14. Plants of moist or wet habitats. 15
    Plants of drier habitats 19
15. Leaf tips boat-shaped; spikelets 2-flowered. . . . . *Poa*
    Leaf tips not boat-shaped; spikelets 1-flowered 16
16. Plants 60 to 150 cm tall; leaves 6 to 15 mm wide . . . . *Phalaris*
    Plants less than 120 cm tall; leaves less than 10 mm wide 17

17. Plants pale-green; panicle composed of several spikes.......... *Beckmannia*
    Plants not as above 18
18. Spikelets sessile, forming a dense spike less than 1 cm thick ...... *Alopecurus*
    Spikelets stalked; spike more than 1 cm thick .... *Calamagrostis*
19. Spikelets 1-flowered 20
    Spikelets 2 or more-flowered.................................... 23
20. Awns more than 5 cm long 21
    Awns less than 5 cm long....................................... 22
21. Awns straight .... *Hordeum*
    Awns bent and twisted .... *Stipa*
22. Panicle cylindrical; awn less than 2 mm long.................... *Phleum*
    Panicle not cylindrical; awn more than 2 mm long .... *Oryzopsis*
23. Leaf tips boat-shaped... *Poa*
    Leaf tips not boat-shaped 24
24. Awns 1 to 1.5 cm long, bent and twisted.................... *Helictotrichon*
    Awns not as above 25
25. Awns bent .... *Trisetum*
    Awns straight 26
26. Panicle with short branches.................................... 27
    Panicle not branched; spikelets sessile 29
27. Spikelets 2-flowered; panicle 10 to 30 cm long................ *Deschampsia*
    Spikelets with 2 or more flowers; panicle 1 to 20 cm long 28
28. Glumes shorter than lemma .... *Festuca*
    Glumes longer than lemma.................................. *Koeleria*
29. Spikelets solitary at each node.......................... *Agropyron*
    Spikelets more than 1 at each node.......................... *Elymus*

Agropyron Gaertn. **WHEAT GRASS**

Wheat grasses are perennial plants, with or without creeping rhizomes. The spikelets are in a single, narrow spike. Each spikelet is several-flowered and sessile, positioned with the flat side towards the rachis of the spike, 1 per node. The glumes are approximately equal in length, 3 to 7-nerved with tips obtuse, acute or short-awned. Lemmas are quite firm, 5 to 7-nerved and acute or awned from the apex.

1. Plants tufted; spike open .... *A. spicatum*
   Plants not tufted; spike closed.................................. 2
2. Leaf margins rolled inward .... *A. dasystachyum*
   Leaves flat, margins not rolled inward .... *A. repens*

Agropyron dasystachyum (Hook.) Scribn.

NORTHERN WHEAT GRASS

As in all wheat grasses, this grass has sessile spikelets forming a very narrow spike. The lemmas may be densely to sparsely pubescent and awnless or short-awned. Densely pubescent lemmas are often good diagnostic features. This grass possesses long, slender rhizomes and grows 4 to 13 dm tall. The leaves are usually involute and glaucous; if flat, they are 2 to 3 mm wide. This plant is common on dry grasslands, south-facing slopes and dry, open woodlands.

Agropyron repens (L.) Beauv.

COUCH GRASS, QUACK GRASS

This grass is a pernicious weed of gardens and lawns. It has long, tough, yellowish rhizomes and is very difficult to eradicate. It grows 4 to 8 dm high and has wide, flat leaves (4 to 8 mm wide). The lemmas may be pointed or may bear an awn that can vary in length from 1 to 10 mm. This plant is found on waste ground and roadsides.

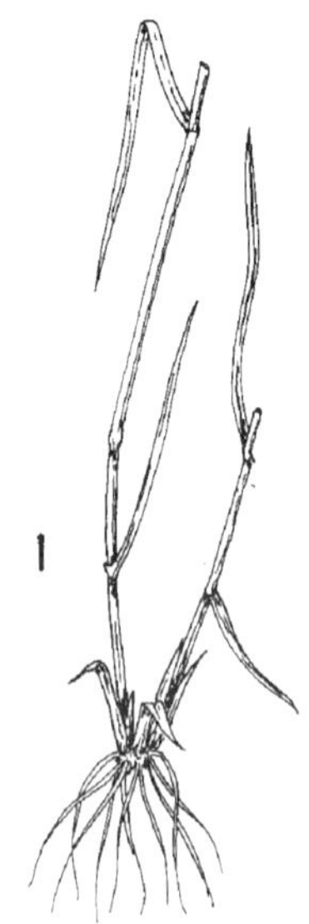

Agropyron spicatum (Pursh) Scribn. & Smith

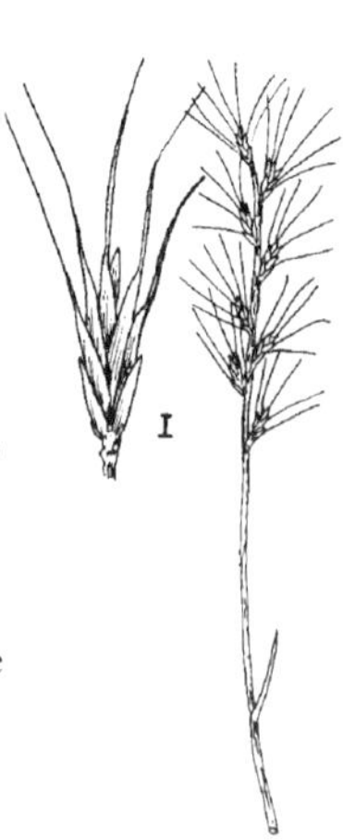

BLUEBUNCH WHEAT GRASS

Unlike northern wheat grass and couch grass, bluebunch wheat grass rarely has rhizomes, but instead tends to grow in dense tufts. Another key difference is the greater space between the spikelets: the spikelets barely overlap. The lemmas may be awnless or have curved, divergent awns (often bending at right angles), 1 to 2 cm long. These long awns give the inflorescence a spiky appearance, hence the species name *spicatum*. The plant grows 3 to 10 dm tall and has slender leaf-blades. It is found in dry, open areas.

Agrostis scabra Willd.

TICKLE GRASS, ROUGH HAIR GRASS

This grass grows in small, dense tufts, 3 to 7 dm tall, with leaves mostly basal and filiform. It is easily distinguished from the other grasses because of its wide-spreading panicle of almost hair-like branches. The panicle is 1 to 2.5 dm long and nearly as wide at maturity. The spikelets are borne near the ends of the branches, are very small (2 to 2.7 mm long) and lack paleas. At maturity, the panicle often breaks away and can be carried by the wind like a tumble-weed. It grows in dry to wet disturbed areas and open woods to the subalpine level.

NOTES: *This very fine panicle can tickle bare legs, hence the common name "tickle grass."*

Alopecurus aequalis Sobol.

WATER FOXTAIL

This grass is a perennial, growing up to 6 dm. It has loosely tufted stems which root at the nodes. Leaf blades are flat, up to 5 mm wide, with ligules 4 to 8 mm long. The flower clusters are soft and dense, and grow in the form of a spike, 2 to 8 cm long and 3 to 5 mm wide. The spikelets are single-flowered. The lemma is hairless, is as long as the glumes and bears a stiff awn. At low elevations, it grows in muddy ground or shallow water, often near beaver ponds.

NOTES: *Water foxtail is often confused with timothy grass,* Phleum pratense, *which also has a spike-like inflorescence.*

Beckmannia syzigachne (Steud.) Fern.

SLOUGH GRASS

Slough grass is a stout, annual grass growing 3 to 10 dm tall, light green in colour with flat leaves, 3 to 10 mm wide. The panicle is the diagnostic feature of this grass. It is narrow and composed of numerous crowded, appressed or ascending spikes. The spikelets are single-flowered, orbicular, flat and 3 mm long; they lie in 2 rows along 1 side of a slender rachis, giving the appearance of shingles arranged in a feather-like fashion. Slough grass is found on low, wet ground.

Bromus L. — BROME GRASS

Brome grasses may be annual or perennial. They possess flat blades and closed sheaths. Spikelets are several-flowered, on stalks that may be upright and lax, or drooping in a contracted or open panicle. The glumes are of unequal length and shorter than the lemmas. Lemmas are 2-toothed at the apex and awnless from between the teeth.

Leaves hairy . . . . *B. tectorum*
Leaves not hairy. . . . . *B. inermis*

Bromus inermis Leyss.

AWNLESS BROME

This large, rhizomatous, perennial grass grows 2 to 15 dm tall and has wide, flat leaf blades (5 to 15 mm wide). The spikelets tend to be large, 1.5 to 3 cm long, on erect or curved branches. The lemmas are awnless or short-awned (despite the common name). The plants can be found in open woods and grassland to subalpine elevations, as well as roadsides, fields and waste ground.

Bromus tectorum L.

DOWNY CHESS

Downy chess is an annual brome grass, much smaller than awnless brome, reaching only 2 to 5 dm tall. The sheaths and leaf-blades have soft hairs, and the spikelets droop down on slender branches. The glumes and lemmas are also soft-hairy; the lemma has slender teeth at the tip and a straight awn, 12 to 15 mm long. This grass is non-native and occurs on roadsides and waste ground, often in great numbers.

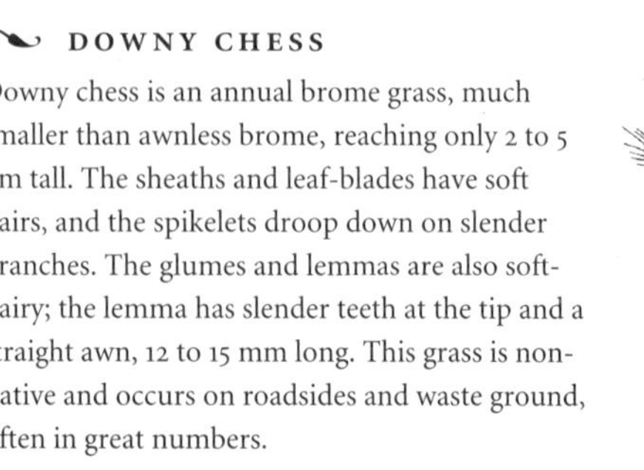

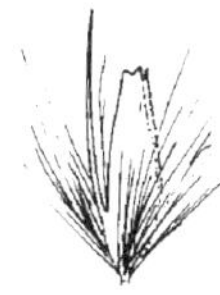

Calamagrostis canadensis (Michx.) Beauv.

MARSH REED GRASS, BLUEJOINT

This tall, woodland grass reproduces by creeping rhizomes and grows 6 to 12 dm high. The many flat, lax leaves are 4 to 8 mm wide, and the ligule is prominent, 3 to 6 mm long. The panicle tends to be large (1 to 2 dm long) and may be broad, open or dense. Spikelets are single-flowered, and the lemmas have a fine, straight awn attached just below the middle. A whorl of fine, straight (callus) hairs arises from the base of the floret, approximately equalling the lemma. This grass may be found in moist woodland, meadows and marshes.

Danthonia californica Boland

CALIFORNIA OAT GRASS

California oat grass grows in tufts up to 9 dm tall. The leaves may be flat or involute and tend to be conspicuously hairy on the lower surface. The ligule is not membranous but is comprised of a tuft of hairs. The spikelets of this grass are its most distinguishing feature. They tend to be large, with glumes of 10 to 18 mm and lemmas of 8 to 12 mm, in a panicle 3 to 9 cm long. The lemmas are prominently 2-toothed at the apex with a twisted awn arising from between the teeth, 5 to 10 mm long. California oat grass is found in dry to moist open grassland and open woodland to the subalpine level.

Deschampsia caespitosa (L.) Beauv.

TUFTED HAIR GRASS

This grass grows in dense tufts, 2 to 2 dm tall, with long, firm leaves mostly arising from the base of the plant. The ligules are prominent and pointed, up to 10 mm long. The panicle is loose, open and nodding but not as broad and fine as in tickle grass. The spikelets are borne near the ends of the branchlets and are 3 to 5 mm long. They glisten and may be pale or purplish. A tiny, hair-like awn arises from near the base of the lemma. Tufted hair grass is found in moist meadows, on river bars and along lakeshores.

Elymus innovatus Beal

HAIRY WILD RYE

Hairy wild rye is similar in appearance to the wheat grasses because its spikelets are sessile and the inflorescence is a spike. The spike tends to be wider and more dense, however, since there are generally 2 spikelets at each node instead of 1. The glumes and lemmas are very hairy; the lemmas are short-awned (1 to 4 mm long), and the spike is often purplish or greyish. The plants grow 4 to 10 dm high from slender rhizomes clothed in brown scales, and can be found in open woodlands, especially pine forests, and grassland.

Festuca L. — FESCUE

Fescue grasses are perennial plants and are often densely tufted. The leaves are narrow and involute or folded. Spikelets are stalked in narrow or spreading panicles. The glumes are narrow, of unequal lengths and shorter than the lemmas. The lemmas are awned from the apex or sharp pointed.

1. Leaf-margins rolled inward 2
   Leaves flat, margins not rolled inward .... *F. scabrella*
2. Leaf-bases reddish brown; spikelets purple-tinged .... *F. rubra*
   Leaf-bases not reddish brown; spikelets not purple-tinged .... *F. ovina*

Festuca ovina L.

SHEEP FESCUE

Sheep fescue grows in dense tufts and reaches heights of 1.5 to 3 dm, occasionally up to 6 dm. Old basal sheaths are obvious, and the leaves are firm, slender and involute. The panicle is stiff, 2 to 10 cm long, with divergent branches when fully mature; the lemmas are short-awned. This grass is non-native and may be found in grasslands and along roadsides.

Festuca rubra L.

RED FESCUE

This grass can be loosely tufted or short-creeping, 1 to 8 dm tall, with brown or reddish basal sheaths that are not persistent. The leaves are narrow, soft and smooth, and are usually folded or involute. The spikelets are often purple-tinged on ascending branches. The lemmas are short-awned. Both introduced and native plants exist in the area, along roadsides and in dry open areas. This grass is occasionally added to lawn-seed mixtures.

Festuca scabrella Torr.

ROUGH FESCUE

This grass grows in large tussocks, 3 to 9 dm high. Unlike red fescue, the leaves are firm, rough and involute, and the old basal sheaths are long persistent. The common and species names reflect the rough (scabrous) nature of the leaves. The panicle is narrow, and the spikelets are elliptic to ovate on appressed or ascending branches. Rough fescue is common in dry grassland, where it is often a dominant component; in fact, fescue grassland is an ecoregion within Alberta.

Glyceria R.Br. — MANNA GRASS

Manna grasses are perennials of aquatic habitats and moist to wet ground. They may be rhizomatous or tufted and have flat or folded leaf-blades tapered to sharp boat-shaped tips. The few to many-flowered spikelets are borne in a large, open panicle that is often nodding at the summit. The glumes are of unequal length, shorter than the lemmas, and scarious or whitish. The lemmas of the species described below are broad and firm, with 7 rib-like nerves running parallel to the sides, and a blunt or obtuse apex.

Plants 9 to 16 dm tall, firm leaf blades, 6 to 15 mm broad . . . . *G. grandis*
Plants 3 to 8 dm tall, partly folded leaf blades 2 to 4 mm wide. . . . . . *G. striata*

Glyceria grandis S. Wats. ex A. Gray

COMMON TALL MANNA GRASS

This grass is a strongly rhizomatous perennial, 9 to 16 dm tall with flat, firm leaf-blades, 6 to 15 mm broad. The large, open panicle is 2 to 3.5 dm long, with numerous spreading branches. Each spikelet has several flowers with purplish, 7-nerved lemmas and lanceolate glumes. The plants grow in sloughs and wet meadows.

Glyceria striata (Lam.) A.S. Hitchc.

FOWL MANNA GRASS

This moisture-loving plant is rhizomatous and often found in large clusters. It grows only 3 to 8 dm tall, and this helps to distinguish it from the common tall manna grass. It is pale-green with flat or folded leaf blades, usually 2 to 4 mm wide. The panicle is open and ovoid, erect or drooping at the top. Spikelets are 3 to 4 mm long and often purplish. Glumes and lemmas are obtuse, and the lemmas are prominently 7-nerved. These plants are common in shallow water and boggy meadows.

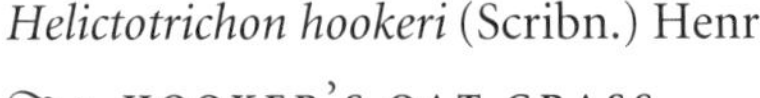

Helictotrichon hookeri (Scribn.) Henr.

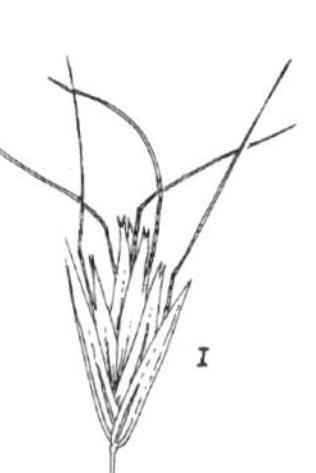

HOOKER'S OAT GRASS

This attractive perennial is easily distinguished by the long, shining, brownish spikelets and the thickened leaf-margins resulting from a whitish nerve on either side. The panicle is narrow with erect branches, each usually bearing a single spikelet. The lemmas are 10 to 12 mm long, firm and brown, with 2 narrow teeth at the apex and awned from slightly below the middle. The awn is 1 to 1.5 cm long, twisted below and bent once or twice. The leaves are flat or folded, usually basal and 2 to 4 mm wide. The plant grows in dense tufts, 2 to 4 dm high, in dry grassland of the foothills.

Hierochloe odorata (L.) Beauv.

SWEET GRASS

This attractive, sweet-smelling grass is notable for its bronze, golden-yellow or purple spikelets in a more or less loose, pyramidal panicle. The spikelets are tulip-shaped, 4 to 7 mm long, with 3 florets. The terminal floret is perfect, and the 2 lateral florets are staminate or neutral. The glumes are broad, ovate and thin, and about the same length as the lemmas. The plant grows from creeping rhizomes and produces sterile leafy shoots. The stems reach 3 to 6 dm tall, are purplish at the base and have 2 or 3 short, broad cauline leaves. Sweet grass is found in moist to dry open areas such as meadows and streambanks.

NOTES: *Sweetgrass was used by the Blackfoot as an incense in many different ways. The leaves were dried and plaited into fillets; these were worn around women's heads or braided into clothing. The leaves were carried in buckskin bags as a perfume, or soaked in water for use as a hair wash. The grass was used in the Sundance ceremony and was burnt on an altar in lodges or tipis.* ~ *The fragrant odour of sweet grass is due to coumarin.*

Hordeum jubatum L.

FOXTAIL BARLEY

This grass is easily identified by its inflorescence, which is a dense, bristly spike, resembling a fox's tail. There are 3 spikelets per node. The 2 lateral spikelets are stalked and reduced to 1 to 3 spreading awns; the middle spikelet is sessile with glumes up to 9 cm long and lemma awns as long as the glumes. These all combine to create a densely bristly spike. The spike is often nodding and is pale-green to purple. The plants grow in tufts, 3 to 10 dm high. They are common in disturbed areas, where they often grow in dense stands, and they are early colonizers of waste ground, including saline areas.

Koeleria macrantha (Ledeb.) J.A. Schultes f.

JUNE GRASS

This plant has a dense, spike-like panicle of short-stalked, appressed spikelets. It is often shiny and may be pale green or purplish. The lemmas are acute or short-awned, the awn arising just below the apex. This perennial plant grows in tufts 2 to 5 dm tall and has flat or involute leaf-blades borne mostly at the base. It is found in dry grassland.

Oryzopsis Michx. — RICE GRASS

Rice grasses are tufted perennials with flat or involute leaves. Spikelets are single-flowered in a narrow panicle. Glumes are equal or nearly equal, quite broad and about as long as the spikelet. The lemmas are convolute and hardened, with a terminal awn that breaks off easily.

Leaf-margins rolled inward; awn 3 to 5 mm long . . . . *O. exigua*
Leaves flat, margins not rolled inward; awn 6 to 8 mm long . . . . *O. asperifolia*

Oryzopsis asperifolia Michx.

WHITE-GRAINED MOUNTAIN RICE GRASS

The hardened (indurate) lemma sets this grass apart from the others. The lemmas bear awns, 5 to 10 mm long, from their tips, which are easily broken off. Spikelets are single-flowered, 6 to 8 mm long, on erect branches in a narrow panicle. The glumes are elliptic to obovate, green with scarious margins and come to an abrupt point at the tip. The plant grows in loose tufts, 2 to 7 dm high. The leaf-blades on the flowering stems are reduced or lacking; the basal leaves are long, flat and glaucous on the lower surface. This grass grows in dry to moist open forests and clearings.

Oryzopsis exigua Thurb.

RICE GRASS

This grass differs from *O. asperifolia* in having smaller spikelets (3 to 5 mm long), lemma awns that are 4 to 6 mm long and geniculate, and leaves that are erect and filiform or involute. Spikelets are again single-flowered and are borne on appressed branches of the narrow panicle. The plant grows in dense tufts on dry, open ground or in open forest.

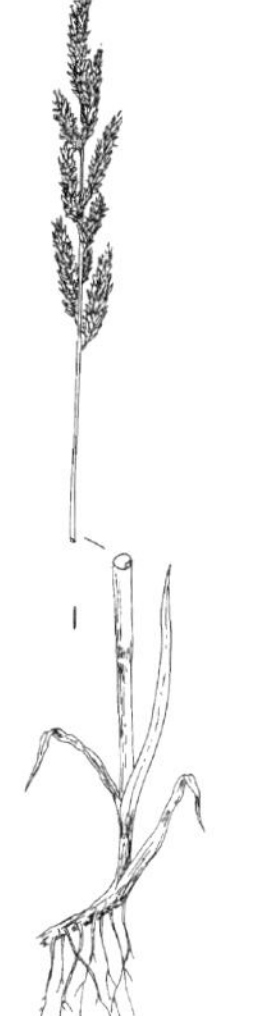

Phalaris arundinacea L.

REED CANARY GRASS

This large, robust perennial of shores and marshes grows 6 to 15 dm tall and possesses long, scaly, pinkish rhizomes. Leaf-blades tend to be long (1 to 3 dm) and wide (5 to 15 mm). The spikelet differs from other grasses, possessing 2 sterile, rudimentary florets which appear as minute, hairy, brownish scales below a single perfect floret. The spikelet is pale in colour, suffused with purple, and the fertile lemma is firm and shiny when in fruit. The panicle tends to be narrow or somewhat open and is comprised of spikes of densely aggregated spikelets. This plant is often found in wet disturbed sites as well as natural wetlands.

Phleum L. — TIMOTHY

Timothy grasses are perennials with solitary or tufted stems and flat leaves. Spikelets are single-flowered and flattened and are densely packed in spike-like, uninterrupted, cylindrical or ovoid panicles. Glumes are equal in length, keeled and short-awned. Lemmas are shorter than the glumes, broadly truncate and hyaline.

Panicle 1 to 3 cm long . . . . *P. commutatum*
Panicle 5 to 10 cm long. . . . . *P. pratense*

Phleum commutatum Gaudin (*Phleum alpinum* L.)

MOUNTAIN TIMOTHY

Timothy grasses are characterized by very dense, spike-like panicles of closely packed 1-flowered spikelets in an uninterrupted cluster forming a cylindrical head. The spikelets are very flat and the lemma is broadly truncate and hyaline. In the case of mountain timothy, the panicle is ovoid or short-cylindric in shape, 1.5 to 3 cm long. The stems are solitary or in small tufts, 1 to 5 dm high, and the upper and middle sheaths are somewhat inflated. This grass is found on the borders of forest and in meadows, up to the alpine zone.

Phleum pratense L.

TIMOTHY

Timothy is a large, non-native grass grown for hay and pasture, but often escaping into natural habitats. It differs from mountain timothy in size (growing 4 to 9 dm tall), in panicle shape (long-cylindric, 5 to 10 cm long), in having an enlarged, bulbous section at the base of the stem, and in having upper sheaths that are not inflated. It can be a common plant in both meadows and woodlands.

Poa L. — BLUEGRASS

The bluegrasses below are perennial species with leaves ending in the characteristic boat-shaped tip. Spikelets are 2 to several-flowered and occur in a dense to open panicle. The glumes and lemmas are keeled and awnless. The lemmas are somewhat acute and pubescent, sometimes with a tuft of crinkly (cobwebby) hairs at the base.

Stem-leaves 1 or 2; alpine or subalpine meadows. . . . . *P. alpina*
Stem-leaves numerous; lower elevations . . . . *P. pratensis*

Poa alpina L.

ALPINE BLUEGRASS

Alpine bluegrass is a small grass, 1 to 3 dm tall, that grows in tufts or mats. The leaves are short and occur mostly on short basal shoots. The panicle is pyramidal or ovoid in shape, and the spikelets are broad and purplish to bronze in colour. The lemmas are very hairy on the keel and marginal nerves. As its name suggests, alpine bluegrass is common at alpine elevations and also occurs at subalpine elevations. It can also be found in meadows and on rocky slopes.

NOTES: *Bluegrasses are characterized by boat-shaped leaf-tips and lemmas that are often sharply keeled on their backs.*

Poa pratensis L.

KENTUCKY BLUEGRASS

Introduced and native Kentucky bluegrass plants are indistinguishable and are commonly found throughout the area. The panicle is pyramidal or ovoid, and dense to somewhat open. The lower branches usually occur in a whorl of 5, all of different lengths. The lemmas have a conspicuous tuft of crinkly (webbed) hairs at their base. This grass is common in grassland, meadows and woodland.

NOTES: *Kentucky bluegrass is a sod-forming grass, spreading quickly by rhizomes which enable it to colonize open areas. Because of this characteristic, it is a common component of lawn-seed mixtures.*

Schizachne purpurascens (Torr.) Swallen

FALSE MELIC

This slender, loosely tufted perennial grows 4 to 8 dm tall and is characterized by an open, lax panicle of more or less drooping branches. The branches occur singly or in pairs and bear 1 or 2 awned spikelets approximately 2 cm long. The spikelets are purplish, bronze or pale in colour. The lemmas are 2-toothed at the apex, awned from just below the teeth and are long-hairy at the base. This grass is usually found in open woods, often at forest edges.

Stipa columbiana Macoun

COLUMBIA NEEDLE GRASS

This perennial bunchgrass derives its common name from the long slender awns (2 to 3.1 cm), usually twice bent, that arise from the apex of the lemma. The branches of the narrow, lax panicle are short and ascending; the spikelets are 1-flowered. The lemma is 4 to 6 mm long, somewhat hardened, rolled inwards along the margin and appressed-hairy. The plant generally grows 4 to 6 dm high, but can reach up to 12 dm, and occurs in dry grassland and open woodland.

Trisetum spicatum (L.) Richt.

SPIKE TRISETUM

This tufted perennial bunchgrass tends to have a dense spike-like panicle, hence its specific name *spicatum*. It differs from June grass in having fine, hair-like awns, 5 to 6 mm long, arising from approximately one-third of the way down from the tip of the lemma. The spikelets vary from pale to purplish, and can be dark purple at higher elevations. The leaves tend to be short and stiff, which often accentuates the colourful spikelets. The plant grows 10 to 50 cm high, generally on dry soils, in open woods, on mountain slopes and alpine tundra.

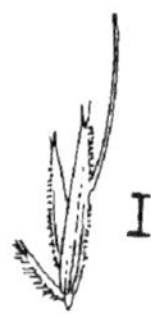

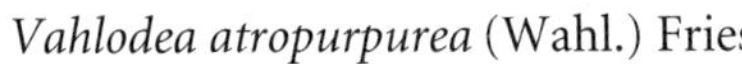

Vahlodea atropurpurea (Wahl.) Fries

MOUNTAIN HAIR GRASS

This grass differs from tufted hair grass in having glumes that extend beyond the upper floret of the spikelet, a more loosely tufted growth habit, and flowering stems that are purplish at the base. It grows 4 to 8 dm high and has flat leaves, 4 to 6 mm wide. The ligules are hairy and 1.5 to 3.5 mm long. The few slender branches of the panicle are drooping, with spikelets that tend to be purplish. Lemma awns may be straight or bent, approximately 2.5 mm long. Mountain hair grass is found in woods and wet meadows, often at high elevations, up to 2700 m.

Cyperaceae SEDGE FAMILY

The sedges are grass-like plants and are often difficult to distinguish from one another. They are usually perennial and arise from rhizomes with fibrous roots. Unlike the stems of grasses, the stems are often 3-sided and are usually solid. The leaves are usually 3-ranked, while those of grasses are 2-ranked. The tiny flowers are in the axils of scales (glumes) and are arranged in spikes or spikelets. There is usually a perianth, a ring of bristles or scales, enclosing 3 stamens and 1 carpel. In the main genus, *Carex*, the carpel is enclosed in a sac called a *perigynium*.

1. Flowers imperfect (flowers with one sex: male or female) 2
   Flowers perfect (flowers with both sexes) 3
2. Fruit enclosed in a flask-shaped bract . . . . *Carex*
   Fruit not enclosed in a flask-shaped bract. . . . . *Kobresia*
3. Perianth (sepals and petals) of numerous silky bristles . . . . *Eriophorum*
   Perianth of 1 to 12 parts 4
4. Spikelets 1; stems leafless. . . . . *Eleocharis*
   Spikelets 1 to many; stems usually leafy . . . . *Scirpus*

Carex L. SEDGE

Sedges are perennial herbs, generally with 3-sided stems. Leaves are 3-ranked with closed sheaths. There may be 1 or several spikes, each with a bract at the base that may be leaf-like or much reduced and scale-like. Plants may be monoecious or dioecious, with male and female flowers in different spikes, in the upper or lower portion of a spike, or on separate plants. Flowers are imperfect (unisexual), lacking a perianth. Male flowers consist of 3 stamens, and female flowers consist of 3 carpels joined together, with 2 or 3 stigmas, and a sac (perigynium) enclosing the ovaries. Each flower is subtended by a scale (glume).

1. Spikes 1 2
   Spikes 2 to 6. 3
2. Leaf-bases purplish red. . . . . *C. scirpoidea*
   Leaf-bases brownish green . . . . *C. filifolia*
3. Stems less than 10 cm tall; fruit orange . . . . *C. aurea*
   Stems more than 20 cm tall; fruit not orange 4
4. Leaf-bases reddish brown; male spikes 1 or 2 . . . . *C. aquatilis*
   Leaf-bases greenish brown; male spikes 2 to 4. . . . . *C. rostrata*

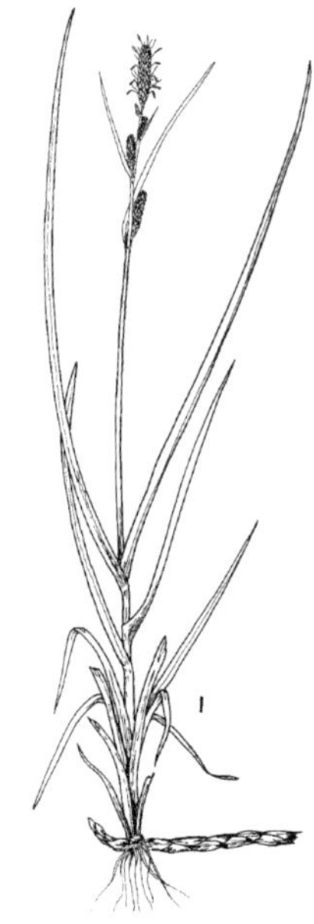

Carex aquatilis Wahlenb.

WATER SEDGE

This densely tufted perennial plant has long rootstocks resembling stolons. It grows 2 to 8 dm high. The stems are reddish at the base and sharply 3-sided at the top. The leaves are 2 to 5 mm wide and often glaucous-green. Male (staminate) and female (carpellary) florets are generally borne on separate spikes. The upper spikes are staminate and narrow, with purplish black to brown scales; the uppermost are stalked. The lower carpellary (pistillate) spikes are cylindric and erect with purplish black scales, often with the midvein extending slightly beyond the tip. The perigynia are light green and strongly flattened, with 2 stigmas. The lower bracts are leaf-like. This is a common sedge of marshy places from low to high elevations.

Carex aurea Nutt.

GOLDEN SEDGE

The distinguishing features of this small (up to 1 dm), stoloniferous sedge are the plump, elliptic-obovoid perigynia that vary from a light glaucous-green to a golden, orange or brownish colour when mature. It is this golden colour that gives rise to the species name: *aurea* means "golden." Staminate and carpellary florets are generally borne on separate spikes. There is a single terminal, primarily staminate, spike and 3 to 5 lateral carpellary spikes. The carpellary scales are pale to reddish brown, widely spreading at maturity, with a broad, green midrib that often extends slightly beyond the tip. There are 2 stigmas. The bracts are long and leaf-like, extending well beyond the inflorescence. This attractive sedge is found in moist meadows and wet, gravelly sites.

NOTES: *Awned sedge* (C. atherodes) *is another wetland sedge, found in wet meadows, ponds, fens, marshes and lakeshores. It grows 3 to 12 dm high from long, slender rhizomes and has webby sheaths at the base. Its dull-green leaves are few to several, usually shorter than the stem. Perigynia are yellowish green to light brown, 7 to 12 mm long and somewhat leathery.*

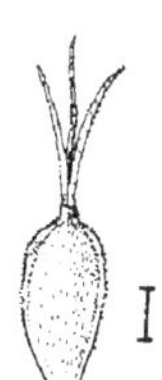

Carex filifolia Nutt.

THREAD-LEAVED SEDGE

This small, densely tufted perennial is characterized by its filiform (thread-like) involute leaves and slender, wiry stems. The old, brownish basal sheaths are conspicuous at the base. The inflorescence is made up of a solitary spike with staminate flowers at the top and carpellary flowers below (androgynous). The pistillate scales are light reddish brown with very broad hyaline margins. The perigynia are obovoid and more or less triangular, with 3 stigmas. This is a sedge of dry grassland and ridges.

Carex rostrata Stokes

BEAKED SEDGE

This large, robust sedge is similar in appearance to water sedge. It grows 5 to 10 dm high and has long creeping rhizomes. The leaves are pale green or glaucous and 2 to 10 mm wide. Staminate and pistillate florets are generally borne on separate spikes. The upper staminate spikes are narrow, have reddish brown scales and are stalked. The carpellary spikes are cylindric and generally lack stalks, although the lower may be stalked. The perigynia are ovoid, yellowish green to straw-coloured and shining, with 3 stigmas; they spread wide at maturity. The scales are narrow and purplish brown. The lower bracts are leaf-like. Beaked sedge prefers perennially wet sites at low elevations.

Carex scirpoidea Michx.

RUSH-LIKE SEDGE, SINGLE-SPIKED SEDGE

This sedge grows 1 to 4 dm high. The rootstocks are stout, scaly and dark reddish purple; the flowering stems are purplish red at the base. Male and female flowers are borne on separate plants (dioecious), and the plants possess only a single spike. Scales are ovate, chocolate-brown, ciliate and hyaline-margined. The perigynia are ovate to lanceolate with 3 stigmas. This sedge is found in moist, open areas to alpine elevations up to 2000 m.

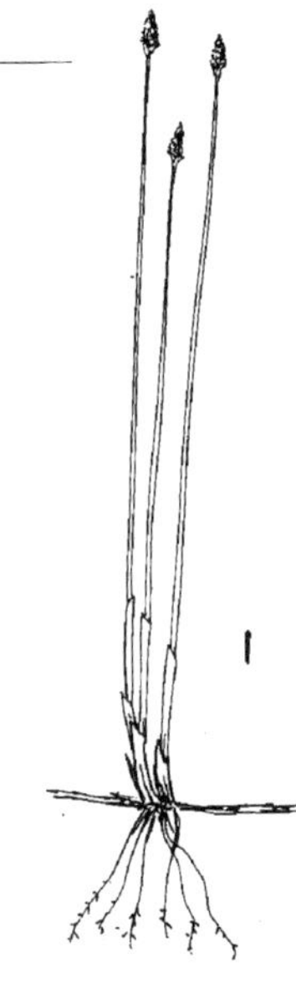

Eleocharis palustris (L.) R. & S.

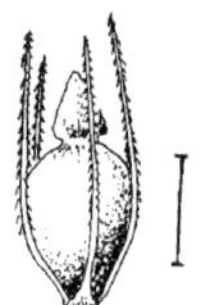

CREEPING SPIKE RUSH

This moisture-loving plant is characterized by single stems with leaves reduced to bladeless sheaths. It grows in tufts from stout creeping rhizomes, 1 to 10 dm tall. The stems are circular in cross-section or somewhat flattened. The sheaths are often reddish below. The solitary, terminal spikelet is 5 to 20 mm long. The 1 to 3 sterile scales at the base of the spikelet are obtuse, whereas the fertile scales above are narrower, usually with a firm midrib extending to the apex. The perianth consists of 4 bristles. The achene is lens-shaped, yellow to brown, and smooth, with the base of the style much enlarged to form a tubercle at the apex of the achene. The spike rush is found in wet places.

Eriophorum viridi-carinatum (Engelm.) Fern.

THIN-LEAVED COTTON GRASS, GREEN-KEELED COTTON GRASS

This plant derives its common name from the perianth, which is cleft to the base into silky bristles. As a result, each spikelet has the appearance of a cotton boll. The spikelets are borne on drooping peduncles, and the bristles can be white to very pale brown. The scales are greenish to blackish with a prominent midrib extending to the tip. Stems are either solitary or in small tufts, 3 to 9 dm tall and 3-sided. These plants occur in boggy woods, meadows and stream margins.

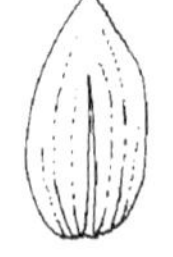

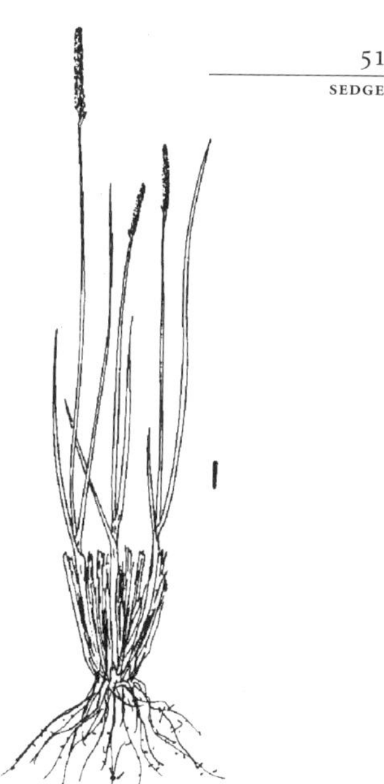

Kobresia myosuroides (Vill.) Fiori & Paol.

BOG SEDGE, BELLARD'S KOBRESIA

Bellard's kobresia grows in compact tussocks, 1 to 3 dm tall, with long, erect, wiry leaves. The spikes are narrow; the upper spikelets are staminate while the lower spikelets each have 1 carpellary and 1 staminate flower (rarely, only the carpellary flower). This plant is easily mistaken for a sedge, but can be distinguished by the conspicuous old sheaths, brown and fibrillose, and the lack of a perigynium surrounding the ovary. In place of the perigynium is a bract-like glume that closely envelopes the ovary; it is light brown and shining. Bellard's kobresia can be found on dry alpine slopes and turfy, exposed alpine areas.

Scirpus L. — BULRUSH

Bulrushes are perennial plants of wet soil. They may be densely tufted or rhizomatous. The species below have stems that are circular in cross-section. Leaf-blades may be grass-like, or they may be much reduced and tapered to a point. There are 1 to several spikelets in an umbellate or spicate inflorescence with a bract at the base. Spikelets are 2 to several-flowered. Flowers are perfect, with a perianth of 1 to 6 short bristles, and surrounded by overlapping scales.

Spikelets 1; 4 to 5 mm long . . . . *S. caespitosus*
Spikelets 1 to 5; 7 to 20 mm long . . . . *S. acutus*

Scirpus acutus Muhl. ex. Bigel.

GREAT BULRUSH

This is a tall plant of marshes and lakeshores, growing 0.5 to 3 m high. It has thick, spongy rhizomes and tends to grow in pure stands. The stems are circular in cross-section and firm. The involucral bract is erect, 1 to 4 cm long, and appears as a continuation of the stem. The inflorescence is composed of several branches, each bearing 1 to 5 spikelets. The branches may be long and the inflorescence open, or short and the inflorescence compact. The scales of the spikelets are pale, often flecked with red or brown, and have laciniate-fimbriate margins.

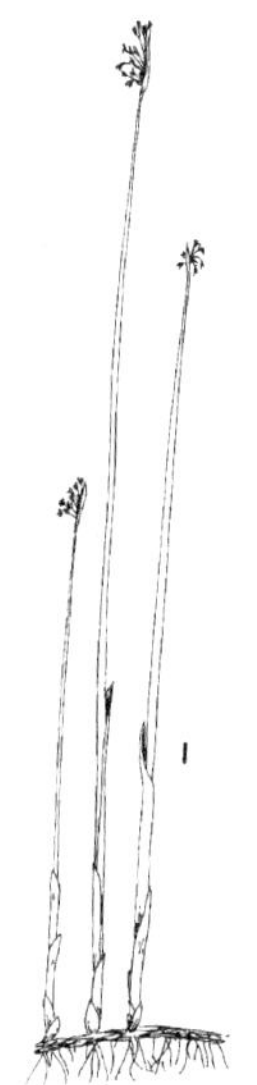

Scirpus caespitosus L.

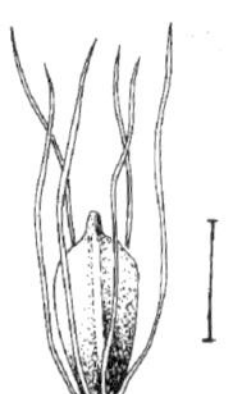

TUFTED BULRUSH

This plant is similar in appearance to spike rush. The stems are circular in cross-section, wiry and smooth, clothed at the base with conspicuous, straw-coloured sheaths. Only the uppermost sheath has a leaf-blade attached; it is very short (5 to 8 mm long). The spikelet is solitary and terminal, 4 to 5 mm long, with chestnut-brown scales. The midvein of the lowest bract is prolonged into a blunt awn shorter than or exceeding the spikelet. The perianth is made up of 6 delicate white bristles surrounding the achene. This plant grows in dense tufts and tussocks in bogs, fens and wet tundra from low elevations to the alpine region.

Lemnaceae DUCKWEED FAMILY

Duckweeds are minute plants, 2 to 10 mm in diameter, found either floating, as a mat of thousands of individual thalli like a green carpet, or submerged in water. Each plant is a flat frond *(thallus)* which is not differentiated into stem and leaf and bears a few tiny rootlets on the undersurface. The flowers are rarely seen; they are unisexual. The male flower consists of a single stamen; the female flower is reduced to a flask-shaped ovary, which, when fertilized, develops into a utricle with 1 to 7 seeds.

Vegetative reproduction is normal in the duckweed family, as sexual reproduction is so problematic: it is done by *budding*. New fronds are produced from the edge or base of the parent frond; these separate to form a new colony. There is also another method of vegetative reproduction. In the fall the fronds form bulblets which sink to the bottom of the pond. In the spring they float to the surface and develop into new fronds. Duckweeds are found in ponds and streams.

Lemna trisulca L.

IVY-LEAVED DUCKWEED

Ivy-leaved duckweed is the most interesting of the duckweeds. When budding takes place, some of the thalli, which have stalks, remain together and form a tiny (6 to 10 mm in diameter) plant which appears to have 3 lobes; hence the specific name *trisulca.* This trait makes it easy to distinguish it from the other species of duckweed. Common duckweed (*L. minor*) has a stalkless thallus only 2 to 5 mm long, which does not appear to be 3-lobed. Ivy-leaved duckweed is found in quiet water; common duckweed is also found in quiet water and sometimes in beaver ponds. Both species often form a green carpet over the water.

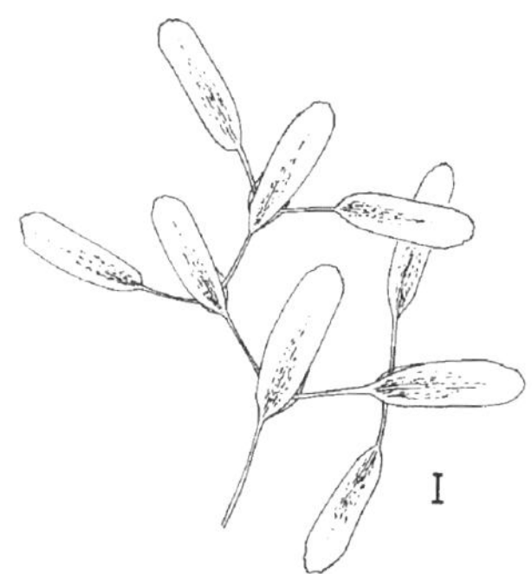

Juncaceae RUSH FAMILY

These plants are perennial, with grass-like leaves, and are either tufted or arise from creeping rhizomes. The flowers are arranged in clusters called *cymes.* The flower parts are either green or brown, and look like tiny brown scales. There are 6 perianth segments, representing the sepals and petals; also 3 or 6 stamens and a superior ovary, divided into 3. The fruit is a capsule that splits into 3 parts.

Leaves flat, edges hairy . . . . *Luzula*
Leaves round in cross-section, edges not hairy . . . . *Juncus*

Juncus L. RUSH

Rushes are smooth, perennial herbs of wet meadows, bogs, shores and moist ground. The terete stems (circular in cross-section) arise singly or in tufts from creeping rhizomes. Leaves may be terete or may be lacking and reduced to sheaths. There are 1 to several heads in a compact or loosely branched cyme of few to several flowers. The involucral bract may be flat or terete and may appear as a continuation of the stem. The flowers are perfect, with 6 perianth segments; these are small, brown and scale-like.

Inflorescence terminal; stem-leaves 1 or 2 . . . . *J. alpinoarticulatus*
Inflorescence on one side of stem; stem-leaves absent . . . . *J. balticus*

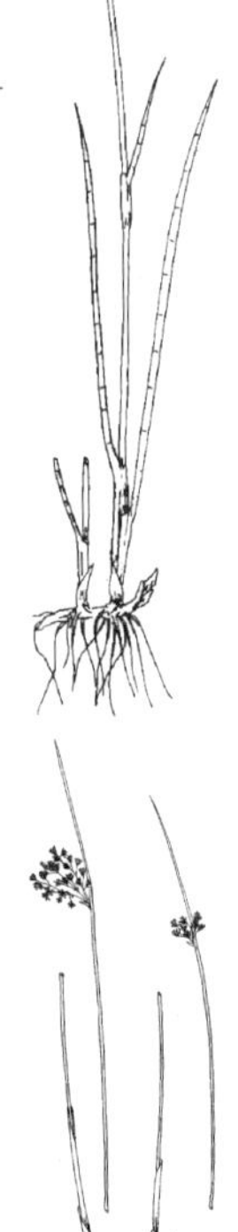

Juncus alpinoarticulatus Chaix (*Juncus alpinus* Vill.)

ALPINE RUSH

Alpine rush is a tufted perennial, 1.5 to 5 dm tall, arising from creeping rhizomes. The plants have 1 or 2 stem-leaves, and the flowering-heads have 2 to 10 flowers, arranged in a cyme. The 6 scales representing sepals and petals are usually purplish brown. The plant is found on the margins of ponds and in wet meadows.

Juncus balticus Willd.

WIRE RUSH

Wire rush is a fairly tall perennial, 2 to 6 dm high, arising from a stout rhizome. The basal leaves are reduced to brownish sheaths. The single inflorescence is characteristic; it appears lateral and is often diffusely branched. This plant is found in wet meadows and bogs, as well as in shallow water or mud.

Luzula DC. — WOOD RUSH

Wood rushes are grass-like perennial herbs. Stems may be single or in tufts, with a cluster of leaves at the base and 1 to several leaves along the stem. Leaf-blades are flat or channelled. The inflorescence may be an open or spike-like cyme, often nodding. The flowers are perfect, each with bractlets at the base and 6 perianth segments. The latter are small, brown and scale-like.

Stem-leaves 1 to 3 . . . . *L. spicata*
Stem-leaves 4 or more. . . . . *L. parviflora*

Luzula parviflora (Ehrh.) Desr.

SMALL-FLOWERED WOOD RUSH

This is a tufted perennial, with stems 3 to 6 dm tall, usually bearing 4 stem-leaves, 4 to 10 mm wide. The inflorescence has slender branches and is characteristically nodding. The small-flowered wood rush is found in moist forests and marshy areas.

Luzula spicata (L.) DC.

SPIKED WOOD RUSH

This is a small tufted alpine plant, with slender stems, 1 to 3 dm tall. The inflorescence is nodding, and the tiny flowers are arranged in a dense spike, hence the species name *spicata*. Spiked wood rush is found on rocky or grassy alpine slopes.

Liliaceae LILY FAMILY

The members of the lily family vary in the arrangement of their flowers (inflorescences) and also in plant-habit. The flowers can be solitary or clustered, but all the flower-parts are in 3s or multiples of 3. The calyx of 3 sepals and the corolla of 3 petals are usually so much alike that their parts are collectively called *tepals*; the lily itself is a good example. There are usually 6 stamens and 3 united carpels in the centre which form a 3-locular structure, with *axile placentation*. The ovary is superior and has 3 styles; the fruit is a capsule or berry.

1. Leaves basal 2
   Leaves alternate 6
2. Leaves with an onion odour . . . . *Allium*
   Leaves without an onion odour 3
3. Flowers yellow or bronze-coloured 4
   Flowers white or greenish white 5
4. Flowers yellow; leaves 2; plants of higher elevations . . . . *Erythronium*
   Flowers bronze; leaves more than 2; montane forests . . . . *Stenanthium*
5. Plants of moist areas; leaves set edgewise to the stem . . . . *Tofieldia*
   Plants of dry open areas and woods; leaves not set edgewise . . . . . . *Zigadenus*
6. Flowers orange or yellow 7
   Flowers white or greenish white 8
7. Flowers orange; fruit a dry capsule . . . . *Lilium*
   Flowers yellow; fruit a red berry . . . . *Disporum*
8. Sepals and petals united; flower stalks twisted . . . . *Streptopus*
   Sepals and petals not united; flower stalks not twisted . . . . *Smilacina*

Allium L. ONION, CHIVES

These are perennial, bulbous plants with a strong odour and slender basal leaves. Each bulb produces a tall scape (flowering stalk) which bears the flowers at the tip in an umbellate cluster. When in bud, flowers are protected by 1 to 3 translucent persistent bracts. There are 6 perianth-leaves, pinkish white or purple, and 6 stamens and 3 carpels, united. The fruit is a capsule with black seeds.

Flowers purple; leaves round in cross-section . . . . *A. schoenoprasum*
Flowers pink or white; leaves flat . . . . *A. cernuum*

Allium cernuum Roth

NODDING ONION

Nodding onion is a low, slender plant, 1 to 5 dm tall, that looks and smells like a small garden onion. The leaves are long, narrow and ridged; they grow from a bulb. The small flowers, pale lavender or white and about 6 mm across, are lily-shaped; there are several in a nodding cluster at the top of the main flowering stalk. The 3 petals and 3 sepals are all alike, 5 to 7 mm long; these are often called tepals. There are 6 stamens and 3 carpels. The latter are joined together in the centre of the flower. When fertilized, they become fruits, small capsules which each split into 3 valves, releasing several black seeds. Nodding onion is found in open, grassy meadows, in thickets, and on open slopes and rocky slides up to 2300 m. NOTES: *The bulbs were eaten by native peoples, and explorers Lewis and Clark found the bulbs a welcome addition to their diet.*

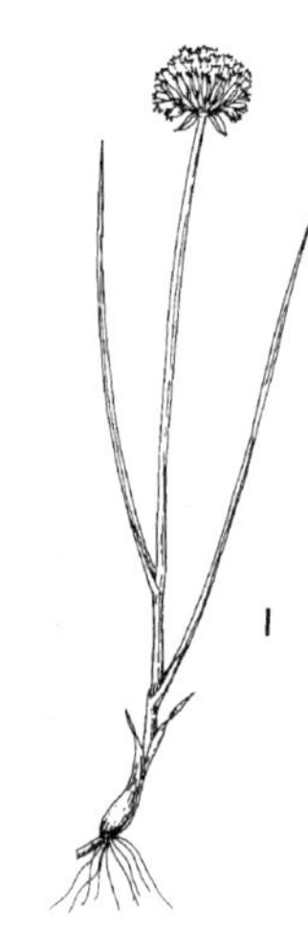

Allium schoenoprasum L.

WILD CHIVES

This plant is similar to nodding onion, but can be distinguished by its hollow leaves and by its flower-head (inflorescence), which is erect and not "nodding." The flowers are also different: instead of pale lavender or white, they are purple. The plant can be seen in damp, open places, such as mountain meadows and lake shores. It can also be found at higher elevations up to 2300 m.

Disporum trachycarpum (S. Wats.) B. & H.

FAIRY-BELLS

This plant grows from a rhizome; the stem is 3 to 8 dm tall. The leaves are ovate and pointed at the tip, with well-marked parallel veins. The flowers are often in pairs at the tips of the branches and are bell-shaped, drooping, whitish or greenish yellow. The flowers are typical of the lily family, with 3 sepals and 3 petals all alike, 6 stamens and 3 carpels joined together in the centre. The fruit is a berry, brilliant red when ripe. Fairy-bells grow in moist woods.

NOTES: *This species has been found north of Wasootch Creek (flowing into the Kananaskis River), on Lusk Creek Ridge in a spruce-aspen-pine forest and in many other places.*

Erythronium grandiflorum Pursh

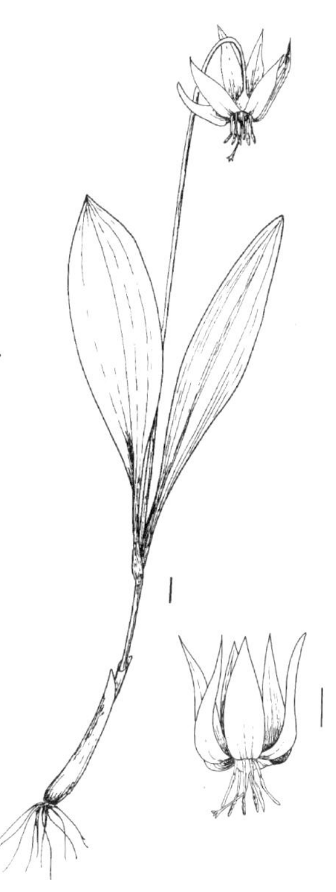

GLACIER LILY, DOG-TOOTH VIOLET

Glacier lily grows 1 to 4 dm high, with 2 leaves arising from the base. These leaves are lanceolate to ovate. The flowers are attractive: large, bright yellow and nodding, with typical lily-like arrangement: 6 tepals all alike, 6 stamens and 3 carpels (joined) in the centre. The tepals are reflexed, so the flower looks something like a cyclamen flower. The fruit is a 3-sided erect capsule. The glacier lily is found in open forests from foothills up to timberline and on grassy slopes. It grows in rich soil and blooms when the snow melts.

NOTES: *This plant has been found in the Kananaskis Valley in a high, moist, alpine meadow between Mount Rae and Mount Arethusa, also north of the Highwood Pass Summit, on the trail to Elbow Lake and in several other places.* ⁓ *The common name "glacier lily" is rather apt: the plant has a lily-like flower and is found in alpine regions where glaciers are not uncommon.* ⁓ *The other common name, dog-tooth violet, is misleading: the plant is nothing like a violet!*

Lilium philadelphicum L.

WESTERN WOOD LILY

Western wood lily is a handsome plant with large, brilliant, orange-red flowers. The colour comes from the 6 tepals which are spotted near the base. There are 6 stamens, free, and 3 carpels, joined, in the centre. These carpels produce the fruit, a capsule with 3 valves. The plant arises from a bulb, often characteristic of the lily family. The western wood lily grows in open, grassy meadows and aspen groves.

NOTES: *This species has been found near the summit of Moose Mountain, also at Sibbald Creek and many other places. Western wood lily is a common spring plant in Bow Valley Provincial Park.*

The specific name philadelphicum *comes from Philadelphia, where Kalm, a pupil of Linnaeus, collected the original specimens.* *The Stoneys collected the bulbs in July and August and ate them raw or in soups.* *This is one of the loveliest flowers in Kananaskis Country—it is outstanding. It has been chosen as the floral emblem of Saskatchewan.*

Smilacina Desf. — SOLOMON'S-SEAL

Members of this genus are perennial plants arising from a branching rhizome. The rhizome produces annual stems with broad to lanceolate leaves, often clasping, and terminating in a cluster of white flowers, a raceme or a panicle. Each flower has 6 tepals (perianth-leaves), 6 stamens and 3 united carpels. The fruit is globose, berry-like.

1. Leaves 3 .... *S. trifolia*
   Leaves more than 3 2
2. Flowering-stalk with numerous branches .... *S. racemosa*
   Flowering-stalk not branched .... *S. stellata*

Smilacina racemosa (L.) Desf.

FALSE SOLOMON'S-SEAL

This plant grows 3 to 10 dm tall from a thick, fleshy rhizome. The stem leaves are usually sessile and lanceolate; they clasp the stem. The plant produces many white flowers, arranged in a pyramidal panicle 5 to 15 cm long. The flowers are small, with the usual lily-like arrangement: 6 tepals, 6 stamens and 3 carpels. The carpels form the fruit, a red berry dotted with purple. False Solomon's-seal is found in moist, shaded habitats at lower and middle elevations.

NOTES: *This species has been recorded from Wasootch Creek, at the west end of Barrier Lake, near the Kananaskis Field Stations, and elsewhere.* *The common name "false Solomon's-seal" is a misnomer: the plant does not look like the true Solomon's-seal.*

Smilacina stellata (L.) Desf.

STAR-FLOWERED SOLOMON'S-SEAL

While false Solomon's-seal is a tall, coarse, plant that can reach 1 m in height, star-flowered Solomon's-seal is a relatively dainty plant, 2 to 6 dm high. The flowers are white and similar to those of false Solomon's-seal, but the plants can be distinguished: star-flowered Solomon's-seal has flowers sparsely arranged in a single raceme, not a panicle, and the individual flowers are larger. It arises from a rhizome and spreads rapidly. The berries of star-flowered Solomon's-seal are distinctive: they are green at first, then striped with deep purple, and finally black. The plant grows in a wide variety of habitats, including prairie grassland, open woods and meadows and higher north slopes.

NOTES: *This species has been found in an alpine meadow on Plateau Mountain at 7600' (2320 m), on a steep, open slope overlooking Cat Creek bridge and many other places.* *The first part of this plant's common name is apt: the flowers are distinctly star-shaped (*stella *is Latin for "star").*

Smilacina trifolia (L.) Desf.

THREE-LEAVED SOLOMON'S-SEAL

This plant can easily be distinguished from the others because it has only 3 leaves and its habitat is different. In three-leaved Solomon's-seal the raceme of flowers has a long, main flowering-stalk (*peduncle*); the peduncle of the star-flowered Solomon's-seal is shorter. Three-leaved Solomon's-seal is found in bogs and wet woods, often in *Sphagnum* moss.

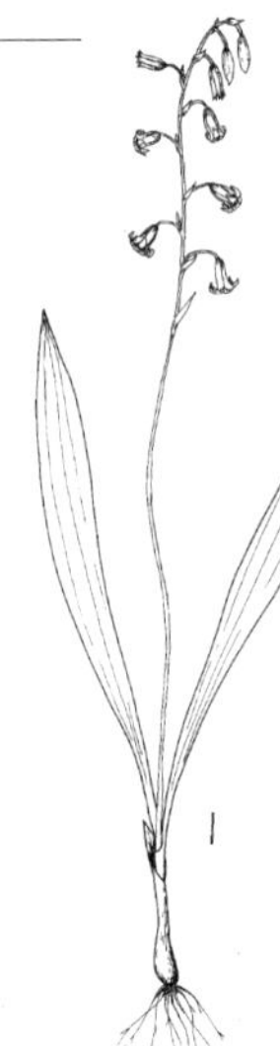

Stenanthium occidentale A. Gray

BRONZE-BELLS

This plant develops from a bulb and has a slender stem bearing several greenish or brownish red flowers in a raceme. The nodding flowers have slender stalks and are greenish bronze and bell-shaped, hence the common name. They have the usual lily-like structure; there are 2 or 3 basal leaves which are linear or linear-lanceolate. The fruit is a capsule. The plant grows in moist woodland.

NOTES: *This species has been found in numerous places, including white spruce woods near Elbow Falls, in* Abies-Picea *krummholz, near the summit of Moose Mountain and elsewhere.* ❧ *The species name* occidentale *means "western."*

Streptopus amplexifolius (L.) DC.

CLASPING-LEAVED TWISTED STALK

Twisted stalk is a perennial plant, growing from a thick rhizome; it reaches 3 to 10 dm, and the stem is often branched. The alternate leaves are characteristic: they are sessile, often clasping the stem, and they have wavy edges. The flowers grow in the axils of the leaves, often hidden by them; each flower is bell-shaped and has a slender, twisted stalk (hence the common name). The petals and sepals are very similar and have up-turned edges. There are 6 stamens and the fruit is a red berry with several seeds; it is **inedible**. Twisted stalk is found in moist, shady woods.

Tofieldia Huds. FALSE ASPHODEL

Perennial plants with short rootstocks and mostly basal leaves, slender. The flower-stalk is tall and slender, crowned with a dense spike-like raceme of greenish white flowers, each with 6 perianth-leaves, 6 stamens and 3 carpels, united. The fruit is a capsule, partly enclosed by a persistent perianth.

Flowering-stem sticky .... *T. glutinosa*
Flowering-stem not sticky .... *T. pusilla*

Tofieldia glutinosa (Michx.) Pers.

STICKY FALSE ASPHODEL

This plant is 1 to 5 dm tall and has a slender, glutinous (sticky) stem, hence the specific name *glutinosa.* There are several linear basal leaves, and the flowers are arranged in a small, dense head at the top of the flower-stalk. They are whitish and have the typical lily-like structure, although some features are very small. The 3 carpels in the centre are joined and are surrounded by the 6 tiny stamens and 6 tiny tepals. The fruit is a capsule, yellowish or red. Sticky false asphodel is found in calcareous marshes.

Tofieldia pusilla (Michx.) Pers.

DWARF FALSE ASPHODEL

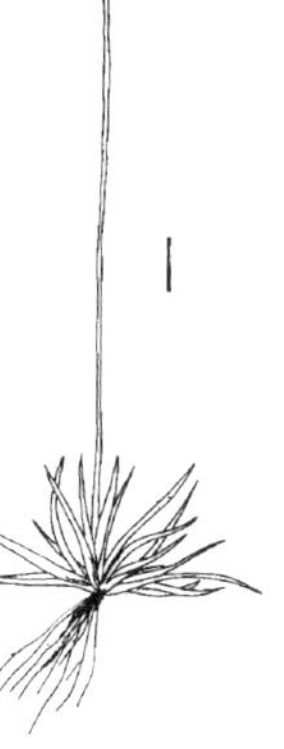

This is a much smaller plant than *T. glutinosa*; it is 0.5 to 2 dm tall, and the linear basal leaves grow in a small tuft. The flower-head is much smaller than that of the other species, and the plant is not glutinous. It is found in calcareous marshes and on wet ledges.

Zigadenus Michx. — CAMAS

Members of this genus are bulbous plants with several grass-like, linear leaves, chiefly basal; the bulbs have blackish coats. The erect flowering-stem rises from the bulb, crowned by a racemose inflorescence, yellowish white, sometimes greenish, with 6 perianth leaves (tepals) with glands at their bases, 6 stamens and 3 united carpels. The fruit is a 3-lobed capsule with persistent styles.

Petals 7 to 10 mm long, greenish white . . . . *Z. elegans*
Petals 4 to 5 mm long, creamy-white. . . . . *Z. venenosus*

Zigadenus elegans Pursh

WHITE CAMAS

This plant is tall (2 to 6 dm) and graceful. It arises from a bulb, and its leaves are linear; they are acute at the apex. The greenish white flowers are arranged in a loose raceme, and all the parts are in 3s or multiples of 3: 6 tepals, 6 stamens, and 3 carpels, which are joined. Each tepal has a dark gland near the base. The fruit is a capsule. The plant is **slightly poisonous** and is found in moist meadows, open woods, and alpine areas.

NOTES: *This plant has been found near the summit of Hailstone Butte, in an alpine meadow near the summit of Moose Mountain, and commonly throughout Kananaskis Country.* *It is an attractive plant and justifies its specific name* elegans.

Zigadenus venenosus S. Wats.

DEATH CAMAS

Death camas is fairly easily distinguished from white camas because the arrangement of the flowers is quite different. In white camas, the flowers are arranged in a loose raceme, but in death camas they are arranged in a short, dense raceme. Death camas flowers are smaller, and their colour is much creamier. Death camas is a smaller plant, found on plains and hillsides.

NOTES: *The common name for this plant is justified: the plant is* ***extremely poisonous.*** *When native peoples were gathering the true camas bulbs* (Camassia) *for food, they had to be careful not to gather death camas bulbs by mistake.*

Iridaceae IRIS FAMILY

These are perennial plants arising from rhizomes, with leaves in 2 ranks. The flower parts are in 3s: 3 petals and 3 sepals (all alike in Alberta species and called "tepals"). There are 3 stamens and 3 carpels, forming a 3-locular ovary, which is inferior. The fruit is a capsule, splitting into 3 parts; there are many seeds. The flower is epigynous: it has an inferior ovary.

The iris family is very similar in some ways to the lily family, but possession of an inferior ovary makes an important difference. The ovary in irises is below the other flower parts, while in the lily family the ovary is superior. Another difference is that the flowers of the lily family have 6 stamens, while those of the iris family have only 3. The species described below look like lilies, but another species, *Iris missouriensis* (not found in Kananaskis Country), looks like the garden iris and its flowers have a more complicated structure.

Sisyrhinchium L. BLUE-EYED GRASS

The blue-eyed grasses are tufted, fibrous-rooted perennials featuring 2-edged stemsand grass-like leaves. The dainty flowers, blue-violet with yellow centres, have 2 ensheathing bracts, 6 perianth leaves, 3 stamens and 3 carpels, united. The fruit is a capsule.

Flowers blue or purple; leaves 1 to 4 mm wide . . . . *S. montanum*
Flowers light purple or white; leaves 1 to 2 mm wide . . . . *S. septentrionale*

Sisyrinchium montanum Greene

COMMON BLUE-EYED GRASS

The stems of this blue-eyed grass are 1 to 5 dm tall. It has attractive blue flowers, with all the parts in 3s. The fruit of the blue-eyed grass is a capsule. The plant is found in moist meadows and also on prairie grassland and dry, grassy hillsides.

NOTES: *The common name "blue-eyed grass" is a misnomer: this plant is a member of the iris family and is* not *a grass, although the leaves are rather grass-like.*

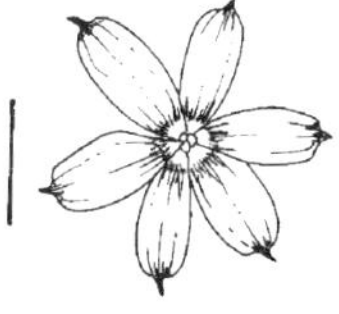

Sisyrinchium septentrionale Bicknell

PALE BLUE-EYED GRASS

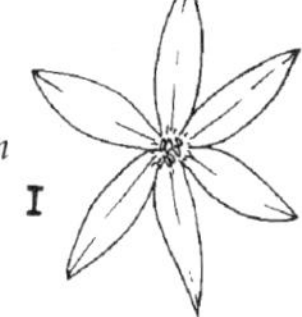

This plant can be distinguished from *S. montanum* because its flowers are a pale blue or white and its tepals are different. In common blue-eyed grass, the tepals have a shallow notch at the tip and a tiny beak; in pale blue-eyed grass, the tepals end in a beak and lack a shallow notch. The plant is found in moist grassy areas.

NOTES: *The specific name* septentrionalis *means "northern"; the name is often seen on old maps.*

Orchidaceae ORCHID FAMILY

Orchidaceae is now recognized by botanists as the most highly developed of all the plant families. Orchids have almost worldwide distribution, and the flower structure is exceedingly complicated, with many structures to aid effective cross-pollination. Pollination is carried out by various types of insects.

The plants arise from bulbs, corms, tubers, or rhizomes, and the leaves are simple, entire and sheathing. The flower structure is as follows: 3 sepals which are often petaloid (coloured) but may be green; 3 petals, white or coloured; 2 of these petals are similar, but the third one, the "lip," is unlike the others and often has a basal spur. The stamens are united with the style and stigma, and form a central *column* (see the Glossary, pp. 311–339). This column is characteristic and is found in no other plant family. The pollen grains are often stuck together in masses called *pollinia.* The style often has a beak called a *rostellum.* The ovary is inferior, and the fruit is a capsule that splits into 3 parts, releasing numerous minute seeds rather like dust. The flowers may be solitary or arranged in racemes or spikes. All orchids are saprophytic to some extent (i.e., they obtain some nourishment from humus and decayed matter in the soil, with the aid of mycorrhizal fungi), but some are only slightly so. *Corallorhiza* species are almost completely saprophytic.

All the members of the orchid family have flowers with the characteristic column, stamens, ovary and stigma, mentioned earlier, but the flowers all have a well-marked inferior ovary, which, being green, can be mistaken for a flower stalk. An excellent textbook describing the complex Orchidaceae family is *The Native Orchids of the United States and Canada, excluding Florida,* by Carlyle A. Luer, 1975, published by The New York Botanical Garden.

NOTE: Habenaria *species have now been revised to the genus* Platanthera. Habenaria viridis *has been revised to* Coeloglossum viride. Orchis rotundifolia *has been revised to* Amerorchis rotundifolia.

1. Leaves reduced to scales; stem appearing leafless . . . *Corallorhiza*
   Leaves green, basal, alternate or opposite . . . 2
2. Leaves basal . . . 3
   Leaves alternate or opposite . . . 5
3. Leaves more than 3 . . . *Goodyera*
   Leaves 2 or less . . . 4
4. Flower with an inflated pouch . . . *Calypso*
   Flower without an inflated pouch . . . *Orchis*
5. Leaves opposite, 2 . . . *Listera*
   Leaves alternate, several . . . 6
6. Flower with an inflated pouch; leaves sheathing . . . *Cypripedium*
   Flower without an inflated pouch . . . 7
7. Flowers with a spur . . . *Habenaria*
   Flowers without a spur; sweet-scented . . . *Spiranthes*

Calypso bulbosa (L.) Oakes

VENUS'-SLIPPER

Venus'-slipper grows from a corm, although the specific name is *bulbosa.* A corm is a solid bulb, a swollen underground storage organ containing starch; it looks like a bulb but has no fleshy leaves. Venus'-slipper has a single, ovate basal leaf with well-marked veins. The flowering stalk is 5 to 15 cm high with a single nodding flower at the apex. The flower has the typical orchid structure: 3 sepals all alike (they may be purple, pink or yellow), 2 lateral petals, similarly coloured, and a third petal modified into a large, distended lip, white with purple markings. In orchid flowers, the stamens and carpels are united in a complex structure called a *column*; the Venus'-slipper is no exception. It grows in coniferous forests and has been found in many pine-forest areas.

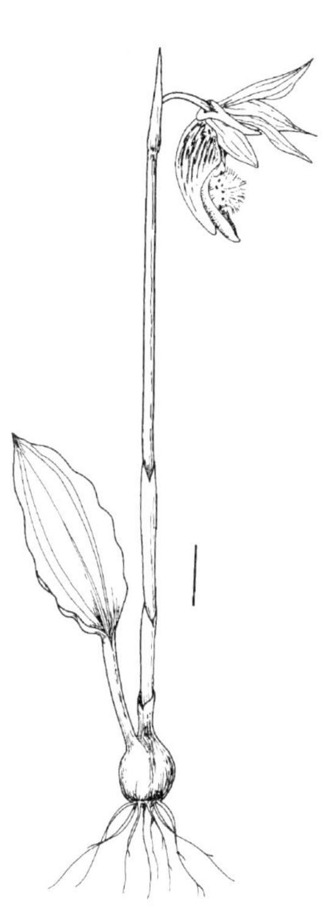

NOTES: *This orchid is one of the loveliest flowers in Kananaskis Country. It is a dainty plant with striking colouration.* ~ *The generic name* Calypso *comes from a Greek sea-nymph Kalypso in Homer's* Odyssey.

Corallorhiza Châtelain — CORAL-ROOT

Members of this genus are saprophytic plants with scales instead of green leaves, arising from branching coralloid rhizomes. The flowers develop in a spike-like raceme with 3 sepals (lateral ones united at base, often spurred), 3 petals, 4 pollinia; the fruit is a capsule. The coral-root species are all characterized by knobbly rhizomes with a superficial resemblance to coral. The genus name *Corallorhiza* stresses this character. They are also easy to distinguish from other orchids because their leaves have been reduced to sheathing scales. Coral-root orchids are saprophytic: they obtain nourishment from decayed matter in the soil with the aid of mycorrhizal fungi, and do not make their food by photosynthesis. This is why the leaves are reduced to scales.

1. Stems yellow-green . . . . *C. trifida*
   Stems purplish brown 2
2. Lower petal with purple stripes; spur absent . . . . *C. striata*
   Lower petal with purple spots; spur present . . . . *C. maculata*

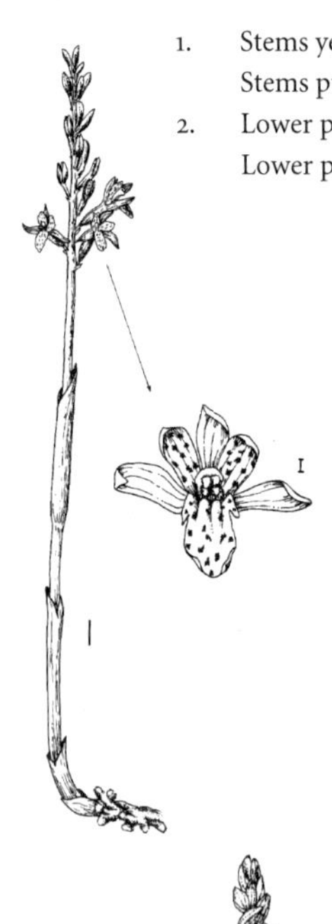

Corallorhiza maculata Raf.

SPOTTED CORAL-ROOT

The flowering-stalk (scape) of spotted coral-root is 2 to 5 dm tall; the flowers are borne in a spike with 10 to 40 whitish flowers spotted with red or purple (*maculata* means "spotted"). The flowers are sessile or have only tiny stalks. The 3 sepals and 2 lateral petals are alike, and the third petal (lip) has 2 lateral lobes and conspicuous spots. This orchid is found in coniferous and deciduous woods.

Corallorhiza striata Lindl.

STRIPED CORAL-ROOT

This plant can readily be distinguished from the spotted coral-root; the petals and sepals are striped (*striata* means "striped") and the lip has no spur. The flowering scape is 1.5 to 4 dm high—shorter than that of the spotted coral-root. The striped coral-root is found in woods.

Corallorhiza trifida Châtelain

PALE CORAL-ROOT

Pale coral-root is only 1 to 3 dm high and is a relatively inconspicuous plant. Its flowers are much smaller than those of striped or spotted coral-root; they are about 5 mm long, yellowish or greenish, sometimes a dull purple. Pale coral-root is found in bogs, thickets, and woods; it is fairly common in Kananaskis Country.

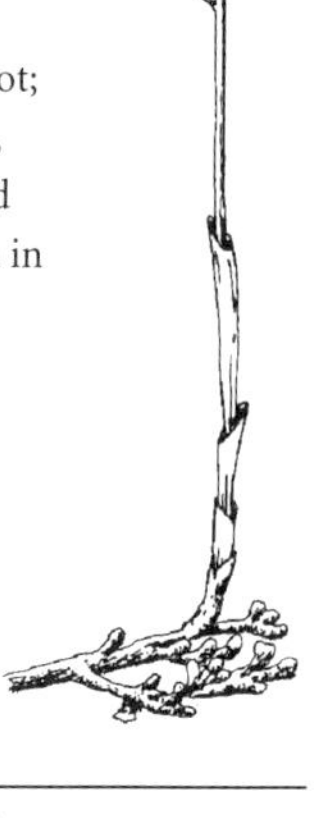

Cypripedium L. — LADY'S-SLIPPER

Members of this genus are rhizomatous perennials with erect stems that bear large leaves, sheathing at the base, with well-marked, parallel veins. The flowers number 1 to 3 and are large and showy, each subtended by a leafy bract, 3 sepals and 3 petals. The lower 2 sepals are united, the upper 2 petals spreading and the lower petal forming the "lip," a conspicuous, pouch-like structure. The central column bears 2 fertile stamens, a single staminode and a broad stigma. The fruit is a 3-valved capsule, with many minute seeds.

Pouch 2 to 4 cm long, yellow . . . . *C. calceolus*
Pouch 1 to 2 cm long, white or pink . . . . *C. passerinum*

Cypripedium calceolus L.

YELLOW LADY'S-SLIPPER

This is a handsome orchid, and the yellow flowers are quite conspicuous. It grows from a rhizome. The stem is 1 to 4 dm tall, usually with 3 or 4 leaves; these leaves are ovate to lanceolate. The 3 sepals are yellowish with purple stripes, and the 2 lateral petals are greenish yellow and spirally twisted. There are usually only 1 or 2 flowers on the flower stalk. The large, yellow flowers are characterized by a pouch-shaped lip 1.5 to 4 cm long: the "lady's slipper". Yellow lady's-slipper is found in moist woodlands.

NOTES: *The common name is a corruption of "Our Lady's slipper"—the species is called* Sabot de la Vierge *in French.*

Cypripedium passerinum Richard.

SPARROW'S EGG ORCHID, NORTHERN LADY'S-SLIPPER

This plant is smaller than yellow lady's-slipper. It grows only 1 to 2.5 dm tall, and the lip of the flower is smaller, only 1.5 cm long. The lip is usually white and pink, with purple spots inside, and is egg-shaped, hence the common name. This orchid is found in moist woodlands and has been collected in spruce woods on the Jumping Pound Road.

NOTES: *The species name* passerinum *refers to sparrows.*

Goodyera repens (L.) R. Br.

LESSER RATTLESNAKE PLANTAIN, CREEPING RATTLESNAKE PLANTAIN

Rattlesnake plantain has a basal cluster of leaves. From the centre of this cluster an unbranched flowering stalk (a spike-like raceme) rises, bearing numerous small flowers which may be white or greenish. In this species, the raceme is 3 to 6 cm long. In a related species, *G. oblongifolia*, the raceme is 6 to 10 cm long. Each flower is "hooded." The hood (5 mm long) consists of the upper sepal, joined to the lateral petals. The lateral sepals are free, and the lip is deeply saccate (sac-like), with recurved or flaring margins. Rattlesnake plantain is found in damp woods.

NOTES: *The common name may have arisen from the native tradition of using the plant to heal rattlesnake bites.* ~ *"Plantain" probably comes from the fact that the leaves resemble those of plantain.* ~ *The generic name* Goodyera *honours John Goodyera (1592-1664), an English botanist from Hampshire who is mentioned in Gerard's well-known* Herbal.

Habenaria Willd. **BOG ORCHID**

The plants belonging to the genus *Habenaria* are usually called bog orchids. They grow in moist woods, bogs and meadows. Bog orchid is identified by a tall flower-stalk (scape) that arises from a bunch of fleshy roots; it has few, sessile leaves and flowers that develop in a terminal, spike-like raceme. Flowers are bracted, white or greenish. The sepals and petals are similar; the lower petal (the "lip") is entire or 3-lobed, prolonged backwardes into a spur. The fruit is a capsule. Many species belonging to the genus *Habenaria* have been revised to species of *Platanthera* L.C. Richard, except *Habenaria viridis*, which has been revised to *Coeloglossum viride*.

1. Flowers white . . . . *H. dilatata*
   Flowers green 2
2. Flowers much shorter than the leaf-like bracts . . . . *H. viridis*
   Flowers about the same length as the bracts . . . . *H. hyperborea*

Habenaria dilatata (Pursh) Hooker

TALL WHITE BOG ORCHID

Tall white bog orchid grows to a height of 2 to 7 dm and is quite a striking plant. There are several leaves, usually lanceolate, and they are reduced upwards. There are several fragrant flowers, growing in a spike, each subtended by bracts. These flowers are normally white and have the usual arrangement seen in orchids: 3 sepals all alike, 2 lateral petals and a lip petal which is broadly lanceolate, much wider at the base, with a conspicuous spur. The tall white bog orchid is found in moist woods, bogs and wet ditches, and has been collected from a bog northwest of the Lower Kananaskis Lake, but it is not common.

NOTES: *This species has been revised to* Platanthera dilatata *(Pursh) Beck.*

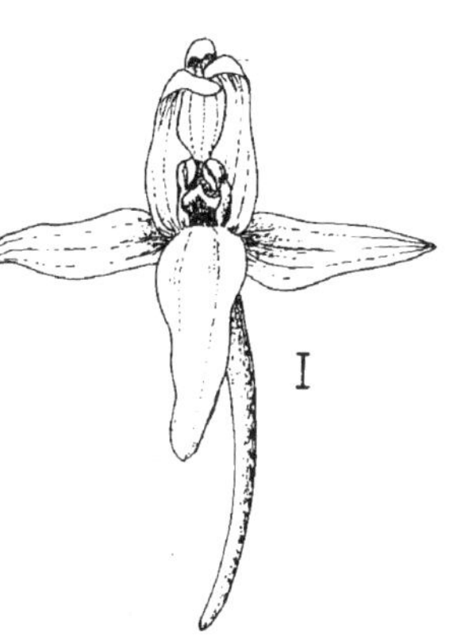

Habenaria hyperborea (L.) R. Brown

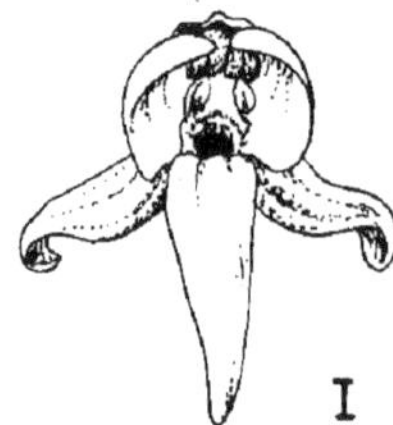

NORTHERN GREEN BOG ORCHID

This plant is similar to the tall white bog orchid but the flowers are greenish, not white, and the 2 lateral petals are shorter. The lip is slightly shorter than that of the tall white bog orchid, and the flower has a spur about as long as the lip. Northern green bog orchid is found in bogs, wet meadows and woods; it is farily common.

NOTES: *Northern green bog orchid has been seen in a bog 2 miles west of Bragg Creek Ranger Station and elsewhere.* *The species name* hyperborea *is a reference to Boreas, the Greek God of the North Wind.* *This species has been revised to* Platanthera hyperborea *(L.) Lindl.*

Habenaria viridis (L.) R. Brown *var. bracteata* (Muhl.) Gray

BRACTED BOG ORCHID

The flowers of bracted bog orchid are greenish and grow in a spike. The bracts underneath the flowers are well developed and quite conspicuous, hence the common name. The shape of the lip helps to distinguish the bracted bog orchid from the other bog orchids, because it is narrowly oblong and the tip is truncate with 2 or 3 teeth; in the other 2 species the lip is entire and the spur is longer. It is found in woods and moist meadows.

NOTES: *The species name* viridis *means "green."* *This species has been revised to* Coeloglossum viride *(L.) Hartman.*

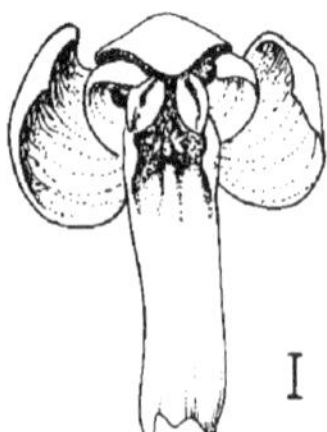

Listera borealis Morong

NORTHERN TWAYBLADE

Northern twayblade is an orchid with inconspicuous flowers; the plant is only 10 to 20 cm high, but the leaves are characteristic and help to identify the plant. There is only a single pair of leaves halfway up the stem, with no leaf stalks. The common name *twayblade* refers to these leaves: twayblade is a corruption of "two blades," i.e., 2 leaf-blades. In northern twayblade the leaves are elliptical, up to 3 by 1.5 cm. The flowers are arranged in a slender raceme and are pale-green to yellow-green. The petals and sepals are 4 to 7 mm long and curved back (reflexed). The lip is broadly oblong, narrowing slightly in the centre and broadening towards the base; the tip is not deeply cut. Northern twayblade is found in boggy spruce-aspen woods, in moist meadows and on mountain slopes.

NOTES: *The generic name Listera honours Dr. Martin Lister (1638-1711), a noted English physician and naturalist.* *The species name* borealis, *which means "northern," comes from Boreas, the Greek God of the North Wind.*

Orchis rotundifolia Banks ex Pursh

ROUND-LEAVED ORCHID

This plant is relatively easy to distinguish from the other orchids because it has a characteristic single oval-round leaf, with well-marked veins; the flowers, arranged in a spike, each have a conspicuous lip, white with purple spots, usually 3-lobed. Venus'-slipper (*Calypso bulbosa*) also has only 1 leaf, but this is different in shape from that of the round-leaved orchid, and it has a solitary flower. Round-leaved orchid grows in woods and has been found in coniferous forest at the west end of Barrier Lake.

NOTES: *Round-leaved orchid is very common near the base of Mount Lorette.* *This species has been revised to* Amerorchis rotundifolia *(Banks) Hulten.*

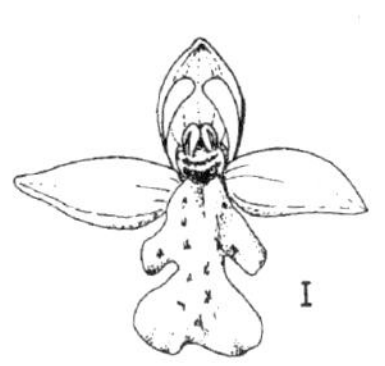

Spiranthes romanzoffiana Chamisso & Schlect.

HOODED LADIES'-TRESSES

Hooded ladies'-tresses is a robust plant with a leafy stem, which grows 1 to 4 dm tall and bears lanceolate-linear leaves. The flower-spike is characteristic: it is tightly twisted, with 3-ranked rows of overlapping, creamy-white, perfumed flowers. The common name ladies'-tresses refers to this spike, which bears a fanciful resemblance to a plait of hair. Each flower has 3 sepals and 3 petals; the sepals and lateral petals are similar and are united, forming a hood. The lip is 9 to 12 mm long and tongue-shaped, with a wavy margin, a constricted middle portion and a rounded terminal lobe. There is no spur. Hooded ladies'-tresses is found on damp grassy slopes, boggy areas, wet meadows and seepage areas.

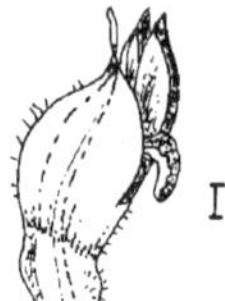

NOTES: *The generic name* Spiranthes *comes from the spiral arrangement of the flowers.* ~ *Chamisso (1781-1838), a German botanist, designated the specific name* romanzoffiana *in honour of his patron Nikolai Rumiantzev, Count Romanzoff (1754-1826). Count Romanzoff, who was a Russian Minister of State, was interested in financing botanical excursions; he financed Chamisso's expedition to Alaska and also Eschscholtz's expedition to California.*

Flowering Plants

DICOTS

Dicots

The dicots can be distinguished by several key characteristics. In dicots, the embryos have two cotyledons (seed-leaves), and the leaves are usually net-veined. The flower parts are in fours or fives, rarely in threes, and the plants are herbaceous or woody.

Most importantly, the possession of cambium cells distinguishes the dicotyledons from the monocotyledons.

KEY TO THE DICOTS

1. Leaves alternate or opposite 2
   Leaves basal or whorled 3
2. Leaves alternate .... Group 1
   Leaves opposite .... Group 2
3. Leaves basal (1 small stem-leaf may be present) ..... Group 3
   Leaves whorled (3 or more leaves originating at 1 node) .... Group 4

GROUP 1: DICOTS WITH ALTERNATE LEAVES

1. Trees, shrubs and woody vines 2
   Herbs (plants with non-woody stems) 10
2. Leaves compound 3
   Leaves simple 4
3. Flowers irregular .... Fabaceae
   Flowers regular .... Rosaceae
4. Flowers borne in catkins 5
   Flowers not borne in catkins 6
5. Leaves with smooth edges; seeds with silky hairs .... Salicaceae
   Leaves with jagged edges; seeds without silky hairs .... Betulaceae
6. Stamens 3; fruit black and berry-like .... Empetraceae
   Stamens 4 or more 7
7. Leaves silver-coloured; flowers yellow within, silvery outside .... Elaeagnaceae
   Leaves green; flowers white, pink or orange 8
8. Leaves "maple leaf-shaped" .... Grossulariaceae
   Leaves not "maple leaf-shaped" 9

9. Leaves mostly evergreen; flowers with fused petals (petals not united in *Ledum*); stamens 5 to 10 . . . . Ericaceae
   Leaves deciduous; flowers with distinct petals; stamens more than 10 . . . . Rosaceae
10. Leaves compound 11
   Leaves simple, sometimes lobed 20
11. Petals distinct 12
   Petals united 16
12. Inflorescence an umbel . . . . Apiaceae
   Inflorescence not an umbel 13
13. Stamens 6 (4 long and 2 short), petals 4 . . . . Brassicaceae
   Stamens 7 or more (sometimes 5) 14
14. Stamens 10; leaflets 3 . . . . Saxifragaceae
   Stamens numerous 15
15. Leaves simple or palmately compound, usually without stipules. . . . . Ranunculaceae
   Leaves once-pinnate, with stipules . . . . Rosaceae
16. Inflorescence a head (capitulum) . . . . Asteraceae
   Inflorescence not a head. 17
17. Flowers irregular . . . . Fabaceae
   Flowers regular 18
18. Leaflets 3; plants of boggy areas . . . . Menyanthaceae
   Leaflets more than 3; drier habitats. 19
19. Stamens shorter than petals. . . . . Polemoniaceae
   Stamens longer than petals. . . . . Hydrophyllaceae
20. Plants white or pink 21
   Plants green. 22
21. Plants white and waxy; saprophytic . . . . Monotropaceae
   Plants pink; parasitic on roots of the species of Asteraceae . . . . Orobanchaceae
22. Leaves lobed 23
   Leaves not lobed 33
23. Plants aquatic; leaves submersed; insectivorous bladders sometimes present . . . . Lentibulariaceae
   Plants not aquatic. 24
24. Inflorescence a head (capitulum) . . . . Asteraceae
   Inflorescence not a head (capitulum). 25
25. Stamens numerous. 26
   Stamens 10 or less 27

26. Stamens united to form a tube; flowers yellowish red . . . . Malvaceae
Stamens not united; flowers white, blue or yellow . . . . Ranunculaceae
27. Stamens 5 or 10 28
Stamens 4 or 6 30
28. Stamens 10 . . . . Saxifragaceae
Stamens 5 29
29. Petals united . . . . Hydrophyllaceae
Petals free . . . . Rosaceae
30. Stamens 4 . . . . Scrophulariaceae
Stamens 6 31
31. Flowers irregular . . . . Fumariaceae
Flowers regular . . . . Brassicaceae
32. Ovary inferior 33
Ovary superior 37
33. Inflorescence a head (capitulum) . . . . Asteraceae
Inflorescence not a head (capitulum) 34
34. Flowers blue; petals united 35
Flowers not blue; petals not united 36
35. Flowers regular, bell-shaped . . . . Campanulaceae
Flowers irregular . . . . Lobeliaceae
36. Sepals 4; petals 4; stamens 8 . . . . Onagraceae
Sepals 5, petals none; stamens 5 . . . . Santalaceae
37. Flowers green, yellowish green or brown in colour, not showy 38
Flowers with some coloured appendages, showy 41
38. Plants with milky juice . . . . Euphorbiaceae
Plants without milky juice 39
39. Sepals 4 to 6; stamens 5, 6 or 8 . . . . Polygonaceae
Sepals 1 to 5; stamens 5 or less 40
40. Fruit 1-seeded; sepals not papery and green . . . . Chenopodiaceae
Fruit several-seeded; sepals papery, not green . . . . Amaranthaceae
41. Stamens 5 or fewer 42
Stamens 6 or more 47
42. Petals free (absent in *Besseya*, calyx 2-lobed) 43
Petals united 45

43. Flowers irregular . . . . Violaceae
Flowers regular. 44
44. Flowers green or yellow . . . . Saxifragaceae
Flowers blue. . . . . Linaceae
45. Flowers irregular. . . . . Scrophulariaceae
Flowers regular. 46
46. Plants densely hairy or bristly . . . . Boraginaceae
Plants not hairy or bristly . . . . Polemoniaceae
47. Stamens 6 (4 long and 2 short) . . . . Brassicaceae
Stamens 8 or more 48
48. Stamens numerous. . . . . Ranunculaceae
Stamens 8 to 10. 49
49. Plants succulent; flowers yellow or red . . . . Crassulaceae
Plants not succulent; flowers white, pink or purple 50
50. Plants slightly woody; leaves evergreen and shiny . . . . Pyrolaceae
Plants with soft stems; leaves not evergreen. 51
51. Flowers irregular. . . . . Polygalaceae
Flowers regular . . . . Saxifragaceae

GROUP 2: DICOTS WITH OPPOSITE LEAVES

1. Trees, shrubs and woody vines 2
Herbs (plants with non-woody stems). 7
2. Woody vines 3
Trees or shrubs. 4
3. Leaves compound with 3 or 5 leaflets; flowers blue or yellow . . . . Ranunculaceae
Leaves simple; flowers orange or yellow . . . . . . Caprifoliaceae
4. Flowers white, orange or pink. 5
Flowers green or brown. 6
5. Bark red in colour; leaves simple; flowers white . . . Cornaceae
Bark not red in colour; leaves simple or compound. . . . . Caprifoliaceae
6. Leaf underside brown-spotted . . . . Elaeagnaceae
Leaf underside not brown-spotted . . . . Aceraceae
7. Inflorescence a head (capitulum). . . . . Asteraceae
Inflorescence not a head (capitulum) 8
8. Plants parasitic, growing on the branches of pines. . . . . Loranthaceae
Plants not parasitic. 9
9. Plants aquatic; floating leaves 3-nerved . . . . . . Callitrichaceae
Plants not aquatic. 10

10. Plants with stinging hairs; leaves toothed .... Urticaceae
    Plants without stinging hairs 11
11. Plants with milky juice................................ 12
    Plants without milky juice 13
12. Leaves compound .... Apocynaceae
    Leaves simple .... Asclepiadaceae
13. Leaves lobed 14
    Leaves not lobed.................................. 16
14. Inflorescence a compact cluster; flowers blue.... Dipsacaceae
    Inflorescence not a compact cluster head; flowers not blue ..
    15
15. Leaves palmately lobed (hand-shaped)......... Geraniaceae
    Leaves pinnately lobed.................... Valerianaceae
16. Flowers regular 17
    Flowers irregular.................................. 22
17. Plants with fleshy stems and leaves..................... 18
    Plants without fleshy stems or leaves 19
18. Sepals 2; flowers white or rose-coloured ...... Portulacaceae
    Sepals 4; flowers yellow .... Crassulaceae
19. Plants often with swollen nodes .... Caryophyllaceae
    Plants without swollen nodes 20
20. Flowers yellow......................... Primulaceae
    Flowers white, blue or purple 21
21. Leaves sharp-pointed; sepals hairy .... Polemoniaceae
    Leaves not sharp-pointed; sepals not hairy..... Gentianaceae
22. Plants with square stems .... Lamiaceae
    Plants with round stems................ Scrophulariaceae

GROUP 3: DICOTS WITH BASAL LEAVES

1. Plants insectivorous; leaves with a sticky upper surface or sticky-tipped hairs; boggy areas 2
   Plants not insectivorous, drier habitats 3
2. Leaves reddish green with sticky-tipped hairs; flowers white
   .... Droseracae
   Leaves yellowish green with a sticky upper surface; flowers purple .... Lentibulariaceae
3. Leaves compound 4
   Leaves simple, sometimes lobed.......................... 6
4. Flowers in 3 to 5 globe-shaped clusters; leaflets 9 to 15.......
   .... Araliaceae
   Flowers not in globe-shaped clusters 5

5. Inflorescence an umbel . . . . Apiaceae
   Inflorescence not an umbel . . . . Rosaceae
6. Inflorescence a head (capitulum) . . . . Asteraceae
   Inflorescence not a head 7
7. Leaves lobed 8
   Leaves not lobed 10
8. Plants with milky juice; stamens numerous . . . . Papaveraceae
   Plants without milky juice; stamens 5 9
9. Petals free . . . . Saxifragaceae
   Petals united; leaves with 5 to 7 rounded lobes . . . . Hydrophyllaceae
10. Leaves evergreen and leathery . . . . Pyrolaceae
    Leaves not evergreen 11
11. Leaf underside strongly ribbed . . . . Plantaginaceae
    Leaf underside not strongly ribbed 12
12. Stamens 4 or 10 . . . . Saxifragaceae
    Stamens 4 to 9 13
13. Stamens 4 to 9; inflorescence an umbel, or flowers in conspicuous clusters; flowers white, pink or yellow . . . . Polygonaceae
    Stamens 5; inflorescence not an umbel, flowers purple 14
14. Flowers irregular, purple . . . . Violaceae
    Flowers regular, white or pink 15
15. Plants with a single stem-leaf; flowers with 5 gland-tipped sterile stamens . . . . Parnassiaceae
    Plants without a stem-leaf; flowers without sterile stamens . . . . Primulaceae

GROUP 4: PLANTS WITH WHORLED LEAVES

1. Plants aquatic or semi-aquatic 2
   Plants not aquatic or semi-aquatic 4
2. Leaves entire and narrow . . . . Hippuridaceae
   Leaves dissected into narrow divisions 3
3. Flowers yellow, irregular; insectivorous bladders present . . . . Lentibulariaceae
   Flowers purplish green, regular; insectivorous bladders absent . . . . Haloragaceae
4. Stems square; leaves narrow, in whorls of 4, 6 or 8 . . . . Rubiaceae
   Stems not square; leaves wide, in an apparent whorl of 4 or 6, but not *Cornus stolonifera* . . . . Cornaceae

Salicaceae WILLOW FAMILY

The members of this family are dioecious trees or shrubs, with alternate leaves that usually have stipules. The tiny flowers are arranged in catkins and have no true perianth; they are either female (carpellary or pistillate) or male (staminate). The flowers are each subtended by a bract.

The bark of mature balsam poplar trees was used by some native peoples for making buckets and other containers, and the aromatic gum from the buds was used as glue. The mass of fluffy hairs from the female catkins when fruiting was used as stuffing for pillows. Willow bark was used for twine, ropes, lines and nets, and the wood was used for fire-drills, snowshoes, and the frames of canoes. Balsam poplars also had ceremonial uses. A balsam poplar tree was used by the Blackfoot as the central support of a Sundance Lodge and was known as "the Tree of Life." Sundance dancers were allowed to drink the sap of a balsam poplar during their 4-day fast.

These trees and shrubs also had medicinal uses. Salicin is present in the sap and bark of poplars, aspens and willows; when ingested into the body, it turns into salicilic acid, the active ingredient of aspirin. A tea made from willow bark was used for reducing fevers, and Okanagan tribes used a brew of aspen branches as bathing water for rheumatism sufferers.

Buds with 1 scale; shrubs or small trees . . . . *Salix*
Buds with 2 or more scales; large trees . . . . *Populus*

Populus L. POPLAR

Trees, sometimes shrubby at high elevations. The bark is smooth at first, becoming furrowed with age; the buds are often resinous. Leaves are broad, ovate, often cordate at base, toothed; stipules fall early. Flowers grow in drooping catkins, expanding with the leaves or just before; they are wind-pollinated. Staminate catkins and carpellary catkins grow on separate trees. Staminate catkins have several to many stamens, with scales; carpellary catkins have 2 carpels (joined) and scales. The fruit is a capsule.

Stalk of leaf round; buds sticky; moist habitats . . . . *P. balsamifera*
Stalk of leaf flat; buds not sticky; drier habitats . . . . *P. tremuloides*

Populus balsamifera L.

BALSAM POPLAR

This tree can grow 20 to 25 m tall. It has deeply furrowed grey bark. The buds are characteristic: they are coated with gum and are sticky. The leaf blades are more or less ovate and gradually taper to a point. The trees are dioecious, that is, they have either male catkins or female catkins, which are wind-pollinated. The male catkins are purplish and produce pollen from the tiny male flowers, while the female catkins bear the carpellary flowers. These are greenish, often 30 to 40 in each catkin; each bears an ovary, which becomes a capsule, with 2 or 3 valves. These burst open to release the familiar fluffy seeds. Balsam poplar is found in riverine forest, boreal forest and alluvial flats. It is a very common tree in Kananaskis Country.

Populus tremuloides Michx.

TREMBLING ASPEN

This tree grows 20 to 25 m high and has a slender trunk; the young bark is light greyish green, but becomes coarsely furrowed and nearly black with age. The leaves are smaller and daintier than those of balsam poplar; they are almost round and narrow to a point. The tree is dioecious, with male and female catkins on separate trees; flowers are wind-pollinated. The male catkins are much smaller than those of balsam poplar; they are hairy-looking and rather inconspicuous. The female catkins are similar to those of balsam poplar, but they are smaller. The catkins are all drooping. Trembling aspen is found in forests and parkland and is a common tree except in the alpine regions.

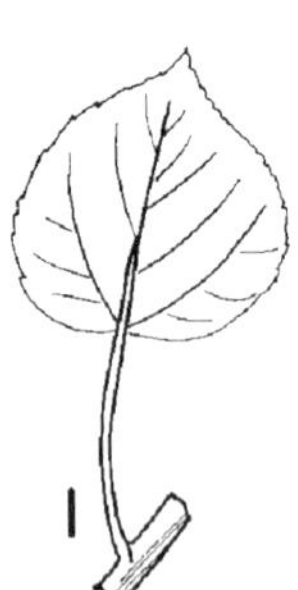

NOTES: *The petioles (leaf stalks) are flattened laterally and are relatively long, so the leaves flutter in a breeze; hence the specific name* tremuloides.

Salix L. **WILLOW**

Willows grow as shrubs, either erect or dwarf (trailing). The winter buds possess a single bud-scale; the leaves are alternate, simple. Flowers appear in catkins (staminate and carpellary on different shrubs), each flower with a bract. Staminate flowers have 1 to 7 stamens; carpellary flowers have 2 carpels, joined. The fruit is a bivalved capsule. The species of *Salix* can be difficult to determine. Usually, mature fruits are needed: the male catkins are not useful for species identification.

1. Prostrate shrub; stems less than 10 cm tall; leaves round . . . . *S. reticulata*
   Erect shrubs 2
2. Leaves leathery, the upper surface deeply veined . . . . *S. vestita*
   Leaves not leathery, the upper surface smooth 3
3. Outer layer of bark shredding; leaves soft-hairy. . . . . *S. glauca*
   Outer layer of bark not shredding; leaves without hairs . . . . *S. bebbiana*

Salix bebbiana Sarg.

BEBB'S WILLOW, BEAKED WILLOW

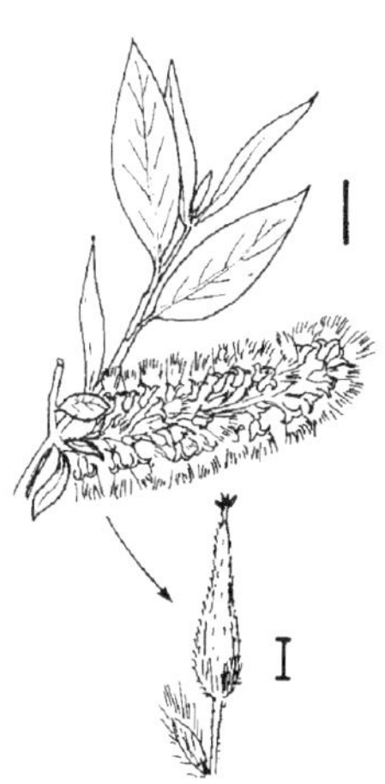

This is a common willow. It grows as a coarse shrub or a small tree 0.5 to 5 m tall, with a narrow crown and ascending branches. The leaves are lance-shaped, 4 to 8 cm long, greyish green below and variably hairy. The shrub is dioecious, with male and female catkins on separate trees. The male catkins are 2 cm long, with many tiny flowers; each has a bract and 2 stamens. These catkins are bright yellow when the pollen is being shed. The female catkins are 3 to 5 cm long and green. The tiny flowers each consist of a bract, with ovary, a tiny style and a stigma. The ovary becomes a white, hairy capsule with 2 valves; it bursts open to release fluffy seeds. Bebb's willow is found in mixed woods.

NOTES: *Bebb's willow has been reported from an area near the Eau Claire campground and in many other places.*

Salix glauca L.

SMOOTH WILLOW

A small, erect shrub 0.5 to 1.5 m tall, with elliptic leaves, rather like those of Bebb's willow, but hairy underneath when young. The catkins appear with the leaves in spring; in the Bebb's willow they often appear before the leaves. The capsules are often white-hairy. This willow is found in alpine and subalpine locations, in moist, coniferous woods and on river flood-plains.

NOTES: *This plant has been found at Hailstone Butte, Cataract Creek and other places.*

Salix reticulata L. *ssp. nivalis* Löre, Löre & Kapoor

DWARF WILLOW

This willow is quite different from Bebb's willow. It is a tiny shrub, growing 1 to 10 cm above the ground and rooting from the stem. The leaves are also quite different; they are much smaller, almost round, and are grey-green underneath. The capsule is grey-hairy. Dwarf willow is found on alpine and subalpine slopes and has been recorded from Plateau Mountain at 2377 m and other places.

Salix vestita Pursh

ROCK WILLOW

These are low shrubs, up to 1.5 m tall, much smaller than those of Bebb's willow. The leaves are ovate, not lanceolate (lance-shaped), and are a lighter green. The catkins are terminal— another difference between the 2 species. This is an alpine or subalpine willow, found in moist habitats up to 2200 m; Bebb's willow is found at lower elevations.

Betulaceae BIRCH FAMILY

The birches are either trees or shrubs, and they are all monoecious, that is, the male and female catkins grow on the same tree. The name is derived from the Greek word *oicos*, meaning "home"—in other words, the catkins of the birches have 1 "home," while the willow catkins have 2 "homes."

The serrate leaves of members of the birch family are entire and alternate. The staminate (male) catkins are drooping, and there are 1 to 3 flowers in the axil of each bract. Each minute flower has a membranous calyx (sometimes absent), which encloses 1 to 10 stamens. The female (carpellary or pistillate) catkins are usually erect, and the flowers are in the axils of bracts. The sepals forming the calyx (if present) are fused with the ovary, which has 2 styles. The fruits are tiny, 1-seeded winged nutlets; the nut may be enclosed by bracts.

Alder wood was used by the Cree to make utensils. It has a low pitch content and so was used, when seasoned, for smoking salmon and deer meat. Both the wood and the bark were used by British Columbian tribes to make dyes for porcupine quills, feathers, and buckskin clothes. The wood of black birch (*Betula occidentalis*) was used by the Stoneys for making bows. The paper birch (*Betula papyrifera*) was another useful tree: when it was mature, the bark was stripped off and used for containers and canoes. The wood was used for fuel and for making snowshoes, bowls and digging sticks.

Female catkins woody (appearing like small cones of conifers), remaining on the branches for several years .... *Alnus*
Female catkins thin, deciduous, never lasting more than 1 season .... *Betula*

Alnus Mill. ALDER

Alders grow either as shrubs with many stems or as small trees; they have smooth bark. The leaves are ovate, petiolate and toothed. Staminate and carpellary catkins appear on the same plant. Staminate catkins are up to 1 dm long, pendulous, growing in clusters at the tips of branches, with peltate bracts bearing 3 flowers and 2 to 4 bractlets; each flower has 4 stamens. Carpellary catkins are erect, with fleshy bracts bearing 2 flowers and 4 bractlets; each flower has 1 or 2 carpels joined. When ripe, the carpellary catkins form little woody "cones" bearing winged nutlets.

Flowers appearing at the same time as the leaves; shrubs of drier habitats ... . *A. crispa*
Flowers appearing before the leaves; tall shrubs or small trees growing along lakeshores and riverbanks................................ *A. tenuifolia*

Alnus crispa (Ait.) Pursh *ssp. sinuata* (Regel) Hult.

GREEN ALDER

Green alder is a sprawling shrub, up to 3 m high; the branchlets are rather glutinous when young. The leaves are shiny, ovate, and sharply toothed. There are 2 kinds of catkins: male (staminate), long and pendulous, in a cluster at the tips of the branches, with 2 stamens in each tiny flower; and female (carpellary), erect, with fleshy bracts, each bearing 2 flowers with ovaries. The ovaries develop into the fruits, little nutlets, with wings.

NOTES: *Green alder has been found in the Lusk Creek area and elsewhere.*

Alnus tenuifolia Nutt.

RIVER ALDER

River alder is much taller than green alder; it grows 2 to 8 m. The leaves of river alder are not shiny like those of green alder, and the little fruits (nutlets) are not winged. Their habitats are rather different: green alder grows in open woods, while river alder, as its name implies, grows on the borders of streams and lakes.

NOTES: *River alder has been found near the fork of Lusk Creek and Stoney Creek. It has also been found along Alder Trail in Bragg Creek Provincial Park and at other places.*

Betula L. — BIRCH

Trees and shrubs with thin bark, conspicuously marked with dark lenticels, elongated horizontally. Leaves alternate, simple and toothed, twigs often glandular. Male and female flowers arranged in separate catkins, both types on the same tree. Male catkins bear bracts; in the axils of the bracts are 3 flowers and 3 bractlets, each flower with 2 stamens. Female catkins have similar bracts with 2 or 3 flowers in their axils, and each flower has 1 to 2 carpels joined; when ripe they form a "samara," a winged achene.

Leaves 1 to 2 cm long; bogs and higher elevations .... *B. glandulosa*
Leaves 2.5 to 6 cm long; streams and lakeshores .... *B. occidentalis*

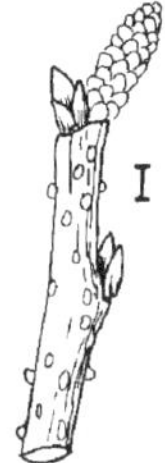

Betula glandulosa Michx.

BOG BIRCH, DWARF BIRCH

As its common name implies, this is a small shrub up to 2 m high. The twigs bear numerous resinous glands, hence the species name *glandulosa*. The leaves are orbicular (rounded) and toothed. There are 2 kinds of catkins: the male (staminate) are up to 2 cm long, and drooping; the female (carpellary) are erect, 1 to 2 cm long. The little fruits (samaras) have narrow wings. Bog birch is found in bogs but also on alpine slopes and in subalpine forest.

NOTES: *This plant has been collected from an old bog near the Kananaskis Field Stations (also the borders of fens in Peter Lougheed Provincial Park) and in other places.*

Betula occidentalis Hook.

WATER BIRCH, BLACK BIRCH

This is a much taller shrub, up to 6 m high. Its leaves differ from those of dwarf birch, which are only 1 to 2 cm long. Water birch leaves are larger, over 2.5 cm long, the blade has a rounded tip, and the margin is bluntly toothed. Both species have numerous glands. Water birch, as its name implies, is found along stream banks and on the shores of lakes.

NOTES: *This species has been collected at the east end of Barrier Lake and is a common shrub throughout the valleys of Kananaskis Country.*

Urticaceae NETTLE FAMILY

Nettles are usually herbs, but some members of the family are shrubs, and all usually have stinging hairs. The leaves may be opposite or alternate, and usually bear stipules. The small greenish flowers are imperfect: either male (staminate) or female (carpellary or pistillate). Each staminate flower has a 3 to 5-lobed calyx, no petals, and 3 to 5 stamens; the carpellary flower has a 3 to 5-lobed calyx, no petals and a single ovary, which becomes the fruit, an achene or a drupe.

Urtica dioica L.
ssp. gracilis (Ait.) Selander

COMMON NETTLE, STINGING NETTLE

(*Urtica gracilis* Ait.) (*Urtica lyallii* Wats.)

Common nettle is a perennial plant, with many unbranched stems arising from strong rootstocks; it reaches 1 to 3 m in height. The leaves are in pairs up the stem, have leafstalks and are coarsely toothed. The tiny flowers are in dense paniculate cymes, which arise from the leaf axils, and are arched. The male (staminate) and female (carpellary or pistillate) flowers are in different clusters. Both sexes are found on the same plant (monoecious), with the female flowers usually on the upper part of the plant. The male flowers have no petals; they have 4 sepals, partly united, and 4 stamens. The female flowers have 4 unequal sepals and a small ovary. The tiny fruit is a 1-seeded achene, enclosed by 2 inner persistent sepals. The plants are found in colonies in moist aspen groves, on roadsides and on mountain slopes.

NOTES: *The plant is covered with stinging hairs. When a hair is touched, the swollen tip breaks off, leaving a needle-like point. The hair is hollow; at its base is a small bulb which contains formic acid, a poisonous fluid. When the hair is touched, the bulb contracts, squeezing the juice through the point at the tip and injecting into whatever touched it. Common nettle is a plant to avoid!*

Santalaceae SANDALWOOD FAMILY

Santalaceae is mainly a tropical family but there are a few temperate genera, e.g., *Comandra.* The plants are semi-parasites and arise from long underground rootstocks; these bear roots with *haustoria* (sucker-like organs), which attach themselves to the roots of other plants, obtaining nourishment from them. Sandalwoods have slender stems with alternate, entire leaves; the flowers are arranged in terminal or axillary clusters. The flowers are inferior (*epigynous*); the hypanthium encloses the conical ovary and forms a flat disc above. 5 petalloid sepals are inserted on the edge of this disc (petals are absent). 5 stamens are also inserted, each associated with a tuft of hairs, which links them to the sepals. In the centre of the disc is a single style with a simple stigma. The fruit is 1-seeded, either succulent or dry.

Flowers greenish bronze; fruit red and berry-like . . . . *Geocaulon*
Flowers white or greenish white; fruit dry . . . . *Comandra*

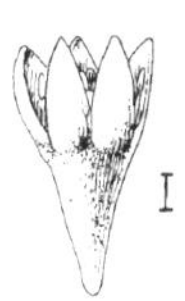

Comandra umbellata (L.) Nutt.

BASTARD TOADFLAX, PALE COMANDRA

(C. pallida D.C.*)*

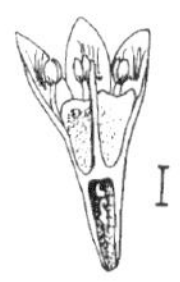

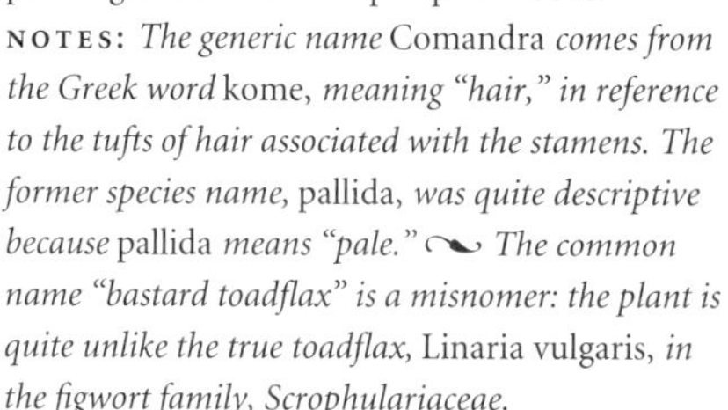

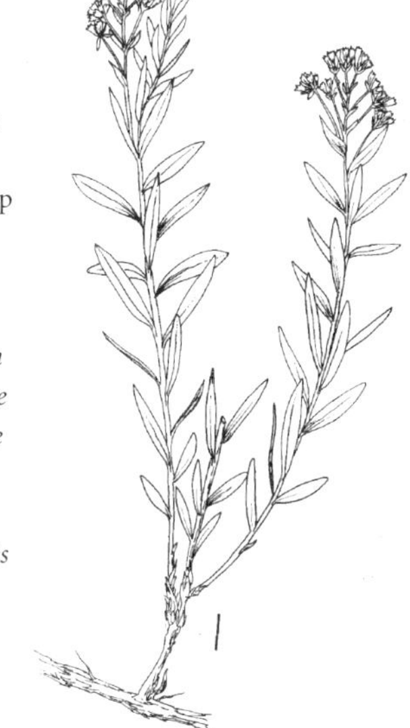

Pale comandra is an inconspicuous, often blue-green plant, 1.5 to 4 dm tall; it grows from rhizomes, which bear tiny suckers on the roots. The leaves are leathery with sharp tips. The plant has small, whitish flowers growing in clusters at the tips of the main stalks. The globular fruit is up to 10 mm across and crowned by the persistent calyx; it is 1-seeded. Pale comandra is found on prairie grassland and in open pine woods.

NOTES: *The generic name* Comandra *comes from the Greek word* kome, *meaning "hair," in reference to the tufts of hair associated with the stamens. The former species name,* pallida, *was quite descriptive because* pallida *means "pale."* *The common name "bastard toadflax" is a misnomer: the plant is quite unlike the true toadflax,* Linaria vulgaris, *in the figwort family, Scrophulariaceae.*

Geocaulon lividum (Richards.) Fern.

NORTHERN BASTARD TOADFLAX

Geocaulon is a perennial plant, growing up to 2 dm high, with numerous alternate leaves arising from spreading rhizomes. The whitish flowers grow in axillary clusters of 3 to 4 flowers each, in the axils of the middle leaves. Each flower has 5 triangular sepals, which are purplish and united; they spread out in a flat ring. The sepals of *Comandra* are more or less erect, and do not spread out. The fruit of *Geocaulon* is round, scarlet, fleshy and drupe-like. *Geocaulon* is found in damp humus and moss, often in shady coniferous woods.

NOTES: *Although* Comandra *and* Geocaulon *are both called "bastard toadflax," they can be distinguished because their leaves are quite different. In* Comandra, *they are bluish green and leathery, with sharp tips, while in* Geocaulon *they are bright green, thin and blunt-tipped. The flowers of* Comandra *are whitish and grow in the upper leaf-axils, but those of* Geocaulon *grow in the axils of the middle leaves and are greenish or bronze. The sepals of* Comandra *are much longer than wide; those of* Geocaulon *are triangular. The fruits of the two plants are also different: those of* Comandra *are greenish and dry, while those of* Geocaulon *are red and fleshy. Both are semi-parasitic plants.*

Loranthaceae (Viscaceae) MISTLETOE FAMILY

Members of this family are true parasites: they obtain their nourishment from trees belonging to Pinaceae, the pine family. Those usually affected are *Pinus banksiana* (Bank's pine) and *P. contorta* (lodgepole pine). Mistletoe stems emerge from the branches of the host tree. They are greenish yellow, slender and much-branched, and the "leaves" have been reduced to tiny scales, with flowers in their axils. These flowers may be either male or female. The fruit is 1-seeded and contains seeds embedded in mucilaginous pulp; this dehisces (opens) explosively, and the seeds are shot out to infect other branches.

Arceuthobium americanum Nutt.

DWARF MISTLETOE

This plant is parasitic on pine species, causing curious outgrowths called "witches' brooms," a tangled mass of branches. The plants emerge from the host-plant, and the slender, greenish yellow, many-branched stems are 2 to 10 cm long. The leaves are reduced to scales, and there are male and female flowers in the axils of the scales. There are no petals; the staminate flowers have a 3-part calyx and 3 stamens, and the carpellary flowers have a 2-part calyx and an inferior ovary. The ovary turns into a berry-like fruit with mucilaginous pulp and a single seed. It bursts open explosively, ejecting the sticky seed, which lands on a neighbouring branch and proceeds to infest it.

NOTES: *The familiar mistletoe of Christmas festivities is quite a different-looking plant and does not grow in Alberta, but it belongs to the same family; its scientific name is* Viscum album.

Polygonaceae BUCKWHEAT FAMILY

The members of the buckwheat family are annuals or perennials with simple leaves. The small flowers of this family often grow in conspicuous clusters. Each flower has 4 to 6 separate sepals, sometimes petalloid, or these are united into a 4 to 6-cleft calyx; there are no petals. There are 4 to 9 stamens and a superior ovary with 2 to 3 styles. The fruits are achenes (dry 1-seeded fruits). An interesting characteristic of this family is the presence of *ocreae* at the bases of the leaves. These are membranous stipules which sheath the stem and continue beyond the insertion of the leaf into a cylinder, which may be cleft or frayed. The members of the *Eriogonum* genus do not possess them, but they are characteristic of the docks (*Rumex* spp.) and knotweeds (*Polygonum* spp.).

1. Stamens 9; sepals 6, coloured . . . . *Eriogonum*
   Stamens 8 or fewer 2
2. Sepals 5; stamens 3 to 8 . . . . *Polygonum*
   Sepals 4 or 6; stamens 6 3
3. Sepals 4; leaves kidney-shaped, sepals 4 . . . . *Oxyria*
   Sepals 6; leaves not kidney-shaped . . . . *Rumex*

Eriogonum Michx. UMBRELLA-PLANTS

These plants are perennials growing from short, branching stems; the leaves are simple, with no ochreae (stipule-sheaths). Flowering stalks bear umbellate clusters of small flowers, often surrounded by an involucre of bracts; there are 6 sepals in 2 whorls, often coloured, and 9 stamens. The fruit is an achene, enclosed by the calyx.

1. Flowers white. . . . . *E. ovalifolium*
   Flowers yellow or greenish white, often tinged with pink 2
2. Flowers with silky hairs . . . . *E. flavum*
   Flowers without silky hairs . . . . *E. umbellatum*

Eriogonum flavum Nutt.

YELLOW UMBRELLA-PLANT

E. flavum is a typical umbrella-plant: it has several flower-heads arranged in an umbel at the top of the flowering stalk, which is covered with whitish hairs. The individual flowers have coloured sepals, no petals and 9 stamens; the ovary is superior and becomes an achene, a dry, 1-seeded indehiscent fruit. Each flower lies in a cup made of joined bracts; this is 5-cleft. The oblanceolate leaves, like those of most *Eriogonum* species, are basal; they are 2.5 to 5 cm long, with slender leaf-stalks. The upper surfaces are green, but the lower surfaces are white-hairy. Yellow umbrella-plant is found on dry prairies and alpine slopes up to 2300 m. **NOTES:** *The* Eriogonum *species are usually called umbrella-plants, because their flowers are often arranged in umbels, like those of the carrot family, Umbelliferae (now Apiaceae).* *The specific name* flavum *means "yellow."*

Eriogonum ovalifolium Nutt.

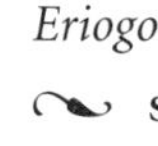

SILVER-PLANT

E. ovalifolium is called silver-plant because its leaves are so densely covered with white hairs on both surfaces that they look silvery. They are densely crowded on top of a stout caudex (root-stock), and they are ovate (oval) with stalks. The leaves are quite different from those of the yellow umbrella-plant: they are smaller (less than 2 cm long), have small leaf-stalks and are more rounded. Silver-plant is a dwarf plant forming a "cushion." Its yellowish flowers form in dense clusters. Each flower lies in a cup, made of bracts joined together. Silver-plant is an alpine species but is also found at lower elevations.

Eriogonum umbellatum Torr.

SUBALPINE UMBRELLA-PLANT

In this plant the umbel arrangement of the flowers is distinct, hence the species name *umbellatum*. The umbels are surrounded by bracts, as in the other species; the individual flowers are pale yellow and turn pinkish when dried. The leaves are spatulate, up to 8 cm long, and white-hairy on the lower surfaces. They differ from those of the yellow umbrella-plant: they are broader and do not have long, tapering leaf stalks. The flowers of this species are much paler than the bright yellow of yellow umbrella-plant. *E. umbellatum* grows in prairie grassland, in higher altitudes, up to 2300 m, and on alpine ridges.

Oxyria digyna (L.) Hill

MOUNTAIN SORREL

Mountain sorrel is a perennial plant, growing from a thick root. The leaves are chiefly basal and kidney-shaped (reniform) with long leaf-stalks; they are characteristic and help to identify the plant. The small, greenish red flowers grow in whorls in a dense raceme. Each flower has 4 sepals in 2 series; the outer ones are reflexed, and the inner ones are erect. In the fruiting stage, they turn bright red and form "wings." The corolla is absent. There are 6 stamens, and the ovary bears 2 reddish stigmas which are plumose. The fruit is an achene, enveloped by a persistent calyx. Mountain sorrel, as its name implies, is found at alpine elevations on moist rocky slopes.

NOTES: *The leaves are edible, rich in vitamins A and C, and can be used in salads. They have a sharp taste, due to the presence of oxalic acid. Oxalic acid is* ***very poisonous*** *if consumed in large quantities, so care is needed.*

Polygonum L. — KNOTWEED

Knotweeds grow as annuals or perennials. The stems have swollen nodes; the leaves have prominent ochreae (stipule-sheaths), and are alternate, entire. Flowers grow in small axillary clusters or in tiny spikes at branch-tips; there are 4 to 6 petalloid sepals (which persist in the fruit) and usually 5 stamens. The fruit is an achene.

1. Leaves heart-shaped; stems twining; flowers green . . . . *P. convolvulus*
   Leaves not heart-shaped; stems not twining 2
2. Stems prostrate; flowers green with white tips. . . . . *P. arenastrum*
   Stems erect; flowers white or pink . . . . *P. viviparum*

Polygonum arenastrum Jord. ex Bor.

COMMON KNOTWEED, YARD KNOTWEED

(P. aviculare L.*)*

This is a common weed in waste places and on roadsides. The plant has prostrate stems, often branched; the main branches are 1 to 8 dm long. The foliage is often bluish green, and the leaves are narrowly lanceolate, sometimes broadly lanceolate (lance-shaped). The ocreae are silvery. The flowers are in axillary clusters of 1 to 5 flowers; they are inconspicuous. The calyx (of sepals) is green, with pink or white margins. There are no petals and usually 5 stamens. The fruit is an achene.

P. convolvulus L.

WILD BUCKWHEAT, BLACK BINDWEED

This plant is easily distinguished from the yard knotweed because its leaves are different: they are much larger (the blades are 2 to 5 cm long) and sagittate (shaped like an arrow-head). The stems are also different, often sprawling or twining, hence the common name "black bindweed." The tiny flowers have a greenish calyx of sepals, usually 5 stamens and the fruit, an achene, is black.

Polygonum viviparum L.

ALPINE BISTORT

Bistort is an erect plant, 1 to 3 dm tall, and has large leaves, lance-shaped, up to 2 dm long. The flower arrangement is different from that of common knotweed: the small flowers are in a spike, dense at the top but more open below. They are greenish white. Alpine bistort grows in moist woods and meadows and is sometimes found on alpine slopes.

NOTES: *The flowers at the base of the spike are modified into little bulblets, which drop off the plant and produce new plants in the soil below. The adjective* viviparous *means "producing live young," which explains the species name* viviparum. *Water smartweed* (P. amphibium) *is a related species found in shallow water and on muddy shores.*

Rumex L. DOCK

Mostly coarse, leafy perennials, docks grow from thick roots, bearing leaves with prominent ochreae, simple and petiolate. The flowers are small, in tall clusters, branched; they are arranged in whorls at the nodes, sometimes bi-sexual. Each flower has a calyx of 6 sepals, in 2 whorls of 3; the outer whorl spreads in fruit, and the inner whorl becomes enlarged and winged, sometimes bearing a tubercle. There are 6 stamens and the fruit is a 3-angled achene.

Members of the genus *Rumex* have distinctive fruits; these are enclosed within the inner sepals, called *valves*. These valves are usually enlarged and have a prominent network of veins; the lower mid-vein often bears a grain-like tubercle.

1. Leaves arrowhead-shaped . . . . *R. acetosa*
   Leaves not arrowhead-shaped 2
2. Leaf margins crinkled . . . . *R. maritimus*
   Leaf margins not crinkled . . . . *R. occidentalis*

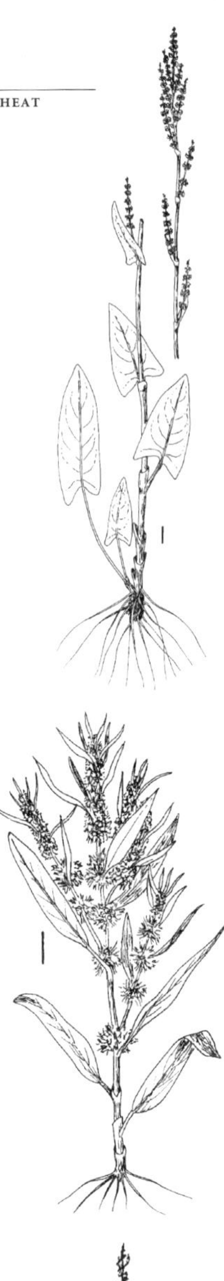

Rumex acetosa L. *ssp. alpestris* (Scop.) Löve

GREEN SORREL

Green sorrel is a tall perennial (up to 1 m high), and the basal leaves are 2 to 10 cm long, with long petioles (leaf-stalks). The leaves become smaller upwards and lose their petioles; the ocreae are usually entire. The flowers are arranged in a panicle (a raceme of racemes), 1 to 2 dm long. Male and female flowers are found on separate plants. The female flowers possess outer sepals nearly 2 mm long; the fruit (an achene) is invested by the inner sepals. The male flowers (staminate) have 6 stamens and sepals 2 to 3 mm long. Green sorrel is found in meadows and also in alpine elevations, hence the sub-species name *alpestris.*

Rumex maritimus L.

GOLDEN DOCK

Golden dock is easily distinguished from green sorrel because the flower clusters and fruits are so different. Those of golden dock are spiny. The panicle is large, freely branched, leafy and golden-brown at maturity, hence the common name. The leaves of green sorrel are sagittate (arrow-shaped), ending in a long point (sometimes cordate), while the leaves of golden dock are not sagittate and do not end in a sharp point—they are lanceolate. Golden dock is found in damp places, on saline shores and also on disturbed ground.

Rumex occidentalis S. Wats.

WESTERN DOCK

Western dock is a single-stemmed perennial, often tinged with red, 5 to 15 dm high. The leaves are lance-shaped, with long petioles, and become smaller up the stem. The flowers grow in narrow clusters. The inner sepals of the flowers form part of the fruit, and are distinctly net-veined. Western dock is smaller than green sorrel, and the leaves are lance-shaped rather than arrow-shaped. Western dock is also different from golden dock: the fruits are not spiny. Western dock grows in willow thickets and other moist, protected places.

Chenopodiaceae GOOSEFOOT FAMILY

The members of this family are often familiar weeds, like lambs'-quarters, pigweed and Russian thistle; some, such as beet and spinach, are useful vegetables. (A weed may be defined as a plant growing in the wrong place; it is certainly not desirable.) They are annual or perennial herbs, and their leaves are often succulent. The small, greenish flowers often grow in dense clusters, sometimes singly. The calyx usually has 5 sepals, and there are no petals. There are 1 to 5 stamens, which arise opposite to the lobes of the calyx. The fruits are small, 1-seeded and often enclosed by the persistent calyx. The family name "goosefoot" has arisen because of the fancied resemblance between the shape of the leaves and that of a goose's foot. The generic name *Chenopodium* comes from two Greek words: *chen* meaning "goose" and *pous* meaning "foot."

1. Stamens 1 . . . . *Monolepis*
   Stamens 2 or more 2
2. Calyx glabrous. . . . . *Chenopodium*
   Calyx hairy . . . . *Kochia*

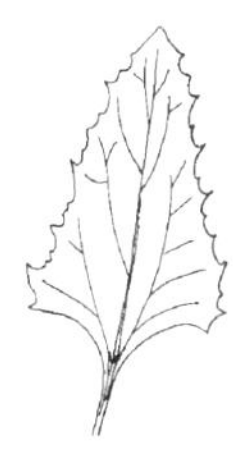

Chenopodium album L.

LAMB'S-QUARTERS, WHITE GOOSEFOOT, PIGWEED

Lamb's-quarters is an annual plant, 4 to 12 dm high, that usually has many branches. The leaves are characteristic of *Chenopodiaceae:* they are more or less arrow-shaped, with the bases tapering into leaf-stalks. They are green above and white-mealy below; the larger leaves are often toothed. The small, greenish flowers are produced in dense clusters and form a spike-like panicle. Each flower has a mealy, 5-lobed calyx, which later encloses the fruit; there are no petals, 3 to 5 stamens and a superior ovary with 2 to 5 stigmas. The 1-seeded fruit is black, 1 mm in diameter and enclosed by the persistent calyx. Lamb's-quarters is a common weed and is found on roadsides, waste places and gardens.

NOTES: *Strawberry blite* (C. capitatum), *a related species, is easily distinguished from lamb's-quarters. Strawberry blite is a branched, yellowish green annual with tiny, bright red flowers and red, juicy fruits.*

Kochia scoparia (L.) Schrad.

BURNING BUSH, SUMMER CYPRESS

Burning bush has a bushy habit, with linear, alternate leaves. The flowers arise from the leaf axils and are either solitary or in small clusters. Each flower has a greenish bract and a calyx with 5 lobes, no petals, and 3 to 5 stamens. The ovary is superior, with 2 stigmas, and turns into a small 1-seeded fruit, enclosed by the persistent, winged calyx.

NOTES: *The common name "summer cypress" seems rather a misnomer since the plant has little in common with a cypress. The other common name, burning bush, does not seem appropriate either, although the plant often turns purplish in the fall.* *Burning bush is a garden plant that has escaped from cultivation and become a common weed.*

Monolepis nuttalliana (Schultes) Greene

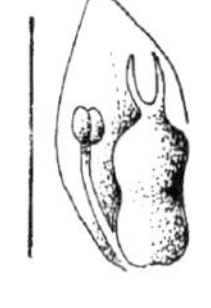

SPEAR-LEAVED GOOSEFOOT

Spear-leaved goosefoot is an annual plant, leafy and densely branched, which grows up to 3 dm in length; the stems are prostrate or ascending. The leaves are alternate and have a characteristic outline because the blade is sagittate (arrow-shaped) and the base tapers into a petiole (leaf-stalk), up to 2.5 cm long. The common name "spear-leaved goosefoot" describes the leaf well. Tiny flowers grow in clusters in the upper leaf-axils. The individual flowers each have a single sepal, which is green and looks more like a bract; they have no petals and only a single stamen. The ovary has a 2-forked style and produces a single-seeded fruit. Spear-leaved goosefoot is a common weed of gardens, roadsides and saline soils.

NOTES: *The species name* nuttallianum *honours Thomas Nuttall (1786-1859), an Englishman who collected many specimens in western North America.*

Amaranthaceae AMARANTH FAMILY

The members of this family are annual herbs and are often weeds. They may be erect or prostrate, with alternate leaves. The plants may be *monoecious* (male and female flowers on the same plant) or *dioecious* (male and female flowers on separate plants). The flowers are small, inconspicuous, greenish or purplish, and each subtended by 3 bracts. Each flower has a 2 to 5-lobed calyx, no corolla, and 2 to 5 stamens, with a superior ovary with 2 or 3 stigmas. The fruit is a 1-seeded capsule which opens transversely.

Stems prostrate, much branched. . . . . *A. graecizans*
Stems erect or ascending . . . . *A. retroflexus*

Amaranthus graecizans L.

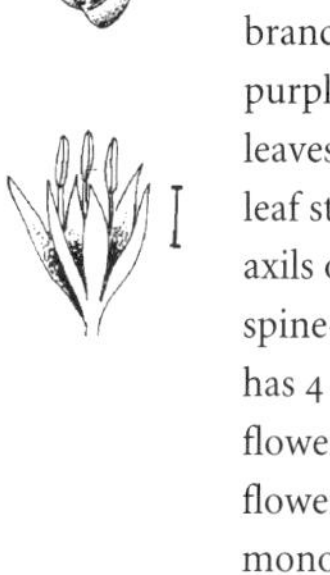

PROSTRATE AMARANTH

As the common name implies, this plant has a prostrate habit and is many-branched; the branches are 3 to 7 dm long. The reddish or purplish stems bear spatulate (spoon-shaped) leaves that are distinctive; they taper into a slender leaf stalk. The small flowers grow in clusters in the axils of the leaves, associated with needle-shaped, spine-tipped bracts up to 3 mm long. Each flower has 4 or 5 sepals (calyx) and no petals. The male flowers have 3 or 4 stamens, and the female flowers have ovaries with 3 styles. The plants are monoecious, with male and female flowers on the same plant. The fruit is a capsule with 2 or 3 beaks. Prostrate amaranth grows as a weed on waste ground and roadsides.

Amaranthus retroflexus L.

RED-ROOT PIGWEED

Red-root pigweed is a sturdy annual plant, 3 to 10 dm tall with erect stems. As the common name implies, the roots are reddish. The stems are many-branched and bear alternate, petiolate leaves which are oval-lanceolate and have prominent white (hairy) veins on the undersurface. The plant is dioecious (male and female flowers on different plants); the flowers are small, inconspicuous and green or purplish. They are grouped into spike-like clusters in the axils of the upper leaves, at the ends of branches, and in a large terminal cluster. The staminate flowers each have bracts surrounding the flower. These bracts are lanceolate and spinose; they are twice as long as the 2 to 5 sepals. There are 2 to 5 stamens opposite the sepals; there are no petals. The carpellary flowers are similar, with spinose bracts and 2 to 5 sepals; they have superior ovaries. The sepals are twice as long as the ovary, which, when ripe, turns into a capsule that splits transversely (it is circumsessile). NOTES: *Red-root pigweed is an introduced plant that grows as a weed in gardens, waste places, roadsides and cultivated fields. It can accumulate high levels of nitrate and has caused poisoning in some livestock.* ~ *Red-root pigweed can be distinguished from prostrate amaranth because it has an erect stem, 3 to 10 dm tall, and its flowers are in spike-like clusters.*

Portulacaceae PURSLANE FAMILY

These are annual or perennial plants with succulent leaves. The flowers each have a calyx of 2 sepals, a corolla of 3 to 6 petals, and 2 to 5 stamens. The fruit is a capsule.

Claytonia L. SPRING BEAUTY

Low, glabrous perennial plants, arising from a corm or a fleshy taproot. The leaves are rather fleshy: in *C. lanceolata* with 2 cauline leaves below the flowering stalk, and in *C. megarhiza* in a basal cluster. The flowers grow in loose terminal racemes and are rather showy, white to pink, with 2 sepals, 5 petals, 5 stamens and 3 joined carpels. The fruit is a 3-valved capsule.

C. lanceolata was used by natives as a useful food supply; it grows from underground corms, 1 to 2 cm in diameter, that are rich in starch. The corms are also dug up and consumed by grizzly bears.

Basal leaves 1 or 2, sometimes absent . . . . *C. lanceolata*
Basal leaves numerous . . . . *C. megarhiza*

Claytonia lanceolata Pursh

WESTERN SPRING BEAUTY

These are attractive little plants that grow from corms, so they appear early in the spring, having a good food supply: hence the common name "spring beauty." They have green-reddish stems, with 2 opposite, lanceolate, succulent leaves; these give rise to the species name *lanceolata*. The attractive, white-pink flowers (3 to 15) grow in a short raceme; they have 2 sepals, 5 petals and 5 stamens. The fruit is a 3-valved capsule, with black, shiny seeds. Western spring beauty is found in meadows, moist woods and alpine slopes.

NOTES: *The generic name* Claytonia *honours John Clayton (b. 1685), who made plant collections in Virginia.*

Claytonia megarhiza (A. Gray) Parry

ALPINE SPRING BEAUTY

This plant is quite different from western spring beauty and is not very attractive. It arises from a fleshy tap-root and numerous leaves grow from this; they are rather crowded and are spoon-shaped (spatulate). The flowers grow in a raceme but are smaller than those of the western spring beauty; they are white or pink. The fruit is a capsule with black seeds. Alpine spring beauty is found on alpine scree slopes and in rock crevices.

NOTES: *The species name* megarhiza *means "large root."*

Caryophyllaceae PINK FAMILY

Pinks are annuals or perennials, with simple leaves borne in pairs. The stem nodes are often swollen. The flowers are often clustered, in cymes; individual flowers have 4 or 5 sepals forming the calyx, 4 or 5 petals forming the corolla—these petals often have a spreading blade and a distinct claw, e.g. *Silene* (campion). There are 8 or 10 stamens, twice as many as the petals, and the ovary has 2 to 5 styles; the fruit is a capsule.

1. Sepals united . . . . *Silene*
   Sepals not united 2
2. Styles 4 or 5 3
   Styles 3 4

3. Petals deeply notched (often appearing as 10) . . . . *Cerastium*
Petals entire . . . . *Sagina*
4. Sepals with 1 nerve. . . . . *Arenaria*
Sepals with 3 or more nerves 5
5. Stems less than 10 cm tall. . . . . *Minuartia*
Stems more than 10 cm tall or long 6
6. Stems leafy; petals notched . . . . *Stellaria*
Stems not leafy; petals not notched. . . . . *Moehringia*

NOTE: *It is easy to distinguish flowers belonging to the genera* Cerastium *and* Stellaria *from those belonging to* Arenaria, *because the latter has flowers with entire petals, while* Cerastium *and* Stellaria *flowers have 2-cleft petals.*

Arenaria capillaris Poir

LINEAR-LEAVED SANDWORT

Sandwort is a densely tufted plant, 1 to 2 dm high; the leaves are paired and needle-like. The flowers are in a terminal cyme and are long-stalked; they are white and the 5 petals are only 5 to 7 mm long. There are 5 sepals and 10 stamens; the fruit is a capsule. Sandwort is found in exposed meadows and rocky slopes from low elevations up to 2700 m.

Cerastium L. CHICKWEED

Hairy annuals or perennials, branching from the base or from a slender rootstock. The leaves are entire. The flowers grow in cymes, with 5 sepals, 5 petals, 10 stamens, and 3 or 5 carpels joined. The fruit is a capsule.

Petals 6 to 8 mm long; alpine elevations. . . . . *C. beeringianum*
Petals 10 to 20 mm long; lower elevations. . . . . *C. arvense*

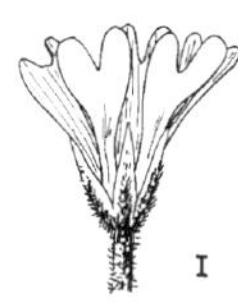

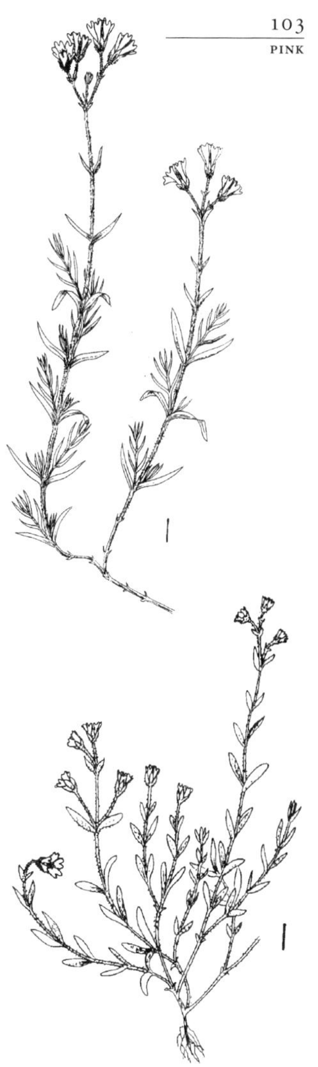

Cerastium arvense L.

FIELD MOUSE-EARED CHICKWEED

Field mouse-eared chickweed gets its rather fanciful name from its leaves which resemble the ears of a mouse. It is a perennial, 1 to 3 dm tall with creeping, branching, delicate stems; the stems are often decumbent at the base and then grow erect. The leaves are narrowly lanceolate, and there are short shoots in their axils. The flowers grow in a loose cymose arrangement; they are white, with 5 sepals, 5 petals, which are bi-lobed, and 10 stamens. The fruit is a capsule. The plant grows in a variety of habitats, from grassy plains to damp spots, and from low elevations to mountain areas. NOTES: *The specific name* arvense *means "of the fields."*

Cerastium beeringianum Cham. & Schlecht.

ALPINE MOUSE-EARED CHICKWEED

This chickweed can be distinguished from the field species because it is a smaller plant, growing up to 1 dm, usually with smaller flowers, and the axillary shoots are usually lacking. It is not a "straggly" plant. The fruit is a capsule, twice as long as the sepals (calyx). It is found on exposed alpine ridges above 2100 m and is definitely an alpine plant.

Minuartia L. SANDWORT

Sandworts grow as annuals or perennials, forming cushions or mats. The leaves are needle-like. The tiny flowers grow in cymes or solitary, with 5 sepals, 5 petals, usually entire and white, 10 stamens, and 3 carpels, joined. The fruit is a capsule.

Petals shorter than the sepals; flowers 1 per stem .... *M. elegans*
Petals longer than the sepals; flowers 1 to 3 per stem .... *M. obtusiloba*

Minuartia elegans (Cham. & Schlecht.) Schischk.

PURPLE ALPINE SANDWORT

(*Arenaria rossii* (R.Br. ex Richards.) Graebn.)

Purple alpine sandwort is a perennial plant, forming a small cushion with the dead leaves remaining on the plant. The leaves are rather needle-like (subulate), and the flowers are small, with 5 sepals, 5 white petals (sometimes absent) and 10 stamens; the ovary has 3 styles, and the fruit is a capsule. Purple alpine sandwort is an alpine plant, found on exposed rocky ledges.

Minuartia obtusiloba (Rydb.) House

ARCTIC SANDWORT

(*Arenaria obtusiloba* (Rydb.) Fern.)

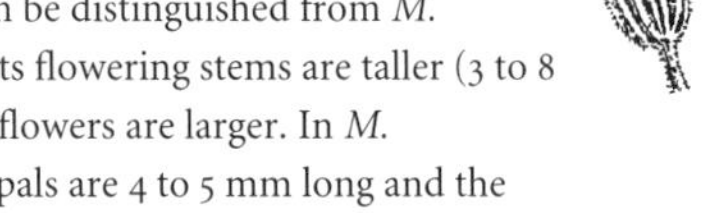

M. obtusiloba can be distinguished from *M. elegans* because its flowering stems are taller (3 to 8 cm tall) and the flowers are larger. In *M. obtusiloba* the sepals are 4 to 5 mm long and the white petals are usually much longer—4 to 8 mm long. Both plants are caespitose, with needle-like leaves. *M. obtusiloba* is found on dry alpine slopes, and is quite common on Plateau Mountain and elsewhere.

Moehringia lateriflora (L.) Fenzl.

BLUNT-LEAVED SANDWORT

(*Arenaria lateriflora* L.)

This plant can easily be distinguished from the *Minuartia* species because its appearance is so different. Instead of a ground-hugging, cushion-like plant, *Moehringia* is a creeping perennial with a slender rhizome, producing slender stems, decumbent at the base and ascending to erect. The leaves are also different from those of *Minuartia:* they are much broader and definitely not needle-like. The white flowers grow on long, slender stalks, and the petals are 3 to 8 mm long. *Moehringia* is found at much lower elevations than *Minuartia,* in moist meadows and woods.

Sagina saginoides (L.) Karst.

MOUNTAIN PEARLWORT

Pearlwort is a very small plant (2 to 8 cm high) and often grows matted. The leaves are in pairs, 5 to 12 mm long; they are needle-like and spine-tipped. The flowers are on very slender stalks, in the leaf-axils, and each has 5 sepals (about 2 mm long, often purplish) and 5 petals (white, entire and shorter than the sepals). There are 10 stamens, and the ovary usually has 5 stigmas; the fruit is a capsule, with 5 valves, which open to release numerous brown seeds. Pearlwort is found on trailsides and damp meadows, and on slopes and ledges at higher elevations.

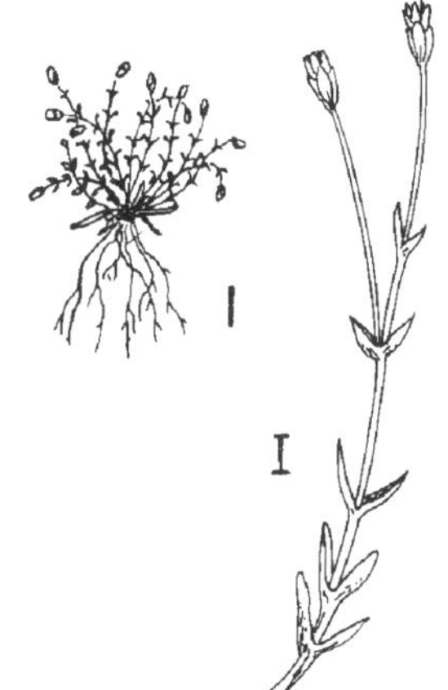

Silene L. — CAMPION

These are perennial plants whose leaves are entire and opposite. The flowers grow in cymes or solitary, each flower with 5 fused sepals forming a tubular calyx, which is many-veined. The corolla has 5 petals (each with a claw and spreading blade), 10 stamens, and 3 to 5 fused carpels (3 to 5 stigmas). The fruit is a many-seeded, toothed capsule.

1. Petals purple or pink; plants forming a dense mat .... *S. acaulis*
   Petals white, sometimes tinged with purple or green 2
2. Flowers more than 1 cm long.............................. *S. menziesii*
   Flowers less than 1 cm long................................... *S. parryi*

Silene acaulis L.

MOSS CAMPION

The common name for this plant is quite descriptive: it has a moss-like habit, and the flower is typically "campion." It has a tubular calyx with 5 teeth, 5 petals and 10 stamens, and the fruit is a capsule. The purple petals each have a spreading blade and a claw. Moss campion is common on alpine slopes and meadows, growing as a dense cushion, often covered with purple flowers; it is an attractive plant, with needle-like leaves.

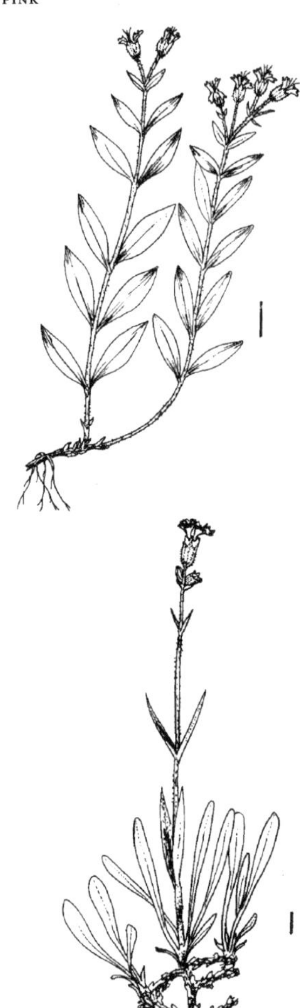

Silene menziesii Hook.

MENZIES' CAMPION, CATCHFLY

Menzies' campion is quite different in growth habit from moss campion. It has several slender stems arising from a slender rootstock; its leaves are much larger and wider (the inflorescence is leafy), and its white flowers are much smaller. Cathchfly is a variable plant: the stems are 1 to 3 dm tall, and the leaves are 2 to 5 cm long. It is not an alpine plant, but is found in woodlands.
NOTES: *The specific name* menziesii *honours Archibald Menzies (1754-1842), a physician-botanist with Captain George Vancouver on his northwest explorations at the end of the 18th century.*

Silene parryi (S. Wats.) C.L. Hitchc. & Maguire

PARRY'S CAMPION

Parry's campion resembles Menzies' campion but differs from it because its inflorescence is not leafy and the main leaves are basal instead of growing up the stem. They are longer (3 to 8 cm long) and are often linear. The calyx has 10 well-marked, purplish veins; this is quite distinctive. Parry's campion is found in meadows and on grassy slopes up to 2600 m.

Stellaria L. — CHICKWEED

Chickweeds grow as annuals or perennials; if perennials, they grow from slender branching rootstocks. The leaves are simple and paired; the flowers are solitary in axils or in terminal cymes. Sepals number 5 and are often sharp-tipped; there are 5 petals, deeply 2-lobed and white, usually 10 stamens, and 3 carpels united. The fruit is a 6-valved capsule.

1. Petals shorter than the sepals, or absent . . . . *S. borealis*
   Petals longer than the sepals 2
2. Stems 4-sided; leaves 2 to 5 cm long . . . . *S. longifolia*
   Stems not 4-sided; leaves 1 to 3 cm long . . . . *S. longipes*

Stellaria borealis Bigel.

NORTHERN STITCHWORT

This chickweed is a dainty plant, with weak stems arising from slender rootstocks; it has a sprawling habit, and the stems break off easily. The leaves are yellowish green, lanceolate and rather wide at the base. The flowers are small, white and inconspicuous. The 5 sepals are only 2 to 4 mm long, and the 5 petals are shorter or absent; the petals, if present, are deeply 2-lobed. The fruits (capsules) are dark. Northern stitchwort grows in moist, shady places.

NOTES: *The specific name* borealis *comes from Boreas, the Greek God of the North Wind.*

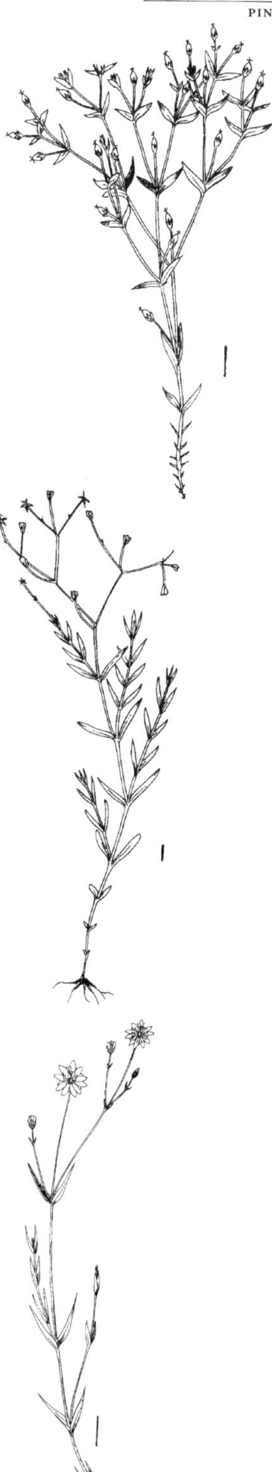

Stellaria longifolia Muhl.

LONG-LEAVED CHICKWEED

Long-leaved chickweed is similar to northern stitchwort: it is a dainty plant with inconspicuous flowers, but not so fragile. Its leaves are narrower and not so wide at the base. The flowers of *S. borealis* are found in the leaf-axils near the base of the plant and spread upwards; in long-leaved chickweed they are chiefly found at the top of the plant. Both plants show cymose arrangement of the flowers. In northern stitchwort the pedicels (flower-stalks) are usually reflexed at maturity, but in long-leaved chickweed they are widely spreading. The fruits (capsules) are dark in *S. borealis* and straw-coloured in *S. longifolia.*

Stellaria longipes Goldie

LONG-STALKED CHICKWEED

The common name is descriptive, because the flower stalks grow up to 8 mm long. There are several differences between this chickweed and the other two. Its leaves are awl-shaped, with a sharp point, and never yellow-green; the flowers are also larger. Although the inflorescence is still cymose, the angle of divergence from the main stem is less in the long-stalked chickweed. The fruits are dark brown to black in colour. Long-stalked chickweed is found on prairies, in woodland and on alpine slopes.

NOTES: *Long-stalked chickweed is a variable plant: the small plants, with single flowers, growing on Plateau Mountain are different from the large plants growing at lower elevations.*

Ranunculaceae BUTTERCUP/CROWFOOT FAMILY

Members of this family are annuals or perennials, sometimes climbing shrubs, with variable leaves. Leaves may be lobed, cleft or toothed, and are sometimes elaborately divided. Many species produce an acrid juice. The flower-structure varies considerably: it is difficult to realize that the common buttercup, with its regular, open flower, belongs to the same family as delphinium, monkshood and columbine, which have complex flowers. Another interesting characteristic of this family is the fact that the petals are often absent and the sepals are petalloid. The basic flower structure involves 3 to 15 sepals, petals present or absent, sometimes with nectar spurs, and numerous stamens. The fruits are either achenes (dry, 1-seeded fruits, often with plumed styles), berries or follicles (pods that open down 1 side only). All the flower parts are free. Some family members are **poisonous**, such as monkshood.

The family name *Ranunculaceae* comes from the word *rana* meaning a frog: *ranunculus* is a "little frog." The plants belonging to the genus *Ranunculus* (buttercups) are often found in damp places.

Some Ranunculaceae plants were used by the Blackfoot. The ripe fluffy seedheads of cut-leaved anemone (*Anemone multifida*) were burned and the smoke inhaled for headaches. The seed and foliage of the western meadow rue (*Thalictrum occidentale*) was kept for its pleasant smell and may have been used as an insect repellant. The berries and roots of red and white baneberry (*Actaea rubra*) are poisonous, but despite this, the Blackfoot brewed the roots to treat coughs and colds.

1. Leaves compound 2
   Leaves simple and lobed 5
2. Woody vines; leaves opposite . . . . . *Clematis*
   Herbs; leaves not opposite 3
3. Flowers white; fruit a red or white berry . . . . . *Actaea*
   Flowers not white; fruit a dry capsule or follicle 4
4. Sepals 5; petals 5, spurred; flowers perfect . . . . . *Aquilegia*
   Sepals 4 or 5; petals none; male and female flowers on different plants . . . . . *Thalictrum*
5. Flowers irregular, blue . . . . . *Delphinium*
   Flowers regular 6
6. Stem-leaves whorled below the flower . . . . . *Anemone*
   Stem-leaves alternate 7
7. Sepals 5, green or purplish; petals 5, yellow or white . . . . . *Ranunculus*
   Sepals 5 to 7, white; petals inconspicuous . . . . . *Trollius*

Actaea rubra L.

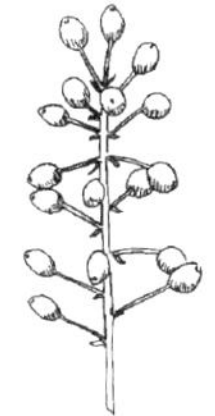

RED AND WHITE BANEBERRY

Baneberry is a large, perennial plant, 3 to 8 dm high, and its large leaves are characteristic: they are compound, with many leaflets. The leaves are "bi-ternate," that is, the leaves are divided into 3 parts, and each of these parts is further divided into 3 parts. The small flowers are white, in a short terminal raceme; there are 5 or 6 whitish sepals and petals, numerous stamens and a superior ovary, in each flower. The fruits are berries, usually red, sometimes white. The plants are found in moist woods.

NOTES: *The common name baneberry suggests one property of the plant: "bane" means "poisonous" or "death." The berries are especially* **poisonous**; *they contain protoanemonine.*

Anemone L. — WINDFLOWER

Windflowers are perennial herbaceous plants with basal, petioled leaves. Leaves on the stem (caudal) grow in an involucre below the flower; the leaf-blades are palmately lobed, each segment divided or lobed again. 5 sepals, petaloid, white or purplish, petals absent, numerous stamens and numerous carpels; fruit an achene.

1. Flowers blue, 2 to 4 cm long . . . . *A. patens*
   Flowers white, yellow or red, less than 2 cm long 2
2. Leaves with 3 leaflets; flowers white; moist habitats. . . . . *A. parviflora*
   Leaves with numerous divisions; dry plains and woods. . . . . *A. multifida*

Anemone multifida Poir.

CUT-LEAVED ANEMONE

Cut-leaved anemone grows 1.5 to 5 dm tall. The common name is appropriate because the leaves have many segments, all sharply pointed and hairy: *multifida* means "much-divided." The flowering stems arise from an involucre of 3 sessile leaves (stalkless), characteristic of anemone plants. The flowers of cut-leaved anemone have no petals (another characteristic of anemones); petals are replaced by coloured sepals, which may be white, cream or reddish purple. There are 5 sepals, numerous stamens and numerous carpels which turn into the fruits (achenes). The plant is found on prairie grasslands.

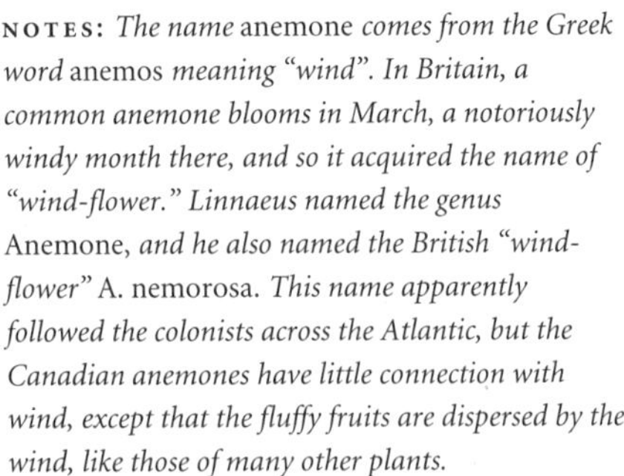

NOTES: *The name* anemone *comes from the Greek word* anemos *meaning "wind". In Britain, a common anemone blooms in March, a notoriously windy month there, and so it acquired the name of "wind-flower." Linnaeus named the genus* Anemone, *and he also named the British "wind-flower"* A. nemorosa. *This name apparently followed the colonists across the Atlantic, but the Canadian anemones have little connection with wind, except that the fluffy fruits are dispersed by the wind, like those of many other plants.*

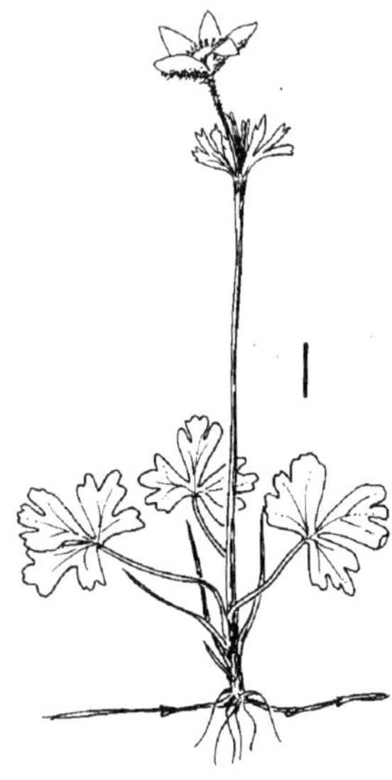

Anemone parviflora Michx.

SMALL WOOD ANEMONE

This plant is much smaller than the cut-leaved anemone (the species name means "small-flowered"), and its leaves are different: they are smaller and only divided into 3 blunt lobes with no sharp points. There are 3 involucral leaves, from which a single flowering stem arises with a single flower; this trait distinguishes it from the cut-leaved anemone. There are no petals: the sepals are white, tinged with blue, or rose. The fruits (achenes) are in a dense head and hairy. The plant is found on moist banks and slopes, in montane and alpine habitats.

Anemone patens

PRAIRIE CROCUS, PASQUE-FLOWER, WIND-FLOWER

The leaves of the prairie crocus are more divided than those of the cut-leaved anemone, and the divisions are pointed. The basal leaves do not expand until the flower has bloomed—the involucral leaves are densely hairy and protect the young flower-bud from the cold winds of early spring. The flower has 5 to 7 sepals, pale blue-purple or white inside, with a deeper colour on the outside. There are numerous stamens surrounding the numerous carpels, which turn into the fruits (achenes). The styles of the ovaries turn into long, fluffy plumes, which aid in fruit dispersal by wind. Prairie crocus is a typical prairie plant, and is also found in dry, open woods.

NOTES: *The common name prairie crocus is a misnomer: it is certainly not a crocus!* ⁓ *The other common name, pasque-flower, is also a misnomer, because* Pasque *means Easter; although the plant does bloom at Easter in Europe, it often blooms after Easter in Alberta.* ⁓ *Whatever name it takes, it is a popular flower, because it blooms soon after the winter snows have melted and is one of the welcome signs of spring.*

Aquilegia L. — COLUMBINE

Columbines are perennial plants rising from a stout rootstock, with large, 3-lobed leaves, with long stalks; the lobes are subdivided again. The flowers have 5 petalloid sepals, 5 erect petals with a long, slender, nectar-filled spur, numerous stamens and 5 carpels which, when ripe, turn into follicles.

Flowers blue . . . . *A. brevistyla*
Flowers yellow . . . . *A. flavescens*

Aquilegia brevistyla Hook.

⁓ **BLUE COLUMBINE**

The columbines differ from the anemones because although their sepals are coloured, they also possess coloured petals. Columbines are easily identified because the petals are spurred, they produce nectar, and the flower is quite distinctive. The attractive basal leaves are also characteristic: they are dainty, long-stemmed and have 3 leaflets, also divided. Blue columbine, as its name implies, has 5 conspicuous blue sepals; the 5 petals are yellowish white, and the spurs are bluish and hooked. There are numerous stamens and 5 carpels, which turn into the fruits (follicles). Follicles are like tiny pea-pods, but they only open down one side. Blue columbine is found in meadows and open woods.

Aquilegia flavescens S. Wats.

YELLOW COLUMBINE

Yellow columbine can be easily distinguished from blue columbine by its colour. The species name *flavescens* comes from the Latin *flavum*, meaning "yellow." The spurs of the flowers are not hooked, but they are more or less curved. The "beaks" of the follicles in yellow columbine are longer than those of blue columbine—they are usually over 5 mm long. Both plants are found in open woods and rocky slopes.

Clematis occidentalis (Hornem.) D.C.

PURPLE CLEMATIS

(C. verticellaris D.C.*)*

Purple clematis is an attractive woodland plant, trailing or climbing, up to 2 m long, with large purple flowers. It climbs by means of its sensitive leaf-stalks. The leaves are divided into 3 leaflets, which are sometimes toothed; the leaves grow opposite each other, unusual in this family. There are 4 large, coloured sepals (blue or purple), no petals and numerous stamens. The fruits are interesting: the numerous achenes turn into a head of "fluffy" fruits, each consisting of a tiny achene and a long, plumed, hairy style. These fruits are characteristic of *Clematis* flowers, and are similar to those of prairie crocus.

NOTES: *The species name* occidentalis *means "western."*

Delphinium L. — LARKSPUR

Larkspurs are erect herbaceous plants with palmately lobed or cleft leaves. The flowers are blue-purple, arranged in a terminal raceme, with 5 petalloid sepals, the upper one forming a long spur, and 4 irregular petals, the upper pair expanding backwards into the sepal-spur. There are numerous stamens and usually 3 carpels, which turn into follicles when ripe.

1. Stems over 1 m tall; flowers numerous . . . . *D. glaucum*
   Stems shorter; flowers fewer than 15 per flower-stalk 2
2. Flowers bluish violet; lower elevations . . . . *D. bicolor*
   Flowers dark blue; higher elevations . . . . *D. nuttallianum*

Delphinium bicolor L.

LOW LARKSPUR

Low larkspur is an attractive plant with a short terminal raceme of large, blue flowers. They have 5 petalloid sepals, coloured blue-purple, with the upper sepal forming a long blue spur; there are also 4 petals, the lower ones dark blue and hairy and the upper ones smaller, creamy-white and hairless, with purple veins. There are numerous stamens and usually 3 carpels; these turn into follicles. The leaves of low larkspur are mostly basal and are divided into 5 leaflets, which are lobed at the tip. It grows in grassland, open moist meadows and woodlands, and can also be found on alpine ridges, up to 2500 m.

NOTES: *This plant is* ***poisonous.***

Delphinium glaucum S. Wats.

TALL LARKSPUR

Tall larkspur can be distinguished by its height; low larkspur is only 2 to 5 dm tall but tall larkspur reaches 1 to 2 m. The raceme of flowers is also taller, up to 2 dm. The leaves of the 2 species are also different: they are more numerous in tall larkspur and the divisions are pointed, not lobed. The stems are hollow—in other larkspur species, the stems are usually solid. Tall larkspur is found in moist woods and meadows, on stream banks and on subalpine slopes.

NOTES: *Delphinium plants contain a* ***poisonous*** *alkaloid, delphinine.*

Delphinium nuttallianum Pritz ex Walp.

NUTTALL'S LARKSPUR

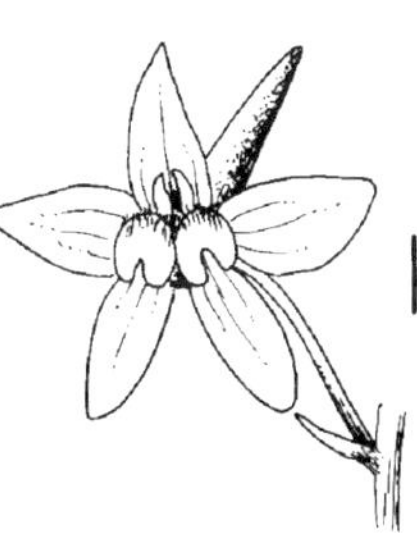

This species can be distinguished from low larkspur because its petals are different: in *D. nuttallianum* the petals are deeply notched. Low larkspur often has fleshy but not tuberous roots, while in Nuttall's larkspur they are often tuberous. Nuttall's larkspur is found on dry mountain slopes.

NOTES: *The species name* nuttallianum *honours Thomas Nuttall (1786-1859), an English botanist who collected plants in western North America.*

Ranunculus L. — BUTTERCUP

Buttercups are perennial, herbaceous plants, sometimes containing acrid juice and often found in moist places. The leaves are variable and may be lobed, toothed or variously divided. The flowers are golden-yellow or white; sepals and petals usually number 5, stamens numerous, carpels 5 to numerous. The petals are often glossy, each with a nectary at its base. The fruit is a head of achenes, sometimes beaked.

1. Leaves lobed or divided into segments 2
   Leaves not lobed or divided into segments; stems often rooting at the nodes . . . . *R. reptans*
2. Basal and stem-leaves different . . . . *R. cardiophyllus*
   Basal and stem-leaves similar 3
3. Plant with leaves 3-lobed or more 4
   Plant not hairy; basal leaves 3-lobed; alpine species . . . . *R. eschscholtzii*
4. Stem-leaves 3-lobed, stalked; petals 8 to 16 mm long . . . . *R. acris*
   Stem-leaves 3 to 7-lobed, not stalked; petals 8 to 10 mm long . . . . *R. pedatifidus*

Ranunculus acris L.

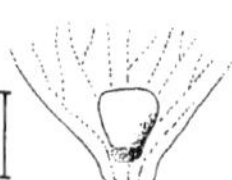

TALL BUTTERCUP

This buttercup deserves its common name, growing to 1 m high. It has no creeping stems. The basal leaves have long leaf-stalks (petioles). The leaves are large (4 to 7 cm wide) and deeply cleft into 3 to 5 parts; these are again divided into narrow, lance-shaped, toothed segments. The leaf is definitely palmate. The few stem (cauline) leaves are usually 3-lobed with short stalks. The flowers are long-stalked and large; each flower has 5 greenish yellow sepals, 5 golden-yellow, shiny petals, numerous stamens and several carpels. These carpels develop into flat achenes with very small beaks. Tall buttercup is a European weed now established on roadsides and moist, open meadows in Kananaskis Country.

Ranunculus cardiophyllus Hook.

HEART-LEAVED BUTTERCUP

Heart-leaved buttercup is well named because it has heart-shaped leaves at its base and linear leaves along its stem. The flowers follow the typical buttercup pattern, with 5 sepals, 5 petals with nectaries at the base of each, numerous stamens and 5 or more carpels, which turn into achenes with short beaks.The plant is found in moist meadows and open woods, at low elevations.

NOTES: *The species name* cardiophyllus *comes from 2 words:* cardio, *meaning "heart," and* phyllos, *meaning "leaf."*

Ranunculus eschschollzii Schlecht.

MOUNTAIN BUTTERCUP

Mountain buttercup grows only 1 to 3 dm high and can be distinguished from heart-leaved buttercup because the basal leaves are quite different: they are split into 3 leaflets, each leaflet deeply lobed. The flowers are yellow and relatively large, but this is a variable plant and size varies with the habitat. The fruiting head of achenes is ovoid to oblong, and the achenes have slender beaks. The habitat of mountain buttercup is quite different from that of the lowland heart-leaved buttercup. As its name suggests, mountain buttercup is found on alpine slopes and meadows.

NOTES: *The rather cumbersome species name honours a Russian botanist, J.F. Eschscholtz (1793-1831), who was a doctor and naturalist on Kotzbue's expeditions, financed by Count Romanzoff.*

Ranunculus pedatifidus J.E. Smith

NORTHERN BUTTERCUP

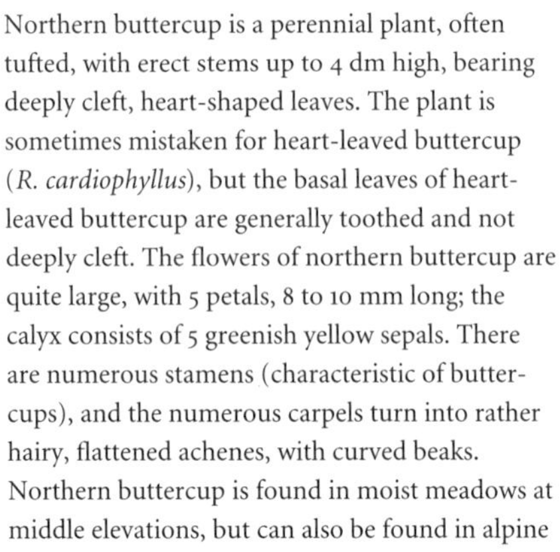

Northern buttercup is a perennial plant, often tufted, with erect stems up to 4 dm high, bearing deeply cleft, heart-shaped leaves. The plant is sometimes mistaken for heart-leaved buttercup (*R. cardiophyllus*), but the basal leaves of heart-leaved buttercup are generally toothed and not deeply cleft. The flowers of northern buttercup are quite large, with 5 petals, 8 to 10 mm long; the calyx consists of 5 greenish yellow sepals. There are numerous stamens (characteristic of buttercups), and the numerous carpels turn into rather hairy, flattened achenes, with curved beaks. Northern buttercup is found in moist meadows at middle elevations, but can also be found in alpine situations.

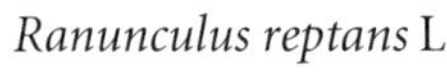

Ranunculus reptans L.

CREEPING SPEARWORT

(R. flammula L.*)*

This plant is easily distinguished from other buttercups because its leaves are filiform or lanceolate. The species name *reptans* means "creeping," and this is appropriate because the stems creep along the ground and root at the nodes (another distinctive trait). Creeping spearwort is found on the borders of ponds and lakes.

Thalictrum L. — MEADOW RUE

Meadow rues are perennial plants that rise from stout rootstocks. The leaves are 2 to 3-ternately compound, basal and cauline, the petioles dilated at the sheathing base. The plants are often dioecious, bearing male flowers with 4 to 5 greenish sepals and numerous stamens, and female flowers with several carpels that ripen into grooved achenes.

Flowering-stalk compact; leaflet underside veiny. . . . . *T. venulosum*
Flowering-stalk wide-spreading; leaflet underside not veiny . . . *T. occidentale*

Thalictrum occidentale A. Gray

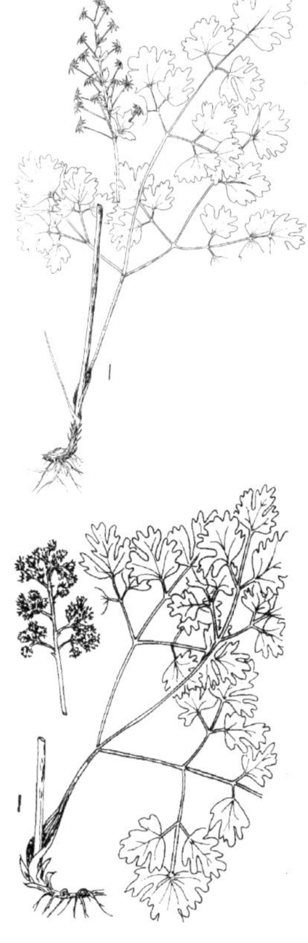

WESTERN MEADOW RUE

Western meadow rue has attractive leaves split into 3 leaflets and each leaflet again split into 3. This is called a bi-ternate leaf. Some meadow rue plants are dioecious. Female flowers have 4 to 5 sepals, no petals and numerous carpels, which turn into achenes 5 to 8 mm long, tapering at both ends. Male flowers have 4 to 5 sepals, no petals and numerous stamens. Western meadow rue grows 5 to 10 dm high and is found in moist woods.

NOTES: *The species name* occidentale *means "western."*

Thalictrum venulosum Trel.

VEINY MEADOW RUE

Veiny meadow rue is a smaller plant than western meadow rue; it grows 2 to 9 dm high and its leaves are smaller, though similar in shape. In western meadow rue the main flower-stalks (peduncles) are wide-spreading and about the same length; in veiny meadow rue they are of varying lengths. Veiny meadow rue has tiny, ascending achenes. It is a prairie flower and is also found in woodlands.

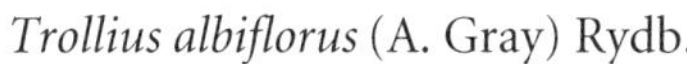

Trollius albiflorus (A. Gray) Rydb.

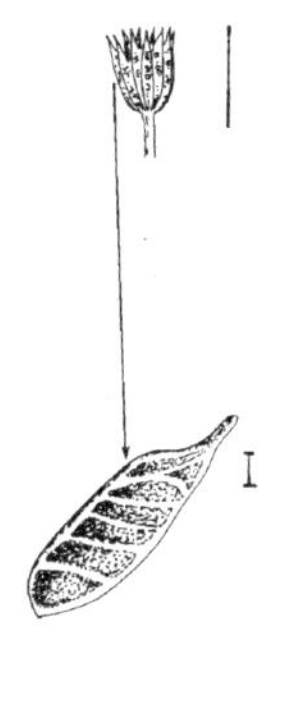

GLOBE-FLOWER

Globe-flower is a glabrous perennial plant with a thick rhizome and erect stems up to 4 dm high. There are a few petioled basal leaves (these have membraneous sheaths), with blades 4 to 8 cm broad, deeply lobed into 5 or 7 parts, which are toothed; there are 1 or 2 stem (cauline) leaves. The large, attractive flowers have 5 to 7 white (sometimes yellow) petaloid sepals and 5 to 8 linear petals which are inconspicuous and surround numerous stamens. There are 10 to 20 carpels in the centre, which develop into follicles. Globe-flowers are alpine or subalpine plants. They are often found in boggy places, flowering just below the edges of retreating snowbanks, from June to the middle of July.

NOTES: *The generic name* Trollius *comes from the German name* Trollblume, *given to a very similar European species,* Trollius europaeus. *The specific name* albiflorus *means "white-flowered."*

Papaveraceae POPPY FAMILY

These are herbs with milky juice. The basal leaves are pinnately dissected, and the flowers are solitary on long flower-stalks. The flower buds are enclosed inside 2 deciduous green sepals which fall off as the bud opens. There are usually 4 petals and numerous stamens; the united ovaries in the centre are characteristic. There are 4 to several carpels joined together; the ovaries are partly divided by little walls (placentae), bearing numerous ovules. The stigmas of the ovaries are united and form a flat crown, a stigmatic disk. There are no styles. This structure, when fertilized, becomes the fruit, a many-seeded capsule. The seeds escape through pores near the edge of the stigmatic disk. The opium poppy (*Papaver somniferum*) is a member of this family, and opium, a milky, reddish brown juice, is found in the unripe fruit when the surface is slit.

Papaver L. POPPY

Poppies are caespitose perennials with milky juice (latex) and basal, pinnately dissected leaves. The flower-buds are nodding on long, leafless stems (scapes); the flowers are large and showy, sulphur-yellow or orange, with 2 deciduous sepals, 4 thin petals, numerous stamens, and 4 to several carpels united into a large, cone-shaped structure which ripens into the fruit. The fruit is a poricidal capsule, with intruding parietal partitions bearing numerous seeds that escape through pores below the stigmatic disk.

Leaves hairy; flowers yellow; alpine slopes . . . . *P. kluanensis*
Leaves not hairy; flowers orange; introduced species . . . . *P. nudicaule*

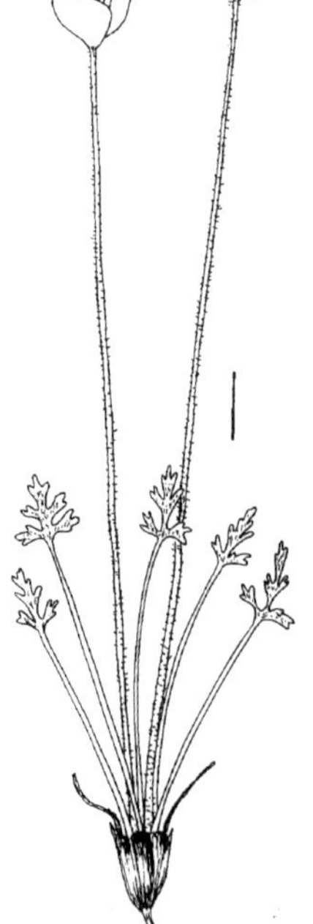

Papaver kluanensis D. Löve

ALPINE POPPY, KLUANE POPPY

Alpine poppy is 6 to 9 cm tall, with numerous densely hairy, basal leaves with long stalks; the blades are 5-lobed. The flower is solitary, and there are 2 deciduous sepals and 4 large, brightly coloured, sulphur-yellow petals, with numerous stamens. The carpels turn into the characteristic poppy-fruit, made up of several carpels joined together into a circular structure, rather pointed at the base, with a flat top with lines radiating from the centre. Inside, there are numerous partitions bearing seeds; the seeds escape through small pores when ripe. The plant contains a milky juice. Alpine poppy is found on alpine slopes.

Papaver nudicaule L.

ICELAND POPPY

The Iceland poppy is taller than alpine poppy, growing 2 to 3 dm tall, but the species are similar in many ways. Iceland poppy has basal leaves with a long flower-stalk rising from them, and a solitary orange or reddish flower. The leaf-shapes are different: those of Iceland poppy are pinnately lobed, while those of alpine poppy are each divided into 5 oval-lanceolate lobes. The species name *nudicaule* means "naked stem," i.e., there are no leaves on the flower-stalk; this is a family characteristic. Iceland poppy has escaped from cultivation and is found on waste ground and roadsides and around Chateau Lake Louise, Alberta. It has been reported from Kananaskis Country.

Fumariaceae FUMITORY FAMILY

These are annual or biennial plants with leafy, glabrous stems growing from a taproot. The numerous stems are often sprawling. The leaves are blue-green with long stalks; they are pinnately divided into leaflets, and the leaflets are pinnately lobed. The flowers are arranged in racemes and are golden-yellow or pink.

Corydalis Medic. FUMITORY

These are dainty, glabrous herbaceous plants with watery juice and dissected leaves. The flowers are irregular, with 2 sepals and 4 petals. The outer pair is sac-shaped and keeled at the apex, one spurred at the base; the inner pair is boat-shaped, adhering at the tip, There are 6 stamens and 2 carpels. The fruit is a long, slender, 2-valved capsule.

Flowers yellow . . . . *C. aurea*
Flowers pink . . . . *C. sempervirens*

Corydalis aurea Willd.

GOLDEN CORYDALIS

Golden corydalis is a dainty plant, 1 to 3 dm high, with decumbent or ascending stems. The leaves have pinnate leaflets, and these are again divided into pinnate lobes. The golden-yellow flowers are attractive. They have 2 sepals and 4 petals, and the petals are irregular in shape. One of the outer petals has a well-marked spur, which is characteristic and helps to identify the flower. There are 6 stamens in 2 series and 2 carpels. The carpels form the fruit, a long, slender, 2-valved capsule. The capsule bursts to release the shiny black seeds. Golden corydalis is found in moist spots along trails, on cliff ledges and in woodlands.

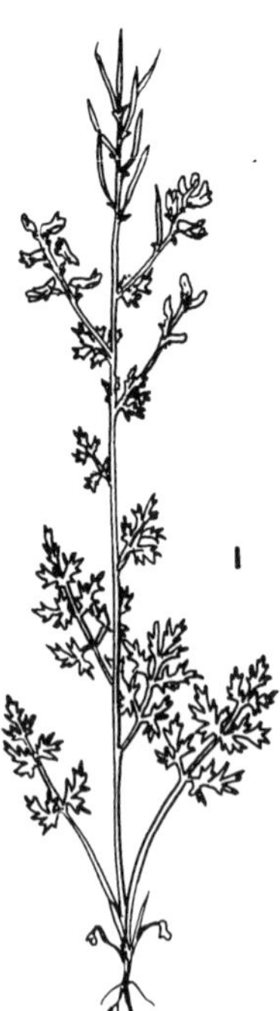

Corydalis sempervirens (L.) Pers.

PINK CORYDALIS

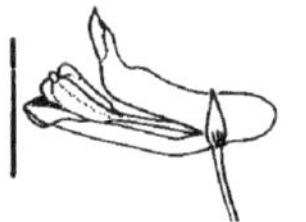

The petals of pink corydalis, as its common name implies, are purplish pink, with yellow tips, and so the plant is easy to distinguish. The fruits (capsules) are also different: in golden corydalis they are spreading or pendent (hanging), while in the pink corydalis they are ascending. Pink corydalis is found in open woods.

Brassicaceae (Cruciferae) MUSTARD FAMILY

Members of the mustard family are herbaceous plants with alternate leaves, often in a basal cluster. The flower structure includes 4 sepals, 4 petals in the shape of a Maltese cross and 6 stamens. The stamens are in 2 series, 2 in the outer series and 4 in the inner one. In the centre of the flower are 2 joined carpels. These form a pod-like fruit with 2 cavities, separated by a thin membrane, to which numerous seeds are attached. This pod may be long and narrow (a silique) or short and broad (a silicle). (The genus *Neslia* is an exception, because it has a ball-like fruit with 1 seed.) It is important to have the mature fruit when attempting to determine the genus and species of a member of this family.

The previous name for the mustard family, Cruciferae, was an appropriate one, because it means the "cross-bearer," referring to the 4 petals in the form of a Maltese cross. It is easy to identify a plant as a member of the mustard family because the flower structure is constant and simple, but as it is a large family, it is not always easy to sort the plants into genera and species. The species of some genera, like *Arabis* and *Draba*, are actually quite difficult to determine.

1. Fruit heart-shaped, ball-shaped or triangular 2
   Fruit pod-like 6
2. Fruit ball-shaped. 3
   Fruit heart-shaped or triangular. 4
3. Fruit inflated, 2-lobed . . . . *Physaria*
   Fruit not inflated, not lobed . . . . *Lesquerella*
4. Fruit triangular. . . . . *Capsella*
   Fruit heart-shaped 5
5. Fruit 10 to 20 mm long; plant strongly scented. *Thlapsi*
   Fruit 2 to 8 mm long; plant not strongly scented . . . . *Lepidium*
6. Fruit flat in cross-section. 7
   Fruit round in cross-section 8
7. Leaves entire or toothed . . . . *Draba*
   Leaves deeply lobed. . . . . *Cardamine*
8. Plants of disturbed areas, often weedy 9
   Plants of undisturbed habitats, not weedy. 10
9. Leaves entire or toothed; plants green; flowers bright yellow. . . . . . *Erysimum*
   Leaves deeply lobed; plants greyish green; flowers pale yellow . . . *Descurainia*
10. Plants of alpine habits. 11
    Plants of lower elevations. 12
11. Flowers creamy-white to yellow . . . . *Smelowskia*
    Flowers purple. . . . . *Arabis*
12. Plants of wet habitats . . . . *Rorippa*
    Plants of dry habitats 13
13. Flowers yellow. . . . . *Erysimum*
    Flowers white to purple 14
14. Low-growing perennial; pods with a bumpy surface . . . . *Braya*
    Erect biennial; pods with a smooth surface . . . . *Arabis*

Arabis L. **ROCK CRESS**

Rock cresses are biennial or perennial herbaceous plants, often pubescent, with branched hairs, petiolate basal leaves (often in a rosette or tufted) and smaller cauline leaves which are sessile, often sagittate or with auricles. The flowers grow in terminal racemes with 4 sepals, 4 white, pink or purple petals, 6 stamens and 2 carpels, united. The fruits are siliques, long and slender.

1. Flowers yellowish white . . . . *A. glabra*
   Flowers pink or purple. 2
2. Plants 5 to 25 cm tall; alpine slopes. . . . . *A. lyallii*
   Plants 40 to 80 cm tall; dry sandy slopes . . . . *A. divaricarpa*

Arabis divaricarpa A. Nels.

PURPLE ROCK CRESS

The rock cresses all have a similar growth-habit: they have a rosette of leaves, more or less flat on the ground, with a flowering-stalk arising from the centre of the rosette. There are several leaves growing on the stem (cauline), which diminish in size upwards. There is a group of small, pale pink to purple, clustered flowers at the top of the main flower stalk. There are 4 sepals, 4 petals, 6 stamens and 2 carpels in each flower. The fruits, called siliques, are long (2 to 8 cm) and thin; they burst open to release 2 rows of seeds. *A. divaricarpa* can be distinguished from other rock cresses, because its long narrow fruits are divergent—that is, they spread out from the main stalk at an angle of 45 degrees. The hairs on the stem are also characteristic: they are 3-rayed, sometimes several-rayed. Purple rock cress is found in damp, grassy places and also on dry sandy slopes.

NOTES: *The species name* divaricarpa *comes from 2 words:* divaricatus, *meaning "divergent," and* carpos, *meaning "fruit."*

Arabis glabra (L.) Bernh.

TOWER MUSTARD

This plant grows 4 to 10 dm high, and its long narrow fruits are erect and closely appressed to the stem. The flowers in *A. divaricarpa* are pale pink to purple, but in *A. glabra* they are yellowish white or have purple tips. The basal leaves in tower mustard are oblong and are usually toothed or cleft. These characteristics are not always reliable for determination, because on the mature plant the leaves are often dry and shrivelled. Tower mustard is found in thickets and fields.

Arabis lyallii S. Wats.

LYALL'S ROCK CRESS

This plant could well be called "alpine rock cress" because it is found on dry alpine slopes and ridges. The plant is smaller than the other two species, only 2 to 25 cm tall, and the flowers are purple, quite a distinctive colour. The basal leaves resemble those of *A. divaricarpa*, but are smaller.

Braya humilis (C.A. Meyer) Robins.

LEAFY BRAYA

Leafy braya is a low-growing perennial plant, with the leaves clustered at the base of the plant. These leaves are 1 to 3 cm long, usually hairy and rather spatulate. There are also some cauline leaves growing alternately up the slender flowering-stem; this trait distinguishes this plant from the species *B. purpurascens*, which has a leafless flowering stem. The flowers grow at the top of the stem in a capitulate inflorescence which elongates considerably in fruit. The flowers are white or lilac, with petals 3 to 4 mm long. The fruits are siliques, 1 to 2 cm long and narrow, and there is a slender style with a coronate stigma. This plant is found in open woods, banks and gravel bars, and can be found in alpine elevations.

NOTES: *The species name* humilis *means "low-growing."*

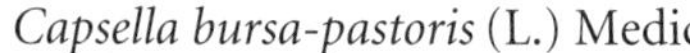

Capsella bursa-pastoris (L.) Medic.

SHEPHERD'S PURSE

Shepherd's purse gets its unusual name from its fruits: they are formed from 2 carpels joined together and are triangular in shape. It requires some imagination to visualize them as "purses," but such an unusual shape makes it easy to identify the plant. If the plant is in flower and not yet in fruit, it is not easy to distinguish it from the rock-cresses at first, because it has a similar habit, with a rosette of leaves, the flowering stalk arising from the centre and a cluster of small white flowers at the top. It grows 1 to 5 dm high. Shepherd's purse is a common annual weed of fields, gardens and lawns.

NOTES: *The species name* bursa-pastoris *literally means the "purse of the shepherd."*

Cardamine umbellata Greene

BITTERCRESS, MOUNTAIN CRESS

(C. oligosperma) Nutt.

Bittercress is a perennial plant, with several erect stems bearing pinnate leaves that are characteristic of this genus. The 3 to 7 leaflets at the base of the leaf are small, but increase in size upwards; the terminal lobe is quite large, up to 2 cm (the leaf is 10 cm long). The leaflets are often round or heart-shaped. The stem-leaves are smaller and variable, but the terminal leaflet is always largest. The flowers are arranged in a short corymb which looks rather like an umbel, hence the specific name *umbellata.* Each flower has 4 sepals, 4 white petals, 6 stamens and 2 united carpels. The carpels develop into the fruit, a long, narrow silique which splits open to release the seeds. Bittercress grows in damp places along trails and along creeks in the mountains.

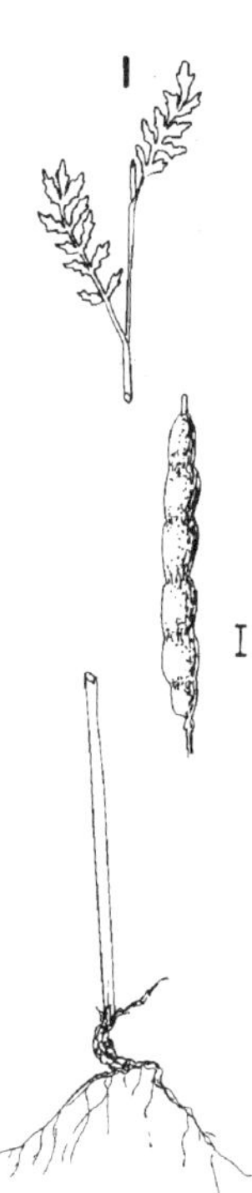

Descurainia richardsonii (Sweet) O.E. Schultz

GREY TANSY MUSTARD

Grey tansy mustard grows from a basal rosette of leaves, which have usually withered by the time the plant is in fruit. It grows 3 to 8 dm high, and the stems are hairy; the stem leaves are pinnate and the leaflets vary, but are not so finely divided as in green tansy mustard (*D. pinata* (Walt.) Britt.), a similar species. Because the leaves are covered with fine soft hairs, the whole plant appears grey, hence the common name. The flowers grow in clusters, each 2 to 4 mm wide, and are pale yellowish with 4 sepals, 4 petals, 6 stamens and 2 united carpels. The 2 carpels form the fruit, a long narrow silique (with a minute beak), which splits open to release the seeds. Grey tansy mustard is found in waste places, roadsides and open areas.

NOTES: *The species name* richardsonii *honours Dr (later Sir) John Richardson (1787-1826), the Arctic explorer who discovered this plant.*

Draba L. — WHITLOW GRASS

Whitlow grasses are low, tufted, herbaceous plants, often mat-forming, with simple entire leaves (often in a rosette). They are pubescent, often with stellate hairs. The flowers grow in a raceme with 4 sepals, 4 white or yellow petals, 6 stamens and 2 united carpels, which ripen into oval fruits, rather pointed at each end.

NOTE: *In order to distinguish different species of* Draba, *it is usually necessary to use a good lens, to examine the hairs on the stems and leaves—the species are notoriously difficult to separate.*

1.	Stems leafless	2
	Stems with leaves	3
2.	Basal leaves less than 1 mm wide	*D. oligosperma*
	Basal leaves more than 1 mm wide	*D. incerta*
3.	Stem-leaves 1 to 2	*D. kananaskis*
	Stem-leaves 3 to 30	*D. aurea*

Draba aurea Vahl.

GOLDEN DRABA, GOLDEN WHITLOW GRASS

Golden draba is a perennial plant, 1 to 5 dm tall. The basal leaves form a rosette, with the flower-stalk arising from the centre, bearing a cluster of small yellow flowers. The flowers have 4 sepals, 4 petals, 6 stamens and 2 carpels. The carpels form the fruit. All drabas have similar fruits: oval in shape with rather pointed ends. In *D. aurea*, the ends are often twisted. The fruit splits open when ripe to release the seeds. Golden draba is found on rocky slopes, open woods and trails.

NOTES: *The species name* aurea *means "golden."*

Draba incerta Payson

WHITLOW GRASS

This is another yellow draba, but it can be distinguished from the golden draba because it has no stem leaves (cauline). The basal leaves of *D. incerta* have many branched hairs—in *D. aurea* these are usually forked, star-like or cruciform. *D. incerta* is found from low elevations up to the 2800 m level.

Draba kananaskis G.A. Mulligan

KANANASKIS WHITLOW GRASS

Kananaskis whitlow grass is a loosely tufted perennial species. It is a small plant with stems only 0.3 to 2 dm tall, with sparse, simple, forked or stellate hairs. The basal leaves are obovate and hairy above and below, with mostly cruciform hairs; the margins are also hairy. There are a few cauline leaves. Each flower has 4 sepals, 4 yellow petals and 6 stamens. There are 2 united carpels which develop into a silicle, a narrowly ovate fruit, with 20 to 24 seeds. It is an alpine plant. Kananaskis whitlow grass is one of the most interesting plants in Kananaskis Country because it is very rare—it is endemic *(see Plants of Special Interest, pp. XLV-XLVIII)*.

Draba oligosperma Hook.

FEW-SEEDED WHITLOW GRASS

This plant is similar to *D. incerta* because it has no stem-leaves (cauline), but the leaves of *D. incerta* are larger, 1.5 to 3.5 mm wide. In *D. oligosperma*, the leaves are 0.7 to 1.7 mm wide, and the hairs are closely appressed to the surface, often comb-like. The pods are 3 to 7 mm long. Few-seeded whitlow grass is found in open rocky places and on scree slopes.

Erysimum inconspicuum (S. Wats.) MacM.

SMALL-FLOWERED ROCKET

Small-flowered rocket is a grey-pubescent perennial plant 3 to 6 dm tall; the stem is usually simple but is sometimes branched, especially later in the season. The leaves are variable; they may be linear or lanceolate, entire or toothed, and they have bifid hairs. The flowers follow the usual pattern of Brassicaceae, with 4 sepals, 4 yellow petals, 6 stamens and 2 carpels (joined). The fruit is a silique, a long, narrow pod, 2 to 5 cm long, which opens to release the seeds. The siliques have persistent stigmas. Small-flowered rocket is found on dry grassland.

NOTES: *The common name "small-flowered rocket" is misleading because a species with which it is often confused, wormseed mustard* (E. cheiranthoides), *has petals only half the length (3 to 5 mm) of the small-flowered rocket. Prairie rocket* (E. asperum) *has much larger petals, 15 to 25 mm long.*

Lepidium ramosissimum A. Nels.

BRANCHED PEPPERWORT

Branched pepperwort is a biennial plant, 2 to 5 dm tall and single-stemmed (much-branched above), covered with dense hairs. The small flowers are arranged in numerous racemes on the branches and are not typical of the mustard family. Each flower has greenish sepals about 1 mm long; the petals are either absent or filiform (thread-like), and there are 2 (sometimes 4) stamens instead of the usual 6. The fruits are silicles, round or broadly ovate. There is a tiny notch at the tip of the silicle, with a persistent stigma, and the silicle is covered with minute hairs; it splits to release the seeds. Branched pepperwort is found on dry slopes and plains.

NOTES: *Branched pepperwort is similar to common peppergrass* (L. densiflorum), *and the 2 species can easily be confused. There are 2 identifying characters: branched pepperwort is a biennial and its silicles are covered with minute hairs, while common peppergrass is an annual and its silicles are glabrous (without hairs).*

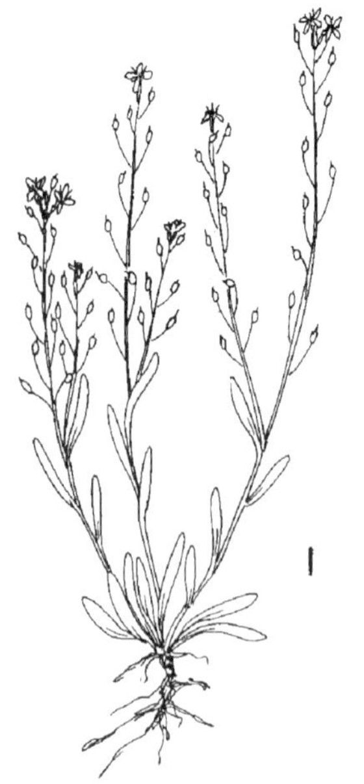

Lesquerella arctica (Wormskj. ex Hornem.) S. Wats.

NORTHERN BLADDERPOD

Northern bladderpod is a tufted perennial with a stout taproot; its basal leaves are 2 to 6 cm long and either spoon-shaped (spatulate) or obovate, tapering into a short leaf-stalk. There are several stems, either decumbent or erect, 5 to 20 cm high; these bear 2 to 14 flowers in racemes. The flowers have 4 sepals, forming the calyx, 4 yellow petals, 6 stamens and 2 carpels, united. The carpels develop into globose pods called silicles, hence the common name "bladderpod." This plant is found on prairie grassland and on dry slopes.

Physaria didymocarpa (Hook.) A. Gray

DOUBLE BLADDERPOD

The basal leaves of double bladderpod are distinctive: spatulate (spoon-shaped) or often fan-shaped, in a rosette, and covered with fine, silvery hairs. The stem leaves are much smaller and lanceolate (lance-shaped). The yellow flowers, in a clustered raceme, have 4 sepals, 4 petals in the form of a cross, 6 stamens and 2 carpels which form the "double bladderpod." The plant is found on dry slopes and plains.

NOTES: *The fruit of double bladderpod consists of 2 rounded, inflated pods which easily distinguish this genus from the other genera of Brassicaceae.*

Rorippa palustris (L) Besser

MARSH YELLOW CRESS

(R. islandica (Oeder) Borbas*)*

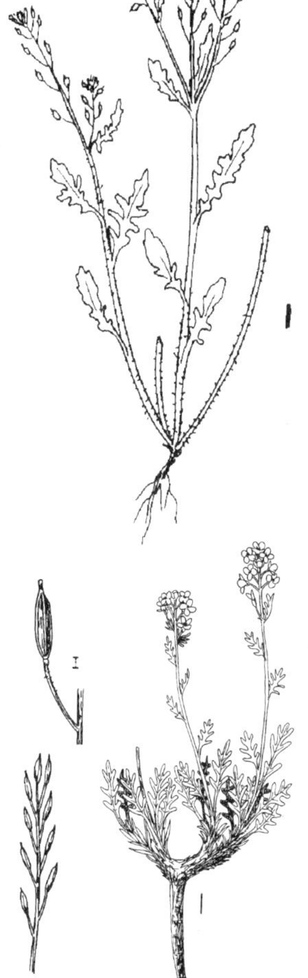

Yellow cress is a bright green annual or perennial plant with several branching stems up to 6 dm high. The basal leaves are clustered, up to 15 cm long, and have leaf-stalks. The stem leaves have no stalks but have auricles (little "ears") at the base, and the leaf-blades are lanceolate, lobed and toothed. The flowers are arranged in long racemes, and each flower has 4 sepals which fall early, 4 yellow petals 2 mm long, 6 stamens and 2 united carpels. The carpels develop into a silicle, which opens to release the seeds. The silicles are almost globose with a notch at the top.

NOTES: *The specific name* palustris *means "of the marshes."* *The former specific name,* islandica, *denotes an Icelandic plant.*

Smelowskia calycina (Steph.) Meyer

SILVER ROCK CRESS

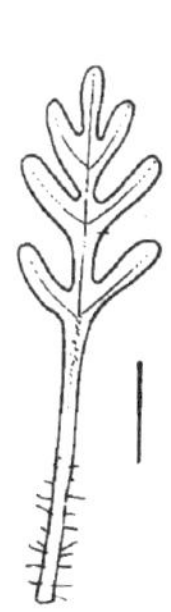

Silver rock cress is a perennial with a branched caudex forming dense tufts on alpine screes. The leaves are pinnately lobed and look grey because they are covered with dense, branched hairs. The flowers are creamy white, often yellowish, and are borne in a simple raceme, like so many flowers in the mustard family. The fruit is oblong, tapering at both ends, and dark-coloured.

NOTES: *The generic name honours a Russian botanist.*

Thlaspi arvense L.

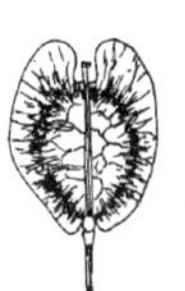

STINKWEED, PENNYCRESS

Stinkweed is an annual, 2 to 5 dm high, with 2 kinds of leaves. The basal ones have leaf stalks (petiolate) and soon drop off, but the stem leaves have no petioles: they clasp the stem and are often toothed. The flowers are small and white—they grow in a raceme on the main flowering stalk, which rapidly elongates after fertilization. The plant turns yellow after flowering. The pods are characteristic: they are heart-shaped, and the stigma is often present. This plant is a familiar weed of gardens and waste places, with an unpleasant smell—hence the common name "stinkweed."

Droseraceae SUNDEW FAMILY

The sundews are insectivorous perennial plants with fibrous roots. Their leaves are arranged in a rosette and are well adapted for trapping insects because they have reddish, sticky, glandular hairs. The plants live in acid swamps and bogs, deficient in essential nutrients, and so have evolved an interesting method of obtaining extra nourishment. The flowers are usually white and are found at the top of a tall scape (a flowering-stalk with no leaves or bracts).

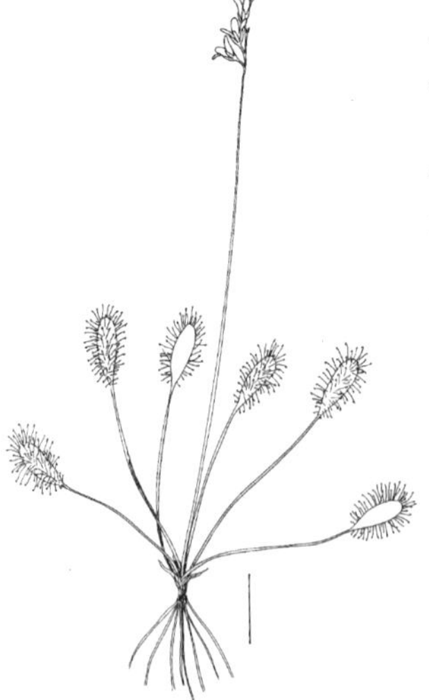

Drosera anglica Hudson

LONG-LEAVED SUNDEW, OBLONG-LEAVED SUNDEW

This sundew is a perennial plant with leaves in a rosette. These leaves are semi-decumbent, partially reclining, with rather long petioles (3 to 4 cm). The leaf-blades are pale green, with bright red, stalked glands covering the upper surface; the blades measure 1 to 3 cm long by 3 to 4 mm wide. The stalked glands each have a "dew-drop" at the tip, and the entire leaf-blade seems to be glistening with dew: hence the common name. The flowers of sundew are arranged in a circinate cyme at the top of a tall flowering-stalk, which arises from the centre of the leaf-rosette. Each flower has 4 to 8 sepals and 4 to 8 petals; the petals are white. There are 4 to 8 stamens and 3 to 5 united carpels in the centre; each ovary has 1 locule. The flower is hypogynous (flower parts below the carpels, which form the gynoecium). The fruit is a many-seeded loculicidal capsule that splits along the middle of the back of each locule. Long-leaved sundew grows in acid swamps and bogs.

NOTES: *This plant has 2 types of glands. Those on the outer margins have longer stalks, and their chief function is to entrap the prey. Those nearer the centre have shorter stalks or are sessile, and their main function is to produce digestive juice, although they aid in entrapment. The prey consists of tiny insects lured by colour and sweet nectar; when an insect crawls over the leaves, it gets stuck in the sticky secretions. The glands on the margins slowly bend towards the centre, securing the prey on top of the glands that secrete digestive enzymes, and the insect is slowly digested.* ~ *The generic name* Drosera *comes from the Greek word* droseros, *which means "glistening." The species name* anglica *was given because this plant is common in England.* ~ *D.* anglica *is very probably a hybrid between* D. rotundifolia *and* D. linearis: *the 3 species often grow together.*

Crassulaceae STONECROP FAMILY

Stonecrops are herbs or shrubs and are usually succulent. The flowers are grouped in terminal cymes, and the flower parts are in whorls of 4 or 5. The fruits are either capsules or clusters of follicles.

Flowers yellow . . . . *Sedum*
Flowers reddish purple . . . . *Tolmachevia*

Sedum lanceolatum Torr.

~ COMMON STONECROP, LANCE-LEAVED STONECROP

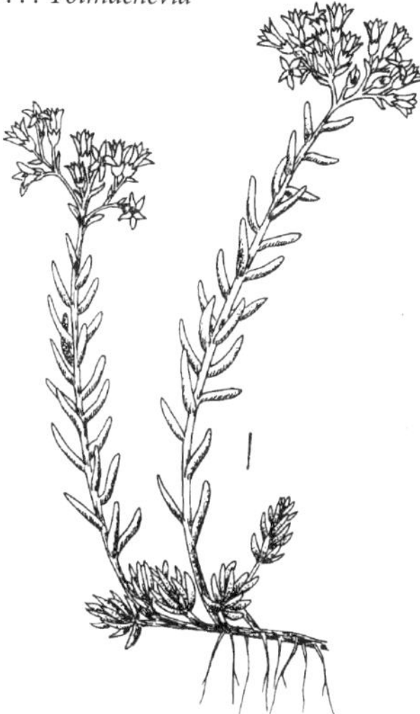

Stonecrops are succulent plants, and lance-leaved stonecrop is no exception: the leaves of lance-leaved stonecrop are fleshy. The basal ones are tufted, and from these rise sterile shoots with fleshy leaves. The flowering shoots also have many leaves, but these drop off early, leaving naked flowering stalks. The small flowers grow in a cluster at the tip of the stalk. Each flower is pale yellow, usually with 5 petals and 5 sepals, 5 stamens and 5 carpels, which form the fruit. The fruit is a tiny cluster of follicles. Stonecrop is found on dry slopes and ridges.

Tolmachevia integrifolia (Raf.) Löve & Löve

ROSEROOT

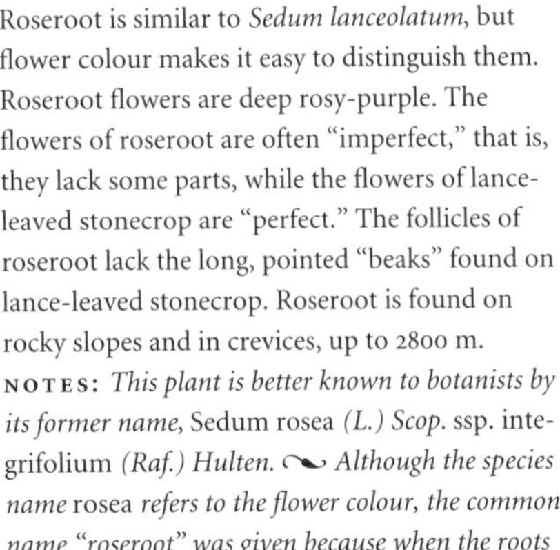

Roseroot is similar to *Sedum lanceolatum*, but flower colour makes it easy to distinguish them. Roseroot flowers are deep rosy-purple. The flowers of roseroot are often "imperfect," that is, they lack some parts, while the flowers of lance-leaved stonecrop are "perfect." The follicles of roseroot lack the long, pointed "beaks" found on lance-leaved stonecrop. Roseroot is found on rocky slopes and in crevices, up to 2800 m.

NOTES: *This plant is better known to botanists by its former name,* Sedum rosea *(L.) Scop.* ssp. integrifolium *(Raf.) Hulten.* ~ *Although the species name* rosea *refers to the flower colour, the common name "roseroot" was given because when the roots were bruised, they were thought to emit the fragrance of roses.*

Saxifragaceae SAXIFRAGE FAMILY

These are perennial herbs with simple leaves; the leaves usually grow in a basal cluster. The flowers are arranged in racemes or cymes, and a few are solitary. The receptacle (hypanthium) is usually well developed and often bears a disc. The flower may be hypogynous, with the ovary on the same level as the other floral parts or above them, or it may be epigynous, with the ovary below the other floral parts. There are 4 to 5 sepals, 4 to 5 petals (sometimes absent), 8 to 10 stamens (twice as many as the petals) and usually 2 carpels. The fruits are either capsules or follicles. The genus *Saxifraga* is sometimes known as the "stone-splitter," because *saxus* means "stone" or "rock" and *frango* means "break."

1. Sepals 4 . . . . *Chrysosplenium*
   Sepals 5 2
2. Leaves alternate . . . . *Leptarrhena*
   Leaves mostly basal 3
3. Stem with 2 or more small leaves . . . . *Lithophragma*
   Stem leafless (1 small leaf may be present) 4
4. Petals 3-lobed or feathery . . . . *Mitella*
   Petals entire 5
5. Stamens 5 . . . . *Heuchera*
   Stamens 10 6
6. Stamens longer than petals . . . . *Tiarella*
   Stamens not longer than petals . . . . *Saxifraga*

Chrysosplenium tetrandrum (Lund) T. Fries

GOLDEN SAXIFRAGE

Golden saxifrage is a dainty, yellowish green plant, 2 to 20 cm tall, which grows from an erect or creeping rhizome and has leafy stolons. The small leaves are characteristic—they are usually orbicular, with rounded lobes—and help to identify the plant. The tiny flowers grow in the axils of the upper leaves; they may be solitary or arranged in a cyme. Each flower has 4 sepals, yellowish green, no petals, 4 stamens, and 2 carpels, united. The fruit is a capsule with numerous seeds. Golden saxifrage grows in moist places and on shady banks and ledges.

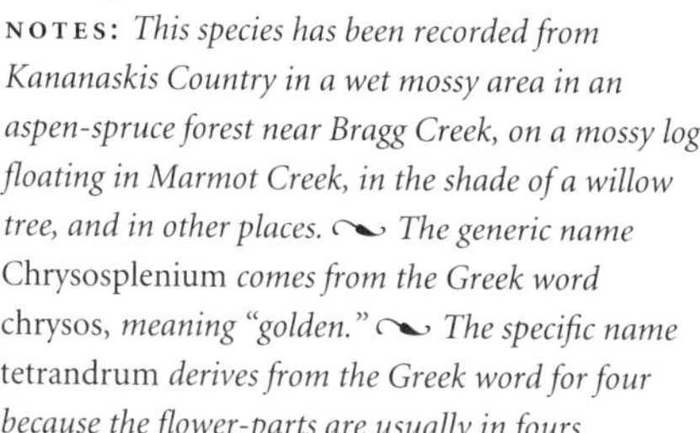

NOTES: *This species has been recorded from Kananaskis Country in a wet mossy area in an aspen-spruce forest near Bragg Creek, on a mossy log floating in Marmot Creek, in the shade of a willow tree, and in other places.* ~ *The generic name* Chrysosplenium *comes from the Greek word* chrysos, *meaning "golden."* ~ *The specific name* tetrandrum *derives from the Greek word for four because the flower-parts are usually in fours.*

Heuchera cylindrica Dougl. ex Hook.

STICKY ALUMROOT

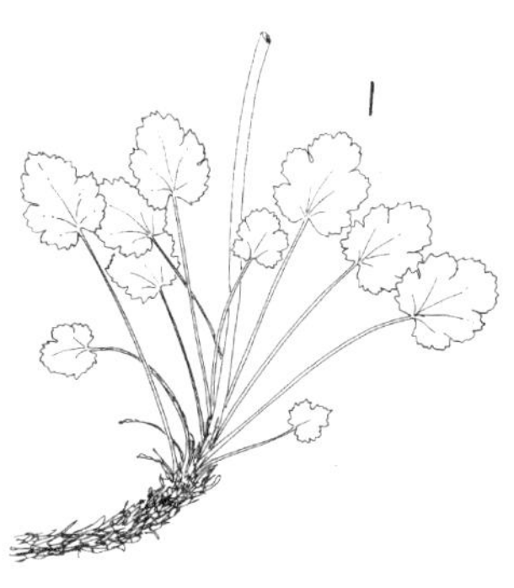

Sticky alumroot grows from a branching rhizome covered with scales, and the flowering-stems are 2 to 5 dm tall. The leaves are basal with long leaf-stalks, are roughly cordate (heart-shaped), with several deep lobes (5 to 7), and they are palmately veined. The tall flowering-stalks are leafless, with the flowers arranged in a cluster at the tip. Each flower has 5 sepals, 5 petals and 5 stamens. There are 2 united carpels in the centre forming the gynoecium, partly inferior; these develop into a 2-beaked capsule. Alumroot is found in open, dry meadows and rocky slopes, from prairie areas to high elevations, up to 2700 m.

NOTES: *The generic name* Heuchera *honours Dr. Johann von Heucher, a professor of Medicine at the University of Wittenberg, Germany.* ~ *The species name* cylindrica *probably comes from the dense, cylindrical panicle of flowers.*

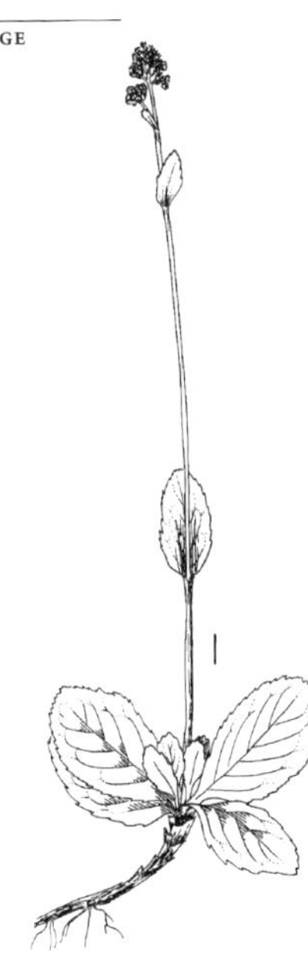

Leptarrhena pyrolifolia (D. Don) R. Br.

LEATHER-LEAVED SAXIFRAGE

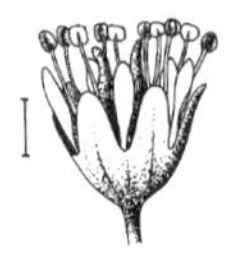

The common name leather-leaved saxifrage highlights one distinctive feature of this plant: the long (3 to 8 cm) evergreen leaves are leathery. They are deep green above and pale or brownish below. The leaves are in a basal rosette, and a long flowering-stalk, 10 to 20 cm high, rises from the centre, bearing a cluster of small flowers at the tip. There are few leaves on the flowering-stalk. The flowers are white or pink, with 4 or 5 sepals, 4 or 5 petals, 10 stamens and 2 carpels, which form the follicle fruits. This saxifrage is found at alpine and subalpine elevations, up to 2300 m, along stream banks and in mossy places.

NOTES: *The species name* pyrolifolia *means "with leaves like wintergreen* (Pyrola)."

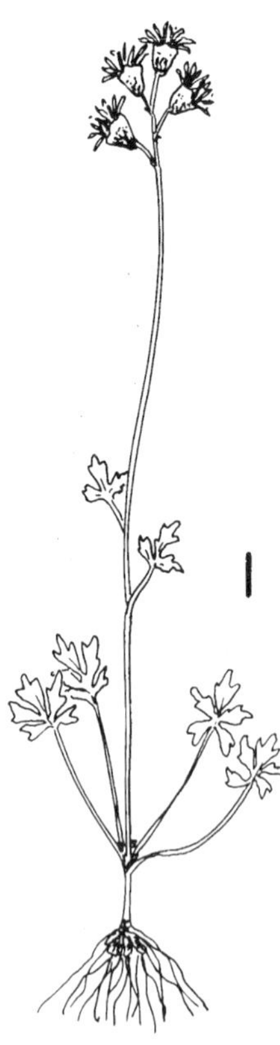

Lithophragma glabrum Nutt. ex T.&G.

ROCKSTAR

(L. bulbifera Rydb.*)*

Rockstar grows only 1 to 2 dm high, with basal leaves twice divided into three. The slender stem is covered with white, glandular, black-headed hairs. The flowers are in short racemes; each flower has 5 sepals, 5 deeply cleft white or pink petals and 10 stamens. There are 3 united carpels in the centre of the flower, forming the gynoecium, the female part of the flower; this ripens into the fruit, a many-seeded capsule, which opens by 3 valves. The flowers are sometimes replaced by bulblets. Rockstar is found on rocky slopes up to 2000 m and in dry montane meadows.

NOTES: *The specific name* glabrum *means glabrous (smooth, lacking hairs); as the stem is covered with glandular hairs, the scientific name seems badly chosen.* *The cauline (stem) leaves bear bulblets in their axils, hence the former specific name,* bulbifera.

Mitella nuda L.

BISHOP'S-CAP, MITREWORT

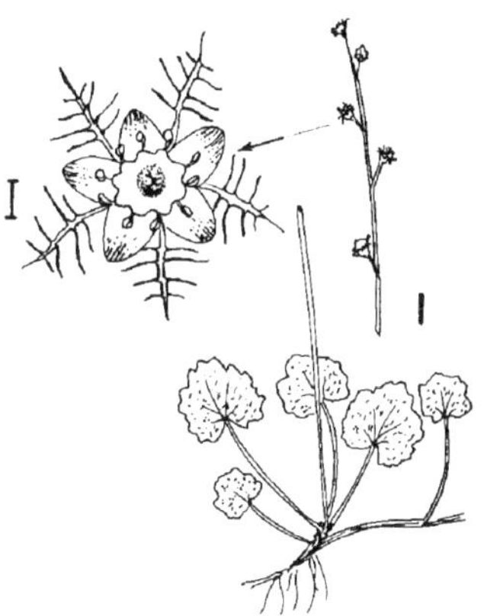

Mitrewort is a dainty plant less than 2 dm high. It rises from slender rhizomes which bear several basal leaves; these are round in shape, with a crenate (scalloped) outline and long leaf-stalks. The tall, flowering-stalks, numbering 1 to 7, are leafless (hence the specific name *nuda*) and bear greenish yellow flowers in a terminal spike-like raceme. Each flower has 5 sepals and 5 feathery petals; there are 10 stamens. There are 2 partly united, partly inferior carpels, each with 2 short styles, which develop into a 2-valved capsule with black or brown seeds. The generic name *Mitella* comes from the Latin word *mitra*, meaning a cap or mitre; the fruit looks like a bishop's mitre. Mitrewort is found in damp places in coniferous woods at lower and middle elevations.

Saxifraga L. — SAXIFRAGE

Saxifrages are perennials, with alternate leaves (opposite in *S. oppositiolia*), usually basal, may be entire, toothed or lobed. The flowers are arranged in cymes; the hypanthium is well developed and joined to the base of the gynoecium. There are 5 sepals, 5 petals, 10 stamens and 2 carpels, united completely or sometimes only at the base. The fruit is a capsule or a pair of follicles.

1. Leaves opposite; flowers purple .... *S. oppositifolia*
   Leaves alternate or basal; flowers white 2
2. Stems leafless; basal leaves fan-shaped, 1 to 4 cm long .... *S. lyallii*
   Stems with leaves; basal leaves 3-lobed or spine-tipped 3
3. Basal leaves 3-lobed; petals white .... *S. caespitosa*
   Basal leaves spine-tipped; petals white with red spots .... *S. bronchialis*

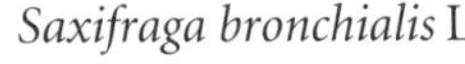

Saxifraga bronchialis L.

SPOTTED SAXIFRAGE, COMMON SAXIFRAGE

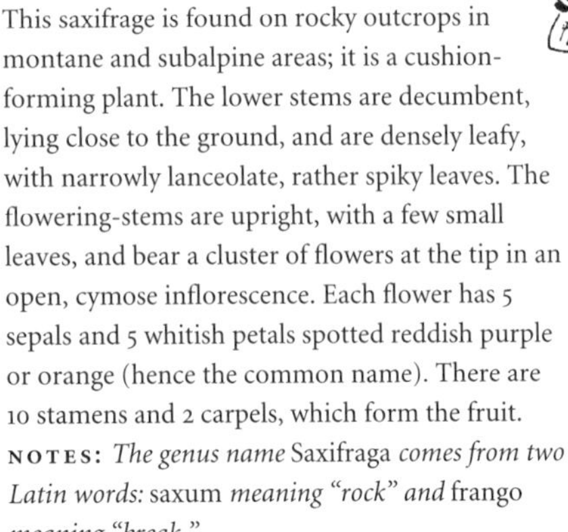

This saxifrage is found on rocky outcrops in montane and subalpine areas; it is a cushion-forming plant. The lower stems are decumbent, lying close to the ground, and are densely leafy, with narrowly lanceolate, rather spiky leaves. The flowering-stems are upright, with a few small leaves, and bear a cluster of flowers at the tip in an open, cymose inflorescence. Each flower has 5 sepals and 5 whitish petals spotted reddish purple or orange (hence the common name). There are 10 stamens and 2 carpels, which form the fruit. NOTES: *The genus name* Saxifraga *comes from two Latin words:* saxum *meaning "rock" and* frango *meaning "break."*

Saxifraga caespitosa L.

TUFTED SAXIFRAGE

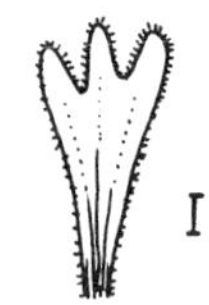

The common name is a descriptive one, because the plants grow in neat little tufts. They are densely leafy but the leaves differ from those of spotted saxifrage because they are 3-lobed at the tip: a distinctive feature. The species name *caespitosa* means "tufted." The flowering stems are erect and bear a few white flowers at the tip, in a loose arrangement. The flower is the usual saxifrage type with 5 sepals, 5 petals, 10 stamens and 2 carpels. The petals are not spotted, which makes the plant easy to distinguish from spotted saxifrage. Tufted saxifrage grows on open, rocky alpine ridges and at middle elevations in moist areas.

Saxifraga lyallii Engler

RED-STEMMED SAXIFRAGE, LYALL'S SAXIFRAGE

Lyall's saxifrage can be distinguished from the other saxifrages already described because of its unusual leaves; they are fan-shaped with a broad blade, have a wedge-shaped base and are sharply toothed. The flowering stems are similar to those of the spotted saxifrage but bear fewer flowers. The petals are not red-spotted; they are white with 2 greenish yellow blotches that soon fade. The fruits are purplish capsules. Lyall's saxifrage is found on stream banks and moist mountain slopes.

NOTES: *The species name of this plant honours David Lyall (1817-1895), a Scottish botanist and surgeon who collected plants in North America while serving with the North American Boundary Commission.*

Saxifraga oppositifolia L.

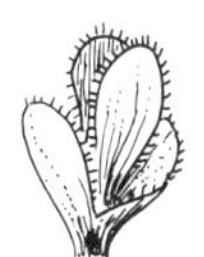

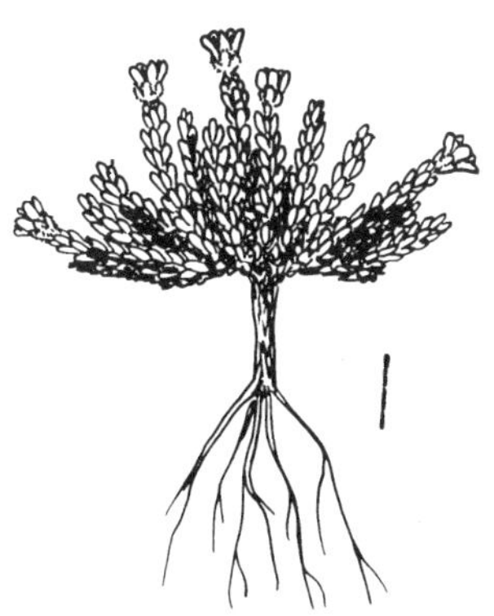

PURPLE SAXIFRAGE

Purple saxifrage is quite different from the other saxifrages. The main difference lies in the flowers, which are large and bright purple: the plant is often covered with these purple blooms. The plant is densely matted, and the prostrate branches crowd together, with leaves overlapping in 4 rows; they have ciliate (hairy) margins. The flower has ciliate (fringed) sepals and 5 large petals, 6 to 8 mm long; the filaments of the stamens and the styles of the carpels are purplish. The plant is found on alpine slopes and ridges.

Tiarella unifoliata Hook.

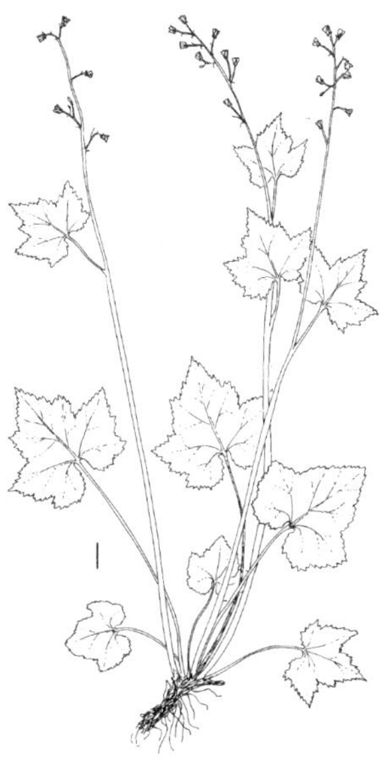

FALSE MITREWORT, SUGAR SCOOP

False mitrewort is a perennial plant with several basal leaves and a few cauline (stem) leaves. The basal leaves are long-stalked, with glandular hairs, and are more or less cordate (heart-shaped) in outline, with 3 well-marked, toothed lobes. The flowering-stems are 1 to 4 dm tall, and the flowers are borne in an elongated panicle. Each flower has the usual saxifrage pattern: 5 white or pink sepals, 5 white petals and 10 stamens. These stamens are exserted, that is, they project from the flower, with long filiform filaments. The petals are 3 times as long as the sepals. There are 2 united carpels, which turn into a capsule with 2 valves. False mitrewort is found in damp, shady woods.

NOTES: *The odd shape of the capsule and its persistent sepals has given rise to the generic name* Tiarella *("little tiara").*

Parnassiaceae GRASS-OF-PARNASSUS FAMILY

These are perennial herbs with entire leaves, usually basal, in a cluster with 1 to several flower-stalks arising from the centre. The flower-stalks bear solitary white flowers. Each flower has 5 sepals and 5 white petals with greenish veins. Opposite the petals are the staminodes, sterile stamens which are tipped with glands, united in 5 scale-like clusters. The 5 stamens alternate with the staminodes. In the centre of the flower is the superior ovary, which forms a 4-valved capsule; this splits open to release the winged seeds.

Parnassia L. GRASS-OF-PARNASSUS

These are perennials, rising from short, erect rootstocks, with basal leaves and tall flowering-stalks (scapes). They usually bear a leafy bract. They produce solitary flowers with 5 sepals, 5 white, often veined, petals, 5 stamens, alternating with 5 clusters of staminodia. United carpels ripen to form a 4-valved capsule.

Petals entire, 5 to 9-nerved . . . . *P. palustris*
Petals frilled, 5-nerved . . . . *P. fimbriata*

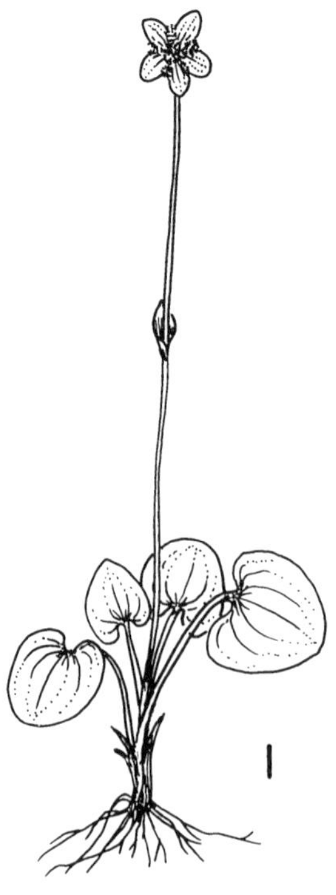

Parnassia fimbriata Konig

FRINGED GRASS-OF-PARNASSUS

The leaves of fringed grass-of-Parnassus are characteristically heart-shaped or kidney-shaped and show parallel veining. They are clustered at the base of the plant and are funnel-shaped when erect. The flowers have 5 sepals, 5 petals (white, with green-yellow veins) and 5 stamens; the fruit is a 4-valved capsule. The staminodes are in 5 clusters. Fringed grass-of-Parnassus is found on the banks of streams and creeks at lower, middle and higher elevations.

NOTES: *Grass-of-Parnassus is linked by its common name with Mount Parnassus in Greece, the legendary abode of Apollo and the Nine Muses.* ~ *"Grass" is a misnomer: the plant does not remotely resemble a true grass.* ~ *The species name* fimbriata *means "fringed" and refers to the bases of the petals.*

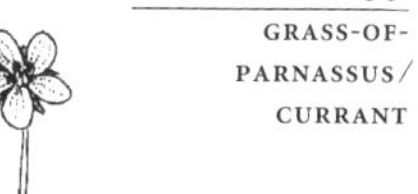

Parnassia palustris L.

NORTHERN GRASS-OF-PARNASSUS, MARSH GRASS-OF-PARNASSUS

(P. montanensis Fern. & Rydb.)

This species is similar to *P. fimbriata* but the petals are not fringed. There are clusters of staminodes opposite the white petals; these staminodes are typically found in grass-of-Parnassus flowers. There are 5 sepals, 5 petals, 5 stamens and 4 united carpels, which turn into a 4-valved capsule. The flowering stalks are 1 to 3 dm high; each bears a heart-shaped, clasping leaf. Fringed grass-of-Parnassus also has a leaf clasping the flowering stem, but it is much smaller and is not a conspicuous feature of the plant. Both *Parnassia* species are found in damp places, on stream banks and in mossy areas.

Grossulariaceae CURRANT/GOOSEBERRY FAMILY

Members of this family are shrubs, and their branches are sometimes prickly. The leaves are alternate, and the veins are definitely palmate. The flowers grow in racemes and cymes; the flower structure is interesting because the hypanthium (receptacle) is joined to the ovary and forms a cup-shaped or tubular structure above it. The flower is epigynous, and the ovary is inferior. The flower-parts are in 5s: 5 sepals, 5 petals and 5 stamens. The ovary is formed from 2 carpels joined together, with 2 styles, and the fruit is a berry.

Ribes L. — GOOSEBERRIES

These are branching shrubs, usually with prickles along the stems and especially at nodes. The leaves are 3 to 5-lobed, palmately veined, and often clustered on short spur-shoots along older stems. Flowers grow in pendent racemes, in pairs or solitary in leaf axils; there are 5 sepals, 5 petals, 5 stamens and 2 united carpels; the ovaries are inferior, and the fruit is a berry.

1. Flowers yellow . . . . *R. aureum*
   Flowers pinkish or whitish 2
2. Fruit black, bristly . . . . *R. lacustre*
   Fruit purple, smooth . . . . *R. oxyacanthoides*

Ribes aureum Pursh

GOLDEN CURRANT

Golden currant is an erect shrub, 1 to 3 m high, bearing distinctive leaves. These are deeply 3-lobed, and the lobes are rounded, not pointed; *Ribes* species tend to have more pointed lobes. The branches are not spiny. The flowers are borne in racemes, and each flower has 5 sepals, 5 petals, 5 stamens, and 2 carpels joined. The petals are much smaller than the sepals and are inserted, erect, at the top of the hypanthium tube. The sepals, petals and hypanthium are all yellow. The fruit is a true berry, black, red or yellow. Golden currant is found on stream banks and on rocky slopes.

NOTES: *The specific name* aureum *means "yellow" or "golden."*

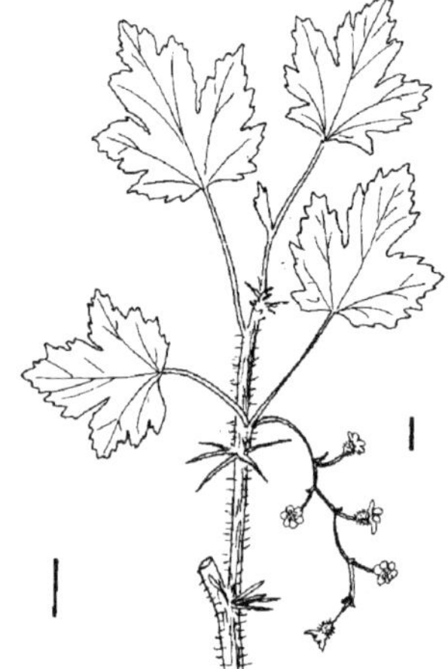

Ribes lacustre (Pers.) Poir

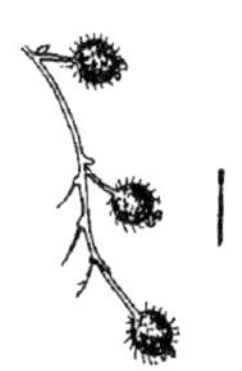

BRISTLY BLACK CURRANT

This shrub has prickly branches. The spines on the nodes are much longer than those on the internodes. The leaves are deeply 3 to 5-lobed, and the lobes are pointed and toothed, so the plant can be distinguished from *R. aureum* . The flowers are arranged in a small, pendent raceme; each flower has 5 sepals, 5 petals smaller than the sepals, 5 stamens and 2 carpels. The carpels are united and form a bristly, dark purple or black berry. The hypanthium (receptacle) is joined to the ovary and is a cup-shaped structure. This shrub is found in moist woods, up to 2300 m.

Ribes oxyacanthoides L.

NORTHERN GOOSEBERRY

Northern gooseberry, a shrub, can be distinguished from other *Ribes* because the flower arrangement is different: in northern gooseberry they grow in pairs in the leaf axils. Currant berries are bristly, while those of gooseberry are smooth. Northern gooseberry is found in moist woods.

NOTES: *The fruits of both gooseberry and currants are true berries—most of the fruits called berries are not really berries at all. For example, the raspberry is a collection of tiny drupes, while the strawberry is an enlarged swollen receptacle and the fruits, the "pips," are tiny achenes scattered all over it. Each "pip" has a seed inside it.*

Rosaceae ROSE FAMILY

The rose family is an interesting one, partly because so many members of the family have attractive flowers (for example, roses, peaches, apples, pears, and hawthorns) and partly because of their particular flower structure. In order to understand this, it is necessary to understand *hypogynous* and *epigynous* flowers. In hypogynous flowers, the carpels in the centre (collectively called the gynoecium) are borne on the hypanthium above the rest of the flower parts. (*Hypo* means "below," so "hypogynous" means the flower parts are below the gynoecium.) Conversely, in epigynous flowers, the carpels (joined) are sunk in the hypanthium, so the flower parts are above the gynoecium (*epi* means "upon"). There is a third type of flower found in this family, perigynous (*peri* means "around"); here the hypanthium is saucer-shaped, with the carpels in the centre and the other flower parts in a ring on its edge. The word *hypanthium* often causes confusion, but it is merely the swollen top of the flower stalk, which supports the sepals, petals, stamens and carpels. See the glossary, pp. 309–338, for detailed illustrations.

In the rose family, the hypanthium can take many forms. For example, the fruit of the rose (the rose-hip) consists of a red, swollen, vase-shaped hypanthium almost completely enclosing the true fruits, and the familiar strawberry fruit consists of a red, fleshy, pyramid-shaped hypanthium with the true fruits (tiny achenes called the "pips") scattered all over it. The hypanthium is often called the "receptacle" but it is seldom in the shape of a receptacle: it is usually flat. The word *hypanthium* comes from 2 Greek words: *hypo* (below) and *anthos* (flower).

Members of the rose family often bear stipulate leaves, that is, leaves with stipules at the base of the leaf-stalk

1. Trees or shrubs 2
   Herbs (plants with non-woody stems) 9
2. Leaves compound 3
   Leaves simple 6
3. Flowers white 4
   Flowers pink or yellow 5
4. Leaflets 3 to 5 .... *Rubus*
   Leaflets more than 7 .... *Sorbus*
5. Flowers pink; leaflets 5 to 9 .... *Rosa*
   Flowers yellow; leaflets 5 .... *Potentilla*
6. Petals 8, yellow or white; prostrate shrubs .... *Dryas*
   Petals 5, pink or white; erect shrubs 7
7. Fruit dry; shrubs up to 2 m tall .... *Spiraea*
   Fruit juicy; shrubs up to 8 m tall 8
8. Fruit red or purple with a single seed (inside a stony endocarp); leaves lance-shaped with a pointed tip .... *Prunus*
   Fruit purple with several seeds; leaves oval .... *Amelanchier*
9. Leaves simple or lobed .... *Rubus*
   Leaves compound 10
10. Stamens 5; leaflets 3; petals yellow; plants of alpine meadows .... *Sibbaldia*
    Stamens many 11
11. Fruit juicy; flowers white or pink 12
    Fruit dry; flowers white or yellow 13

12. Leaves basal; leaflets 3 .... *Fragaria*
    Leaves alternate; leaflets 3 to 5 .... *Rubus*
13. Fruit feathery or with hooked prickles........ *Geum*
    Fruit not as above........ *Potentilla*

Amelanchier alnifolia Nutt.

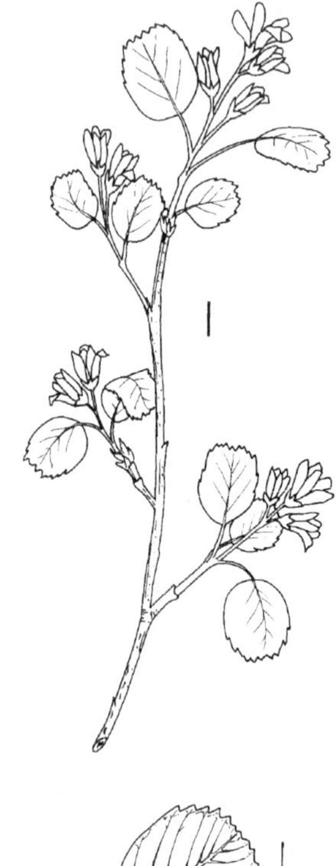

~ SASKATOON, JUNE-BERRY, SERVICE-BERRY

Saskatoon is a shrub or small tree found in open woodlands; it is common in Alberta. It grows 1 to 6 m high and spreads by stolons, so that it forms colonies or clones. The branches are smooth and the leaves are alternate, simple and toothed. The flowers are white and grow in a dense raceme. The calyx (of sepals) is 5-cleft; there are 5 petals and about 20 stamens. The carpels are inferior (below the rest of the flower) and joined together, enclosed by the hypanthium. The fruit is a blue-purple pome (apple-like) and is sweet; it makes delicious pies.

NOTES: *The specific name* alnifolia *means it has leaves like those of the alder,* Alnus. ~ *The Saskatoon shrub was used in many ways by the Natives, especially the Blackfoot. Its name is derived from the Cree word* misaskwatomin, *meaning "a tree with much wood and berry."* ~ *The berries can be eaten raw but are usually cooked; they can be dried and made into flat cakes for use in the winter-time and when walking on trails. The berries were mixed with soups and stews and helped to flavour pemmican. The fruit juice was used for making black and purple dyes, and the wood was used for making arrows and pipe stems.* ~ *Saskatoon also had medicinal uses: the Blackfoot used the berries to make a brew to treat liver problems and stomach ache. The berry juice helped to soothe bloodshot eyes.*

Dryas L. — MOUNTAIN AVENS

Mountain avens are creeping, prostrate shrubs bearing evergreen leaves with scalloped, rolled margins, or entire. The flowers are solitary on leafless, hairy scapes; sepals number 8 to 10 and are narrow; petals number 8 to 10 and are white or pale yellow. There are numerous stamens and carpels; the carpels turn into achenes.

1. Flowers yellow . . . . *D. drummondii*
   Flowers white 2
2. Leaf-edges with rounded teeth; leaf underside with a prominent midrib . . . . *D. octopetala*
   Leaf-edges smooth; leaf underside without a prominent midrib . . . . *D. integrifolia*

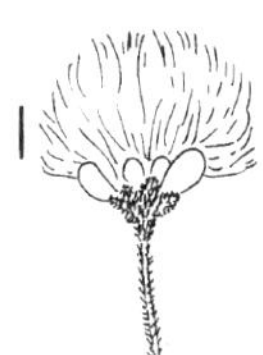

Dryas drummondii Richards.

YELLOW DRYAD, YELLOW MOUNTAIN AVENS

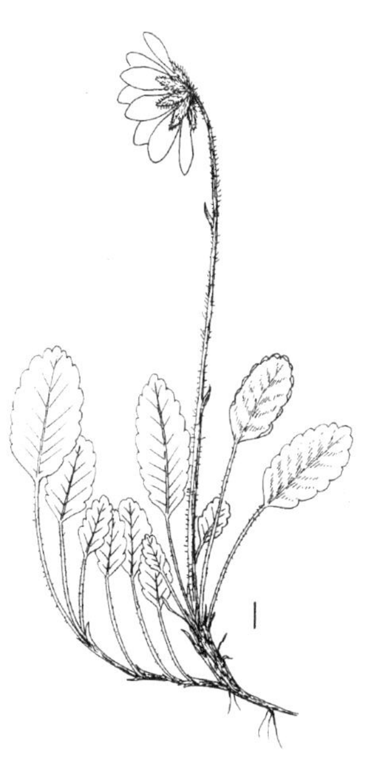

The 3 *Dryas* species are low, prostrate shrubs; the horizontal stems form dense mats and are freely branched. The leaves of yellow dryad are distinctive: they have scalloped edges and are whitish underneath, because the undersurface is tomentose, that is, it is covered with a dense mass of hairs. The flowering stems are 5 to 25 cm high, each bearing a single yellow flower with 8 to 10 sepals, 8 to 10 petals, and numerous stamens and carpels. The carpels turn into plumed achenes like those of old man's whiskers (*Geum triflorum*). Yellow dryad is common on gravelly slopes and river bars.

NOTES: *The species name* drummondii *honours Scottish botanist Thomas Drummond (1780-1836).*

Dryas integrifolia M. Vahl.

WHITE DRYAD, NORTHERN WHITE MOUNTAIN AVENS

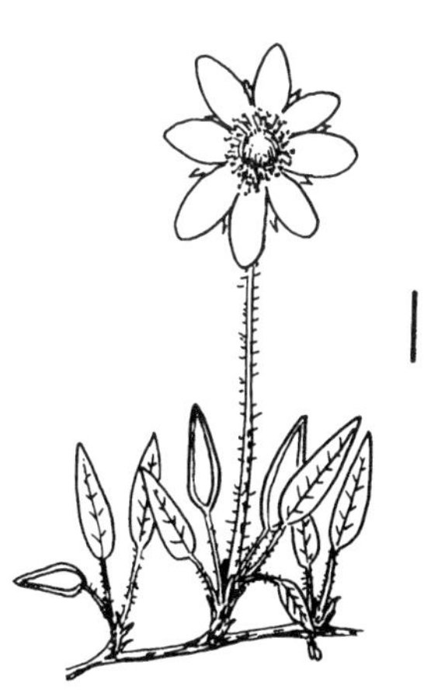

This species is similar to yellow dryad, but there are 3 differences: white dryad is a smaller plant, its flowers are white, not yellow and its leaves have a smooth outline and are pointed at the tip. As the name northern white mountain avens suggests, this plant is found on alpine slopes.

NOTES: *The species name* integrifolia *means "entire leaves," that is, leaves with a smooth outline.*

Dryas octopetala L.

WHITE DRYAD, WHITE MOUNTAIN AVENS

(D. hookeriana Juz.*)*

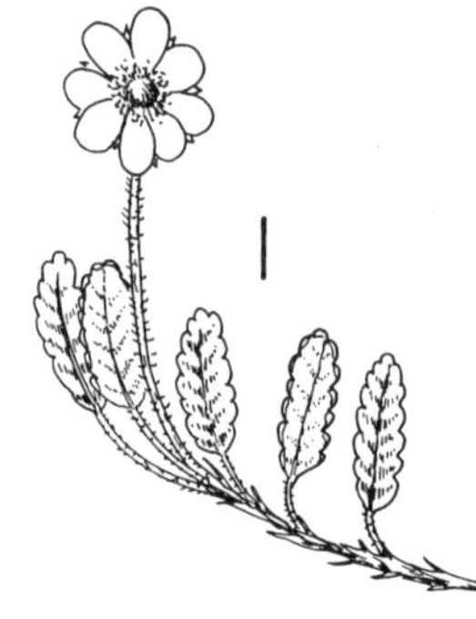

This plant is similar to yellow dryad, but the 2 can be distinguished by flower colour; also, the flowers of yellow dryad are often nodding and those of white dryad are not. The leaves of *D. octopetala* are 0.8 to 2.5 cm long and are usually subcordate or truncate (cut off sharply). The undersurface of the leaves in all *Dryas* species is whitish, due to the dense mat of hairs. This plant is found on alpine slopes. White mountain avens can be distinguished from the northern white mountain avens because the leaves of the former are crenate and glandular, while those of the latter are more or less entire and not glandular.

NOTES: *The species name* octopetala *means "eight petals."*

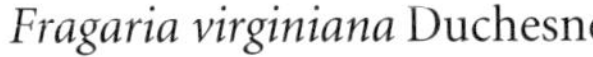

Fragaria virginiana Duchesne

WILD STRAWBERRY

The wild strawberry is a familiar plant, easy to recognize because it is so similar to the garden strawberry. It has several long-stalked leaves, each with 3 lobes; all the lobes are toothed. A flowering-stalk arises from the centre of the plant, bearing white flowers in a cluster at the top. These flowers have 5 sepals, 5 petals and numerous stamens (about 20). The fruits are small versions of the garden strawberry. They each consist of a fleshy, swollen receptacle or hypanthium, covered with tiny "pips." Each "pip" is really a tiny achene with a seed inside it. Wild strawberry is found in woods, and on plains, up to subalpine elevations. It spreads by means of stolons, long, thin stems close to the ground, which produce young plants at the nodes.

NOTES: *The species name* virginianum *indicates that the plant was first described in Virginia.*

Geum L. — AVENS

Avens are hairy perennials that rise from stout rootstocks. The leaves are pinnate, petiolate, alternate and chiefly basal. Flowers grow in small diffuse clusters; they are formed from 5 joined sepals, with 5 alternate bractlets, 5 petals, numerous stamens and numerous carpels that turn into a cluster of achenes, each with a plumose or hooked style.

1.	Flowers reddish purple; stem-leaves 2, opposite	*G. triflorum*
	Flowers yellow; stem-leaves numerous	2
2.	Terminal leaflet 3-lobed	*G. macrophyllum*
	Terminal leaflet not 3-lobed	*G. aleppicum*

Geum aleppicum Jacq.

YELLOW AVENS

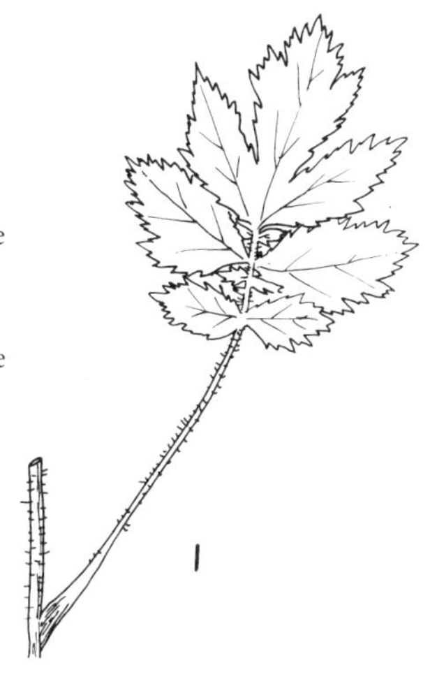

Yellow avens is a coarse plant, 4 to 10 dm high, with a bunch of basal leaves; these are pinnately compound, with leaflets of different sizes, with the terminal one larger than the others. These terminal leaflets are wedge-shaped (cuneate) at the base. The flowering stalks arise from the centre and are often leafy, with a cluster of flowers at the tip. These have 5 sepals, with 5 small bracts in between, 5 yellow petals and numerous stamens. In the centre of the flower are numerous carpels which produce a head of hooked achenes. These hooks are twice as long as the achene itself; and help in fruit dispersal. Yellow avens is found in moist woods and meadows.

Geum macrophyllum Willd.

LARGE-LEAVED YELLOW AVENS

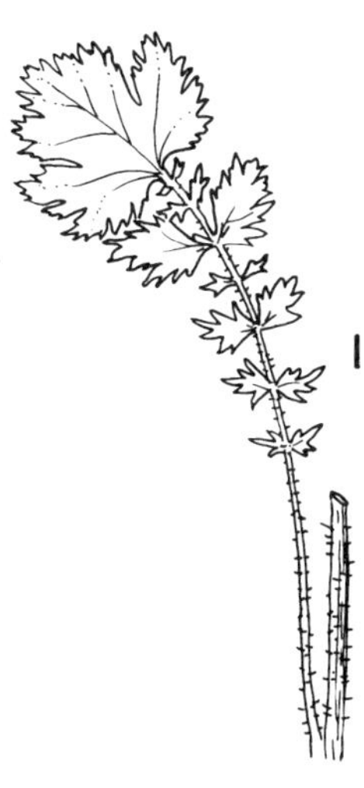

There are 2 species of *Geum* with the common name yellow avens, and they are not easily distinguished. The terminal leaflet of *G. macrophyllum* is cordate (heart-shaped) to truncate, rather than wedge-shaped, as in *G. aleppicum*, and is usually much larger than the other leaflets. The species name *macrophyllum* means "large leaf" and probably refers to this feature, although it is a leaflet, not a leaf. The flowers and fruits are similar to those of *G. aleppicum* but there is a tiny difference: in *G. aleppicum* the achenes are hairy, but glabrous along the sides, whereas in *G. macrophyllum* the achenes are hairy along the sides. *G. macrophyllum* and *G. aleppicum* grow in the same habitats.

Geum triflorum Pursh

OLD MAN'S WHISKERS, PRAIRIE SMOKE, THREE-FLOWERED AVENS

This plant is fairly easy to distinguish from the other *Geum* species because the fruits are quite different. The flowers are also quite different because they are usually in a group of three at the top of the flowering stem and are often drooping; the sepals are reddish purple and so are the bractlets between them. The petals are yellow-white, sometimes pink, and are often hidden by the sepals. The carpels develop into plumed achenes because the style elongates and becomes feathery. Old man's whiskers is a typical prairie plant.

NOTES: *The common name "old man's whiskers" arose because when a group of these plants is in fruit, the long feathery plumes look like whiskers. The other common name, prairie smoke, arose in the same way.*

Potentilla L. — CINQUEFOIL

Shrub or perennial, leaves alternate, stipulate and compound, sepals 5, alternating with 5 bractlets, petals 5, yellow, stamens few to many, carpels numerous, forming a head of achenes.

Potentillas are often called "cinquefoils" because their leaves often have 5 leaflets, but this is not always true: those of the Pennsylvanian cinquefoil (*P. pensylvanica*) are usually pinnate, but two of the species considered here have leaves with 5 leaflets.

1. Shrub . . . . . . *P. fruticosa*
   Herbs 2
2. Flowers 10 to 15 mm across; plant densely hairy; alpine species . . . . . . *P. nivea*
   Flowers 15 to 20 mm across; plant not hairy; lower elevations . . . . . . *P. gracilis*

Potentilla fruticosa L.

SHRUBBY CINQUEFOIL

Shrubby cinquefoil is freely branched, 3 to 10 dm high, with numerous leaves, usually with 5 leaflets. It bears many yellow flowers with 5 sepals, 5 petals, 5 bractlets and numerous stamens. It has densely hairy carpels in the centre. The shrub is found on plains and in open woods.

NOTES: *Shrubby cinquefoil is the only member of the* Potentilla *group in Kananaskis Country with a shrubby habit, so it is easy to distinguish from the rest of the group.* ~ *The Blackfoot used the dry bark, which flakes easily, for making friction fires.* ~ Potentilla *species possess 5 bractlets between the sepals, which help to distinguish cinquefoils from buttercups.*

Potentilla gracilis Dougl. ex Hook.

GRACEFUL CINQUEFOIL, SLENDER CINQUEFOIL

P. gracilis deserves the description "graceful," with tall stems, 3 to 7 dm high, bearing yellow flowers. The stems are usually delicate, but the plant can become rather coarse. The leaves, with 5 leaflets, are often densely tomentose (hairy) underneath, so that they look white; they vary greatly in size in different plants. The flowers are arranged in a cyme and have 5 sepals, 5 bractlets, 5 petals and several stamens; the fruits are achenes. Graceful cinquefoil is found on prairie grassland and in open woods.

Potentilla nivea L.

SNOW CINQUEFOIL

P. nivea arises from a stout caudex, with branches, and the stems are only 0.5 to 1.5 dm tall. This "cinquefoil" has leaves with only 3 leaflets. The stems and undersurfaces of the leaves are densely hairy. The yellow flowers are in a cluster of 1 to 5 and have the usual *Potentilla* structure. The 5 bractlets alternating with the sepals often give the impression that the flower has 10 sepals. This plant is found on alpine slopes.

NOTES: *There are a few other* Potentilla *species found on alpine slopes, e.g.,* P. hookeriana; *these would also qualify for the name "snow cinquefoil."*

Prunus L. — CHERRY

Cherries grow as shrubs or small trees. The leaves are simple, alternate and toothed; the flowers grow in racemes or in umbellate clusters. The hypanthium is cup-shaped; the flowers are perigynous. There are 5 sepals, 5 white petals, and about 20 stamens, often as long as the petals; the fruit a 1-seeded drupe.

Flowers in clusters of 2 to 8; petals 6 mm long. . . . . *P. pensylvanica*
Flowers in a raceme of 10 or more; petals 4 mm long . . . . *P. virginiana*

Prunus pensylvanica L.f.

PIN CHERRY

Pin cherry is a slender shrub or small tree that grows up to 8 m high; it bears lanceolate or ovate leaves with small teeth. The flowers are usually borne in the leaf-axils and are arranged in an umbel-like cluster. Each flower has 5 sepals, 5 white petals and numerous stamens; the fruit is a 1-seeded globose, reddish drupe. Pin cherry grows in woods and clearings; it is generally found in dry habitats.

Prunus virginiana L.

CHOKE CHERRY

Choke cherry is a slender shrub or small tree similar to pin cherry, but can be distinguished from it because the leaves of choke cherry are not so broad and are drawn out into a pointed end. The flower arrangement is also different: the flowers of choke cherry are arranged in a pendent raceme, 5 to 15 cm long. The fruits are also different—those of choke cherry are reddish purple or black, with an astringent taste. Choke cherry is found in woods and clearings.

NOTES: *Choke cherries were an important food-source for the native tribes. The berries were crushed and formed into cakes, then grease was added and they were left in the sun to dry. The cakes could be soaked in the winter to make a drink.* ~ *The shrub also had medicinal uses. For example, some native tribes made a strong, astringent, black tea from small choke cherry twigs, used for relieving fever and as a tonic for new mothers; the twigs had all the leaves stripped off because they contain small amounts of cyanide. The dried choke cherry roots were chewed and placed on wounds to stop bleeding.*

The Blackfoot used the berries to make a grey dye and used the wood for bows, arrows and pipe stems. Today the berries are used in jams and jellies, and to make wine and sauces. The seeds of choke cherry berries contain small quantities of ***cyanide****; this* ***dangerous poison*** *is also found in peach pits, apple pips and apricot pits, but probably only in trace amounts, and is given off through drying or cooking.*

Rosa L. — ROSE

Rose plants are prickly shrubs with pinnate leaves (5 to 11 leaflets), conspicuously stipitate. The flowers are large, solitary and perigynous, with a globose hypanthium, contracted at the mouth. The structure includes 5 sepals, 5 petals and numerous stamens, all arising from the "rim," and numerous carpels inside. The hypanthium turns red and fleshy, and the carpels inside turn into achenes.

Leaflets 3 to 7, each more than 25 mm long; fruit pear-shaped . . . *R. acicularis*
Leaflets 5 to 9, each less than 25 mm long; fruit globe-shaped . . . . . *R. woodsii*

Rosa acicularis Lindl.

PRICKLY ROSE

This shrub grows 0.5 to 1.5 m high and is freely branched; the stems are covered with thorns. The leaves are pinnate, with 3 to 7 toothed leaflets; the leaflets are usually more than 25 mm long. There are stipules at the bases of the leaves. The pink flowers are very attractive: prickly rose has been chosen as the floral emblem of Alberta. Each flower has 5 sepals, 5 petals, numerous stamens and numerous carpels. The fruits, called "hips," are bright red and pear-shaped or globose (round). Their structure is interesting: the red, fleshy part is really the swollen receptacle which has grown into a cup-shape and practically encloses the true fruits, the achenes. Prickly rose is found in woody areas and on banks, at low elevations.

NOTES: *The common name "prickly rose" is explanatory: the plant is certainly prickly. The species name* acicularis *also means "prickly."* *Rose hips are very high in Vitamin C, and in World War II, when orange juice was not available, school children were encouraged to collect rose hips from the hedges and parklands all over Britain; these were made into rose-hip syrup, richer in Vitamin C than orange juice.*

Rosa woodsii Lindl.

COMMON WILD ROSE, WOODS' ROSE

Common wild rose is a bushy shrub, 0.5 to 1.5 m high, and is similar to prickly rose. The two plants can be distinguished because the thorny stems are different: in prickly rose, the thorns are more or less the same size and are found all along the stem, but in common wild rose there are large, tough thorns at the nodes and the stem prickles are not nearly so dense. In prickly rose, the leaves bear 3 to 7 leaflets usually more than 25 mm long; in common wild rose, the leaves bear 5 to 9 leaflets usually less than 25 mm long. But as the 2 species hybridize (they grow in similar habitats), correct determination is always difficult.

Rubus L. — RASPBERRY

Raspberries are perennial shrubs or herbaceous plants, often with prickly stems. The leaves are compound. The flowers may be solitary, in small open clusters or in racemes, with 5 sepals, 5 white or rose-coloured petals, numerous stamens and numerous carpels; these become drupelets, which together form an aggregate fruit on the enlarged receptacle.

1. Stems bristly; leaves with 3 to 5 leaflets; flowers white . . . . *R. idaeus*
   Stems not bristly; leaves with 3 leaflets; flowers white or pink 2
2. Flowers pink, 10 to 15 mm long; leaves 2 to 4 . . . . *R. arcticus*
   Flowers white, 4 to 8 mm long; leaves numerous . . . . *R. pubescens*

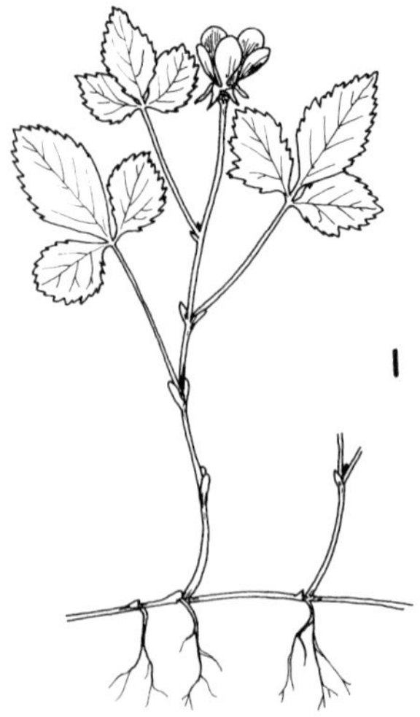

Rubus arcticus L. ssp. *acaulis* (Michx.) Focke

DWARF RASPBERRY

Dwarf raspberry, as its name implies, is a low perennial plant with shoots growing from a slender creeping rootstock; the stem is slender and bears 2 to 4 leaves, each with 3-toothed leaflets. The plant is not thorny. The flowers are purplish rose-coloured, and each flower has the "rose pattern," with 5 sepals, 5 petals, numerous stamens and numerous carpels. The fruit is a typical "raspberry," its carpels turned into tiny drupes called drupelets, each one like a tiny plum, with a seed inside a tough shell, called the stony endoderm, surrounded by a fleshy layer. Raspberry pips are *not* seeds; the seed is inside the pip.

Rubus idaeus L.

WILD RED RASPBERRY

Wild red raspberry can easily be distinguished from dwarf raspberry because it is an erect shrub with canes up to 2 m tall. Its stems are thorny. The leaves are compound with 3 leaflets, each toothed. The flowers of wild red raspberry are white and the fruits are red, although similar to those of dwarf raspberry. Wild red raspberry looks similar to the garden raspberry and grows in open woods.

Rubus pubescens Raf.

RUNNING RASPBERRY, DEWBERRY

Running raspberry is a perennial with slender, trailing stems that root at the tips, hence the common name. This feature distinguishes it from wild red raspberry. Running raspberry has white flowers, which distinguishes it from dwarf raspberry, although the species sometimes hybridize. Running raspberry is found in moist woods.

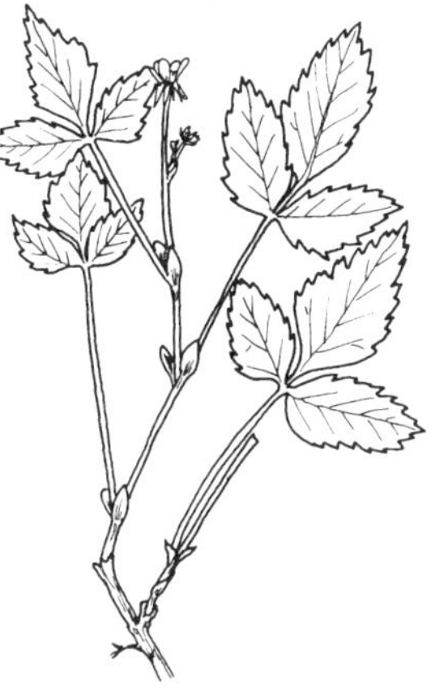

Sibbaldia procumbens L.

SIBBALDIA

Sibbaldia is a low, creeping alpine plant, less than 1 dm high, with creeping rootstocks. It is a densely tufted, perennial plant with ternate leaves. The leaflets are 1 to 3 cm long with wedge-shaped bases and 3 terminal teeth. The flowers help to identify this plant because the green sepals, which are narrow and pointed, are larger and more conspicuous than the yellow petals; the flower looks yellowish green. There are 5 sepals with 5 bractlets, 5 petals and only 5 stamens—stamens are usually numerous in the rose family. The 5 to 15 carpels develop into small achenes. The flowers are arranged in small cymes. Sibbaldia is found on stable, moist alpine slopes and in alpine meadows.
NOTES: *This species has been reported from the alpine meadow at Highwood Pass, and other places in the Kananaskis Country.* ~ Sibbaldia *honours Sir Robert Sibbald (1641-1722), the first Professor of Medicine at Edinburgh University. Sibbald Creek, in Kananaskis Country, is named for Frank Sibbald, a pioneer rancher.* ~ *The specific name* procumbens *means a plant that trails or lies on the ground and does not root at the nodes.*

Sorbus L. — MOUNTAIN ASH

Mountain ash grow as shrubs or small trees with large, pinnate leaves, with 9 to 17 serrate leaflets. The hypanthium is urn-shaped and the flowers grow in compound cymes; flowers have 5 sepals, 5 white petals, numerous (15 to 25) stamens and 3 to 5 carpels. The fruit is a red, berry-like pome.

Leaflets 9 to 11; stems brown . . . . *S. sitchensis*
Leaflets 11 to 13; stems reddish brown . . . . *S. scopulina*

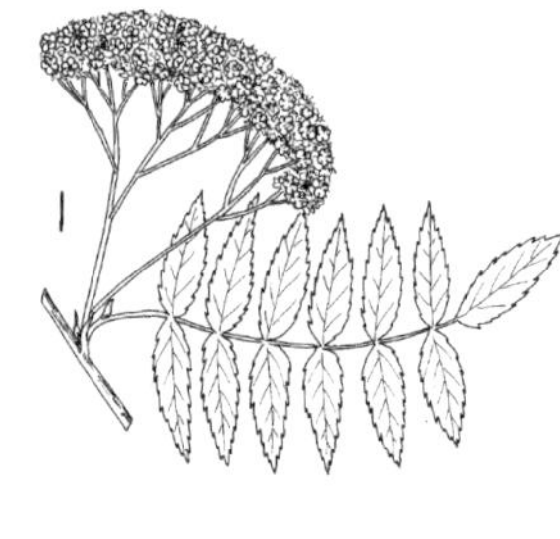

Sorbus scopulina Greene

WESTERN MOUNTAIN ASH

Mountain ash is a shrub 1 to 4 m tall. The leaves are compound, with 11 to 13 leaflets, and are toothed to near the base of the leaflet. The flowers are arranged in a dense, rounded cluster; each flower has 5 sepals, 5 white petals, numerous stamens and 3 to 5 carpels. The carpels turn into fleshy fruits that look like berries, but are really pomes (like apples). The shrubs are found in woods and on moist open slopes.

NOTES: *The common name "ash" refers to the leaves, which resemble those of the ash tree, but the ash,* Fraxinus, *belongs to a different family, Oleaceae.*

Sorbus sitchensis Roemer

SITKA MOUNTAIN ASH

This plant is a shrub, 1 to 3 m tall, but can be distinguished from *S. scopulina* because the leaves are different. In *S. scopulina*, the leaflets (11 to 13) are pointed at the tip and toothed down to the base; in *S. sitchensis* the leaflets are fewer (9 to 11) and the teeth do not extend down to the base; almost half of the leaflet is toothless (entire). The twigs are also different: in *S. scopulina* the young twigs are covered with white hairs, but in *S. sitchensis* the twigs have rust-coloured hairs. Sitka mountain ash is found in montane woods.

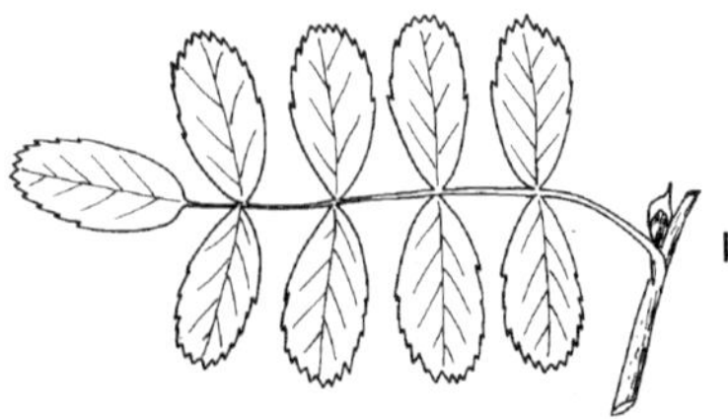

Spiraea L. **MEADOWSWEET**

Meadowsweets are small shrubs with entire, toothed leaves, lacking stipules. The flowers are small, in a flat-topped terminal inflorescence; flowers have 5 sepals, 5 pink or white petals, 15 or more prominent stamens and 5 carpels, inserted at the bottom of the cup-shaped hypanthium. The carpels develop into follicles.

Petals white . . . . *S. betulifolia*
Petals pink. . . . . *S. densiflora*

Spiraea betulifolia Pallas

SPIRAEA, WHITE MEADOWSWEET

(S. lucida Dougl.*)*

White meadowsweet is a low shrub, 2 to 8 dm high, found on open slopes and in woods. It grows from a woody rootstock, rather like a rhizome, with many roots growing from it. The species name *betulifolia* means "leaves like those of the birch," and the leaves are certainly birch-like, oval and coarsely toothed; they grow alternately on the twigs. The white flowers grow in dense clusters at the ends of the twigs (terminal), and each flower has 5 sepals, 5 petals, numerous stamens and usually 5 carpels. The carpels turn into follicles, like pea pods but only opening down 1 side.

Spiraea densiflora Nutt.

PINK SPIRAEA, PINK MEADOWSWEET

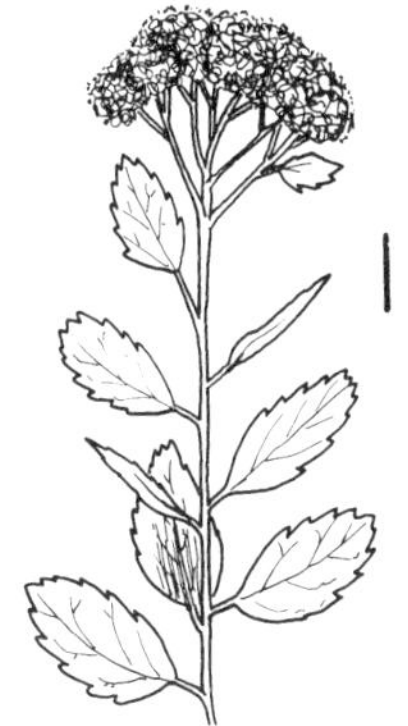

Pink meadowsweet is a slender shrub, 0.5 to 1.5 m high, taller than white meadowsweet, although the leaves of the 2 species are similar. The flowers are similar to those of white meadowsweet, but can be distinguished because of their pink or rose colour. The common name is appropriate, not only because it is sweet-smelling but because it grows in meadows and moist montane woods.

Fabaceae (Leguminosae) PEA FAMILY

Fabaceae is better known by its former name, Leguminosae, the legume family, because the characteristic fruit of this family is the legume. It is a large family, of worldwide distribution, and has been split into 3 sub-families: Mimosoideae, Caesalpinioideae and Papilionoideae. The plants of Kananaskis Country all belong to the sub-family Papilionoideae, so this is the only subfamily described.

The members of this sub-family are perennials, either herbaceous or woody, and usually have compound leaves, either pinnate or palmate. These leaves possess stipules. The roots often bear nodules which contain nitrogen-fixing bacteria; these can turn free nitrogen in the air of the soil into nitrates and other nitrogenous compounds. The flower-structure is unusual and is characteristic of the whole sub-family, except for *Petalostemon* (purple prairie clover). There are 5 sepals, forming a cup-shaped calyx, with 5 prominent teeth, and 5 petals that are arranged as follows: a large one at the back of the flower, called the "standard," 2 smaller ones on either side, called the "wings," and 2 joined together, forming the "keel." Inside the keel are 10 stamens; usually 9 of these are partly joined by their filaments, leaving 1 stamen free. Inside the keel is also the single carpel which turns into the fruit, a *legume*, a pod that opens down both sides to release the seeds. This elaborate arrangement of flower parts is useful for ensuring cross-pollination. In the genus *Hedysarum* the legume is constricted into 1-seeded portions; this type of fruit is called a *loment.*

1.	Leaflets 3	2
	Leaflets 5 or more	5
2.	Flowers in globe-shaped clusters	3
	Flowers not in globe-shaped clusters	4
3.	Pods straight	*Trifolium*
	Pods coiled or curved	*Medicago*
4.	Flowers more than 10 mm long	*Thermopsis*
	Flowers less than 10 mm long	*Melilotus*
5.	Climbing plants with tendrils	6
	Erect plants, tendrils absent	7
6.	Flowers purple, leaflets 8 to 20	*Vicia*
	Flowers yellow, leaflets 6 to 8	*Lathyrus*
7.	Leaves basal	*Oxytropis*
	Leaves alternate	8
8.	Leaves palmate (hand-shaped)	*Lupinus*
	Leaves pinnate	9
9.	Fruit a loment	*Hedysarum*
	Fruit a legume	*Astragalus*

NOTE: *The genera* Lathyrus *(vetchling) and* Vicia *(vetch) differ from the other plants of the family Fabaceae (Leguminosae) because their leaves possess* tendrils. *These are leaflets that have been adapted for climbing purposes and are always found at the tips of the leaves, replacing the terminal leaflet or leaflets.*

Astragalus L. — MILK VETCH

These are perennial species with rhizomes or taproots. They have pinnate leaves with more than 5 leaflets. The flowers are arranged in axillary or terminal racemes, with a calyx of 5 united sepals and 5 petals (a standard petal, 2 "wings," and 2 united and rounded keel petals), 10 stamens (9 united, 1 free) and 1 carpel that forms a legume.

1. Flowers yellow; pods inflated . . . . *A. americanus*
   Flowers purple; pods not inflated 2
2. Flowers 7 to 12 mm long . . . . *A. alpinus*
   Flowers 15 to 18 mm long . . . . *A. striatus*

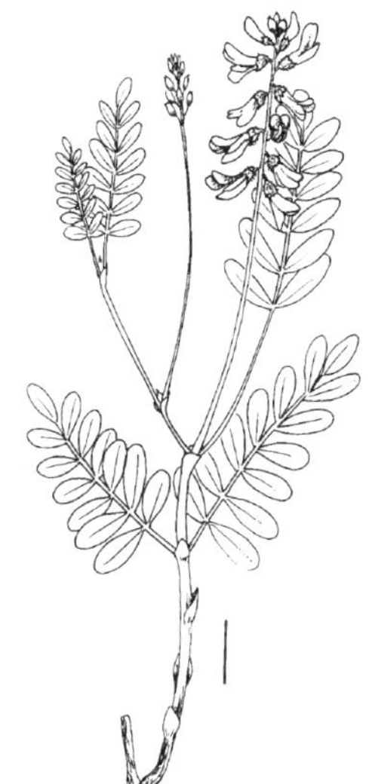

Astragalus alpinus L.

ALPINE MILK VETCH

A low plant which forms mats with creeping rhizomes. The leaves have 13 to 25 leaflets, up to 12 mm long, and the flowers are borne in a short raceme, 1 to 4 cm long. The calyx is covered with black hairs, and the corolla is purplish white. The fruits are brown pods, reflexed downwards, covered with black hairs; each is 1 cm long. Alpine milk vetch is a dainty plant, found on rocky slopes.

Astragalus americanus (Hook.) M.E. Jones

AMERICAN MILK VETCH

(*A. frigidus* (L.) A. Gray var. *americanus* (Hook.) S. Wats.)

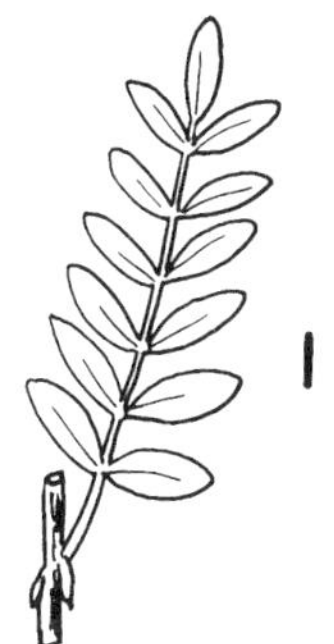

American milk vetch is easily distinguished from alpine milk vetch: it is a much larger plant, 5 to 10 dm high, its leaves are larger, there are only 7 to 15 leaflets and the plant is more erect. Leaflets of the American milk vetch are each 2 to 4 cm long. The flowers of the 2 species are also different: those of American milk vetch are white, not purplish, turning yellowish later. The plant grows in moist woods and on stream banks.

Astragalus striatus Nutt.

ASCENDING PURPLE MILK VETCH

The flowers of this milk vetch are purple and are arranged in a dense raceme at the end of a long flower stalk, like a purple ball. *A. striatus* grows 2 to 4 dm high, but is not such a large, erect, spreading plant as American milk vetch. Ascending purple milk vetch is found on prairie grassland.

Hedysarum L. — SWEET VETCH

Sweet vetches are all erect or ascending perennials. The leaves are stipulate with several pinnate leaflets. The flowers bloom in axillary racemes, usually well above the leaves, with 5 united sepals and 5 petals; the keel petals are much longer than the wings and standard, and the keel is truncate (square-shaped). The characteristic fruit is a loment, a legume with constrictions between the seeds.

1. Flowers yellow . . . . . *H. sulphurescens*
   Flowers reddish purple 2
2. Leaflets 9 to 13 . . . . . *H. boreale*
   Leaflets 15 to 21 . . . . . *H. alpinum*

Hedysarum alpinum L.

ALPINE HEDYSARUM

Hedysarum alpinum is a bushy plant, 2 to 7 dm high, and the leaves have 15 to 21 leaflets with prominent veins. The flowers are arranged in tall racemes; together with the flowering-stalk they are up to 25 cm tall. The flowers are pink to reddish purple, nodding (reflexed) and wing-margined. They produce the distinctive loments, which are often reflexed. Alpine hedysarum is found in moist, open woods and slopes and on gravelly banks and prairies.

NOTES: *The species name* alpinum *is not well chosen because the plant is not a true alpine plant.*

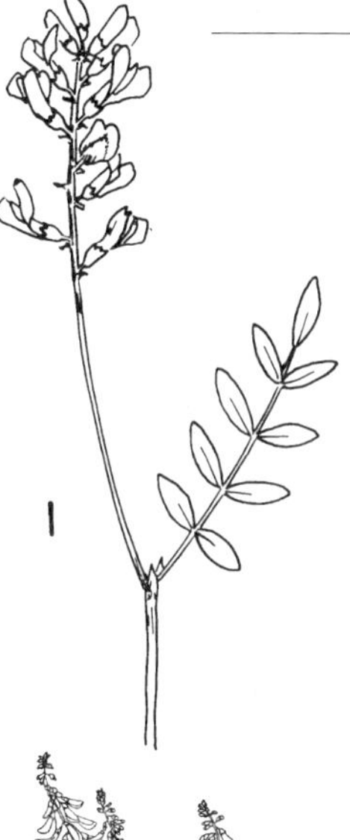

Hedysarum boreale Nutt. var. *mackenzii* (Richards.) C.L. Hichc.

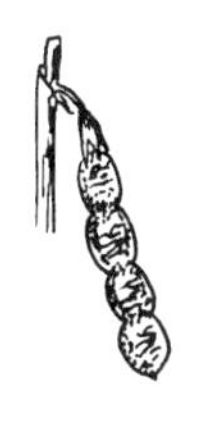

MACKENZIE'S HEDYSARUM, NORTHERN HEDYSARUM

(H. mackenzii Richards.*)*

This species grows 2 to 6 dm tall but can be distinguished from *H. alpinum* because the flowering raceme is not as tall and the variety *mackenzii* has deep purple flowers. The flowers produce loments, but these are not wing-margined (those of *H. alpinum* are). The leaves of *H. boreale* have 9 to 13 leaflets. The plant is found on plains and open slopes in the mountains, where it sometimes provides a splendid show of colour.

Hedysarum sulphurescens Rydb.

YELLOW HEDYSARUM

The flowers of this plant are sulphur-yellow, hence the species name *sulphurescens*, and this makes it very easy to separate it from the other hedysarum species. The flowers are arranged in a tall raceme, and each flower is 15 to 18 mm long; the flowers produce loments which are wing-margined. The leaves have 9 to 17 leaflets, and the plant is found in prairie grassland in woodlands at middle elevations and mountain woods.

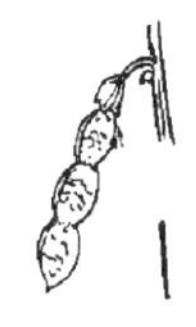

Lathyrus ochroleucus Hook.

CREAM-COLOURED VETCHLING, PEA VINE

Vetchling, or pea vine, usually has 3 to 5 tendrils at the tip of each leaf. There are 4 to 6 leaflets, and each leaf has 2 large stipules at the base; these stipules are leaflike and characteristic of the pea family. Pea vine has stems up to 1 m long and climbs by means of its tendrils. The large flowers are yellowish white, and the plant can be easily distinguished from purple vetch (*L. venosus*). It is found in moist woods and clearings.

Lupinus sericeus Pursh

SILKY PERENNIAL LUPINE

Lupines can often be distinguished from other plants in this family by their leaves because these are palmate, that is, the leaflets radiate from the centre. There are 5 to 10 leaflets in this perennial lupine. The plant grows erect, up to 4 to 8 dm tall. The blue flowers grow in a dense raceme and are quite striking; a colony of lupines is often a sheet of blue. The legume is 2 to 3 cm long and has 4 to 6 seeds. The plant is found on gravelly outwash plains, in aspen groves and in open montane woods.

NOTES: *The species name* sericeus *means "silky" and refers to the dense covering of long, white hairs on the leaves and stems.*

Medicago L. — MEDICK

Medicago species are perennials or annuals with several erect stems, although they may be prostrate. The leaves have 3 leaflets. The flowers are small, in racemes or axillary clusters. The calyx has 5 sepals joined; the corolla has 5 yellow or bluish purple petals, 10 stamens and 1 carpel, which forms a curved or spirally coiled legume. *Medicago* can be distinguished from other genera in the pea family if the fruits are present, because these are curiously coiled or curved structures, quite unlike the familiar legume.

Flowers yellow; pod kidney-shaped . . . . *M. lupulina*
Flowers blue or purple; pod coiled . . . . *M. sativa*

Medicago lupulina L.

BLACK MEDICK

Black medick is a prostrate or decumbent introduced annual, with branches radiating from the base, 2 to 8 dm long. Each leaf usually has 3 leaflets. The flowers are small, yellow and arranged in a short, dense raceme. The black legumes are tightly curved, with conspicuous veins, and they are not spirally twisted. The plant is found on waste ground.

Medicago sativa L.

ALFALFA, LUCERNE

The well-known alfalfa is an introduced forage plant. It is found in fields and waste ground, and also along roadsides. It grows up to 1 m tall, is often bushy and, with its racemes of blue flowers, can be a striking plant. The flower colour can vary from deep purple to purplish blue to pale blue, sometimes almost white. The fruiting pods are characteristic because they are coiled 1 to 3 times.
NOTES: *Although there is also a yellow-flowered form of alfalfa, it is easy to distinguish it from black medick because the growth-habits of the 2 plants are so different.*

Melilotus Mill. SWEET CLOVER

Sweet clovers are annuals or biennials growing from a strong taproot; the plants are leafy and branching. The leaves are stipulate, with 3 leaflets; the flowers grow in tall, loose racemes. Sweet clovers have a calyx of 5 sepals, united, 5 white or yellow petals, 10 stamens and 1 carpel forming a spindle-shaped legume.

Flowers white . . . . *M. alba*
Flowers yellow . . . . *M. officinalis*

Melilotus alba Desr.

WHITE SWEET CLOVER

White sweet clover grows 5 to 25 dm tall. The leaves have 3 leaflets; they are trifoliate like many members of the pea family. The white flowers are arranged in tall, terminal racemes, and they are often pendent; the pods are rounded. White sweet clover is an introduced plant, found in waste places and on roadsides.
NOTES: *This species is called "sweet clover" because it contains coumarin and is fragrant when dried.*

Melilotus officinalis (L.) Lam.

YELLOW SWEET CLOVER

Apart from the flower-colour, it is difficult to distinguish the sweet clovers, but there are 3 small differences. In white sweet clover, the standard petal at the back of the flower is longer than the wing petals, the pod is not wrinkled, and the seeds are yellow-green. In yellow sweet clover, the standard petal is about the same size as the wing petals, the pod is wrinkled, and the seeds are olive-green or purple spotted. Yellow sweet clover is an introduced plant and is found in waste places and along roadsides—the same habitats as white sweet clover; in fact, the two plants often grow side by side.

Oxytropis D.C. — LOCOWEED

Locoweeds are tufted perennials, apparently stemless, arising from a strong taproot and root-crown. Leaves are pinnate, with many leaflets, usually with grey or silvery hairs. Flowers are yellow or bluish purple and grow in spicate racemes. The calyx has 5 fused sepals and a corolla of 5 petals (the 2 keel petals fused, with a distinctive beak), 10 stamens and 1 carpel, forming an oval legume with a deep groove.

1. Flowers yellow . . . . *O. sericea*
   Flowers blue or purple 2
2. Leaves 1 to 5 cm long; stem with 1 to 3 flowers; pod inflated . . . . *O. podocarpa*
   Leaves 7 to 26 cm long; stems with 12 to 35 flowers . . . . *O. splendens*

Oxytropis podocarpa A. Gray

INFLATED OXYTROPE

Inflated oxytrope is a low caespitose plant, with small leaves, 1 to 5 cm long, and with 1 to 3 purple or blue flowers growing in racemes. Its common name comes from the fruiting legumes, which become inflated when ripe. The plant is found on grassy alpine hillsides and high exposed ridges.

NOTES: Oxytropis *flowers can be distinguished from* Astragalus *flowers because the keels have "beaks" and from* Hedysarum *flowers because the keels are not truncate.* Oxytropis *comes from two Greek words:* oxys *(sharp) and* tropis *(keel). The common name "locoweed" was given to this genus because many contain the mineral selenium. When this is ingested by cattle, it affects their nervous systems and they become dazed and disoriented. "Locoweed" derives from the Spanish word* loco, *meaning "crazy."*

Oxytropis sericea Nutt.

EARLY YELLOW LOCOWEED

Early yellow locoweed blooms in spring and is quite an attractive plant with tall racemes of yellow flowers. It is easy to distinguish it from inflated oxytrope because early yellow locoweed is a much taller plant and has yellow flowers. The fruits, legumes, are not inflated. Early yellow locoweed grows on prairie grassland and dry hillsides.

Oxtropis splendens Dougl. ex Hook.

SHOWY LOCOWEED

Showy locoweed is a silver-hairy plant with dense racemes of blue-purple flowers.The racemes elongate to 10 cm in fruit. It grows on open, grassy slopes, on roadsides, in woodlands and on gravelly banks.

NOTES: *Showy locoweed has a well-chosen common name because it certainly is showy, quite striking when a colony is in full bloom. The combination of the purple flowers and numerous silver hairs makes this plant so attractive.*

Thermopsis rhombifolia (Nutt.) Richards.

GOLDEN BEAN

The familiar golden bean appears in the spring; it is an erect plant, 1 to 4 dm high, with alternate leaves and large stipules. Each leaf has 3 leaflets—it is trifoliate. The flowers are arranged in a short raceme and are golden-yellow. The plants spread by creeping rootstocks and form large colonies, creating a striking swath of golden-yellow. After fertilization the fruits appear; they are legumes (pods) and are strongly curved into a semi-circle. The seeds are **poisonous**.

NOTES: *The Blackfoot called golden bean "buffaloflower" because when it bloomed the buffalo bulls left their winter range for summer grazing on the prairie.* *The species name* rhombifolia *means "rhombic-shaped leaves," but this adjective actually applies to the stipules, not the leaves: the large stipules are usually rhombic-shaped.*

Trifolium L. — CLOVER

Clovers are biennials or perennials that may be erect or creeping. Their leaves are stipulate with 3 leaflets; the flowers grow in globose clusters or short racemes. They have a calyx of 5 united sepals, 5 petals, 10 stamens and 1 carpel, which forms a tiny legume, 1 to 6 seeded, often not opening and obscured by persistent sepals.

1. Leaflets 1 to 2 cm long; flowers white . . . . *T. repens*
   Leaflets 2 to 5 cm long; flowers pink. 2
2. Flowers 8 to 9 mm long . . . . *T. hybridum*
   Flowers 12 to 20 mm long . . . . *T. pratense*

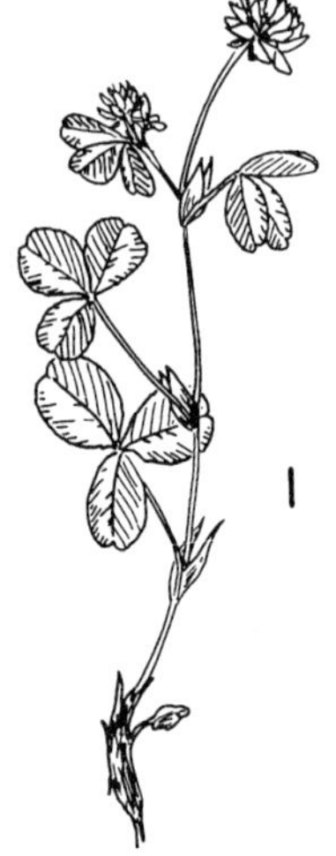

Trifolium hybridum L.

ALSIKE CLOVER

Alsike clover is an introduced plant whose stems can be 5 dm long. The genus name *Trifolium* indicates that the leaves have 3 leaflets—they are trifoliate. The flower-heads arise from the leaf axils on long slender stalks and are 8 to 9 mm long. The petals vary in colour from nearly white to pink to nearly red. It has a weedy growth-habit in Kananaskis Country and is found along roadside ditches.

NOTES: Alsike *is a Norwegian word meaning "clover."*

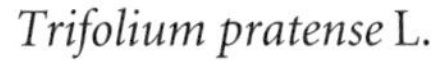

Trifolium pratense L.

RED CLOVER

Red clover is an introduced plant and can grow up to 6 dm high. It can be distinguished from alsike clover because the flower-heads are larger and the petals are usually dark red, although the colour can vary. A more definite difference is that the flower-head is subtended by broad bracts, and the heads are terminal, not axillary. Red clover was introduced for forage and has escaped; it is quite common in moist places, along roadsides and on waste ground.

Trifolium repens L.

WHITE CLOVER, DUTCH CLOVER

Dutch clover is a slender, creeping perennial, rooting from the nodes. The petals are creamy-white, sometimes tinged with pink. The flowers in the flower-heads tend to droop. It grows as a weed in gardens, on lawns and along roadsides.
Notes: The species name *repens* means "creeping," a trait that separates it from other clovers.

Vicia L. — VETCH

Vetches are slender climbing plants whose leaves are pinnate, stipulate and have 3 or more tendrils at the tips replacing leaflets. The flowers grow in axillary, sometimes crowded, racemes, featuring a calyx of 5 united sepals, often swollen at the base, 5 petals, 10 stamens and 1 carpel forming a legume up to 3.5 cm long, flat and pointed at both ends.

Flowering-stalk with 2 to 9 flowers; leaflets 8 to 14 . . . . *V. americana*
Flowering-stalk with 15 to 40 flowers; leaflets 10 to 20 . . . . *V. cracca*

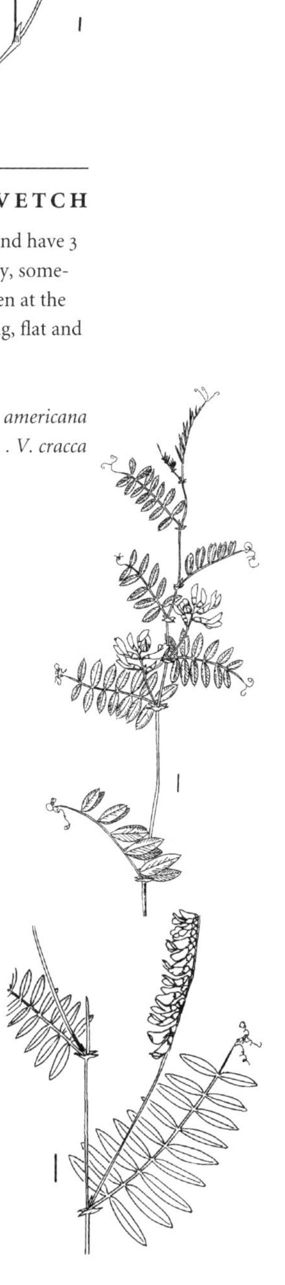

Vicia americana Muhl.

WILD VETCH

Wild vetch has stems 3 to 10 dm long; the leaves have 8 to 14 leaflets. The reddish purple flowers are arranged in a short raceme; there are usually 3 to 9 flowers. The pod, or legume, is pointed at both ends and grows up to 3.5 cm long. The plant grows in meadows and open woods.

NOTES: V. americana *was considered a separate species from* V. sparsifolia T. & G., *but there are so many gradations between the species that* V. sparsifolia *is now included in* V. americana.

Vicia cracca L.

TUFTED VETCH

This species of introduced vetch can easily be distinguished from *V. americana* because the flower arrangements in the two species are quite different. *V. americana* has flowers in a short raceme with only 3 to 9 flowers, but in *V. cracca* the purple flowers are in a tall raceme, with 15 to 40 flowers, and the racemes are single-sided. It is found along roadsides and on waste ground.

Geraniaceae GERANIUM FAMILY

Geraniums are herbaceous, leafy, hairy plants and may be annuals, biennials or perennials. The stipulate leaves are opposite; they are usually cleft and toothed. The flowers usually have long stalks, and are arranged in terminal clusters. Each flower has 5 sepals and 5 petals, which may be white or pink-purple. There may be 5 or 10 stamens, and there are 5 carpels; these are unusual, because the styles are united in a central column, and the 5 carpels each form a separate, 1-seeded fruit, clustered at the base of the styles. When ripe, the separate little fruits spring apart. The long "beak" formed by the styles is a conspicuous feature of the fruit and has given rise to the common name "cranesbill"—*geranos* is the Latin word for "crane."

Flowers rose-purple; leaves and stem sticky .... *G. viscosissimum*
Flowers white; leaves not sticky .... *G. richardsonii*

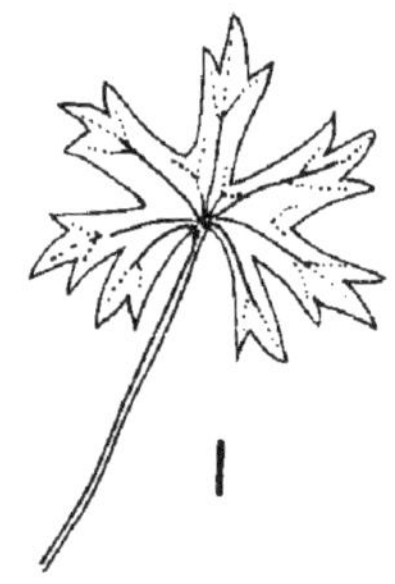

Geranium richardsonii Fisch. & Trautv.

WILD WHITE GERANIUM

White geranium grows to a height of 4 to 8 dm and bears typical geranium leaves, with 3 to 7 deep divisions. The flowers have white petals, usually pink or purple veined, and the sepals are purple-tipped. The fruit is the typical "beaked fruit" of the geranium family: there are 5 carpels and the styles are united to a central column. The ripe fruit splits from below into 5 parts, each containing 1 seed. This type of fruit is called a *schizocarp* or split-fruit, which breaks up into small fruits called *mericarps*. The plant is found in moist thickets and in open woods.

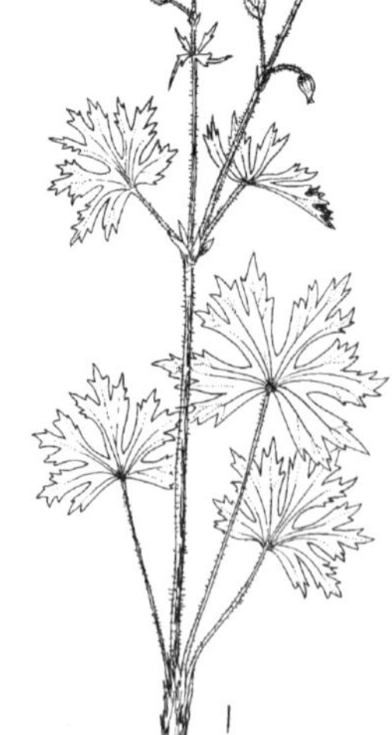

Geranium viscosissimum Fisch. & Meyer

STICKY PURPLE GERANIUM

Sticky purple geranium grows 2 to 9 dm high and is a leafy plant. The leaves are palmately divided into 5 to 7 divisions; they are often glandular and densely hairy. The common name describes it well because the flower stalks, sepals and carpels are glandular ("sticky"), and the petals are a deep rose-purple. Both these characteristics make it easy to distinguish it from white geranium. The plant is found in moist coulees and on grassland slopes.

NOTES: *The species name* viscosissimum *is Latin for "very sticky."*

Linaceae FLAX FAMILY

These are herbaceous plants, either annual or perennial, with narrow leaves that are alternate, simple, sessile and entire. The flowers are found on the upper portions of the stem and are blue or yellow; they are borne on slender stalks. Each flower has 5 sepals and 5 petals (these quickly fall), and there are 5 stamens. In the centre of the flower are the 5 united carpels, which turn into a rounded capsule. This capsule has 5 locules (or divisions) and these are often divided by "false" partitions to make 10 locules. When ripe, the capsule splits from the top to release the seeds.

Linum lewisii Pursh

WILD BLUE FLAX

Wild blue flax is a dainty plant, 2 to 7 dm tall, with slender stems bearing linear leaves. The flowers are found on the upper, branched stems and are a delicate shade of blue, borne on slender flower stalks. There are 5 sepals, 5 petals, 5 stamens and 5 carpels. The petals are deciduous; they fall early. The fruit is a globose capsule that splits when ripe to release the seeds. Wild flax is found on grassy prairies and in dry, open woods.

NOTES: *The species name* lewisii *honours Captain Merriwether Lewis of the Lewis and Clarke expedition.*

Polygalaceae MILKWORT FAMILY

These are herbaceous perennial plants, with alternate leaves and usually several flowering stems. The flowers are in dense racemes at the tips of the flowering stalks, and the flower-structure is characteristic and unmistakable. There are 5 sepals: the 2 inner ones are much larger, wing-like, often petalloid. There are 3 petals (unusual in a dicotyledon flower) and the inner one is boat-shaped, with a fringed crest. The filaments of the 8 stamens are united into a split sheath, and in the centre is the superior ovary. This ovary becomes the fruit, a capsule that splits lengthwise to release 2 seeds.

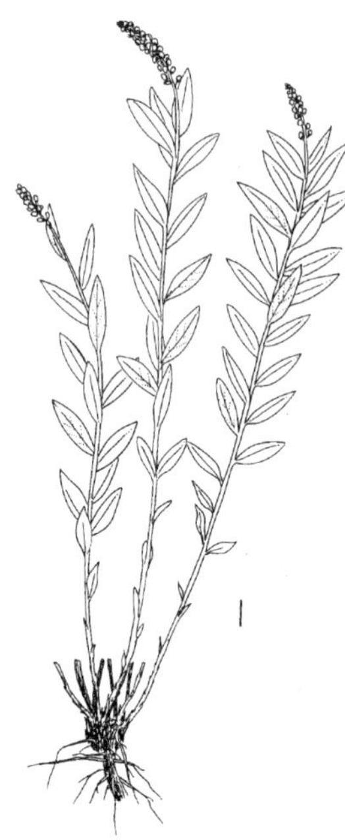

Polygala senega L.

SENECA SNAKEROOT

Seneca snakeroot is a perennial plant with several stems branching from the base, 1 to 5 dm high. The flowers grow in a dense raceme, 1 to 4 cm long, at the top of the flowering stalks and are greenish white, sometimes pink. The flower-structure is notable because the 5 sepals are all free: 2 of them form wings and are sometimes petaloid. The 3 petals are unusual, too, because the middle petal is boat-shaped and has a fringed crest. The fruit is a small capsule. Seneca snakeroot is a prairie plant and is found in open woods.

NOTES: *The common name "Seneca snakeroot" has an interesting history. The rootstock has a twisted, coiled snake-like appearance, and some native tribes believed it was good for snake-bite. Seneca snakeroot is associated with the Seneca people, who used to collect the root in the fall. It is* not *good for snake-bite, but is a useful medicine for lung diseases and was used by the Blackfoot for this purpose.*

Euphorbiaceae SPURGE FAMILY

The spurge family is distinctive because the male flowers (staminate) and the female flowers (carpellary or pistillate) are developed inside a *cyathium*, a cup-like structure, usually with a single carpellary flower and several staminate flowers. Another distinctive feature is that the cyathia often have curiously shaped glands bearing nectar on their margins. These cyathia are borne in cymes.

Members of the spurge family are annual or perennial herbs with milky juice, and the leaves are variable: they can be simple and alternate, opposite or whorled. The carpellary flower has a 3-loculed ovary, with 3 styles and no calyx or corolla. The male flower is a single stamen with a minute bract. The female flower develops into a capsule.

Euphorbia L. SPURGE

Spurges are annuals or perennials containing milky juice. They have whorled, alternate or opposite leaves. The flowers are monoecious and are borne in cyathia, cup-shaped structures with 1 carpellary flower and several staminate flowers. The cyathia are arranged in cymes. The carpellary flower has no calyx or corolla, 3 carpels joined, with 3 styles, which becomes a 3-lobed capsule; staminate flowers in bundles, each with 1 stamen.

Perennial, with branched leafy, erect stem . . . . *E. esula*
Annual, with leafy prostrate stem . . . . *E. serpyllifolia*

Euphorbia esula L.

LEAFY SPURGE

Leafy spurge is a perennial plant with long, thin, creeping rhizomes and alternate leaves. The flower-stalk (peduncle) bears a primary umbel with 7 to 12 rays; there is a whorl of leaves under the umbel. The cyathia are 2 to 3 mm high with 4 crescent-shaped glands (sometimes horned) on their margins. The fruit is a capsule. Leafy spurge is a common weed found on roadsides, in fields and on river flats.

Euphorbia serpyllifolia Pers.

SPURGE

This spurge is an annual with prostrate stems, which are smooth, not hairy, and freely branched; they spread over the ground and form mats. These stems are very slender and bear small, oblong, finely toothed, opposite leaves. The flowers develop inside a cup-like cyathium; the cyathia are arranged in cymes. Each cyathium has a single carpellary (female) flower and a few (1 to 5) staminate flowers; the glands on the margin are elliptic, with white appendages. The fruit is a 3-lobed capsule which splits lengthwise. Spurge is found on waste ground, along railroads and on sandy soil.

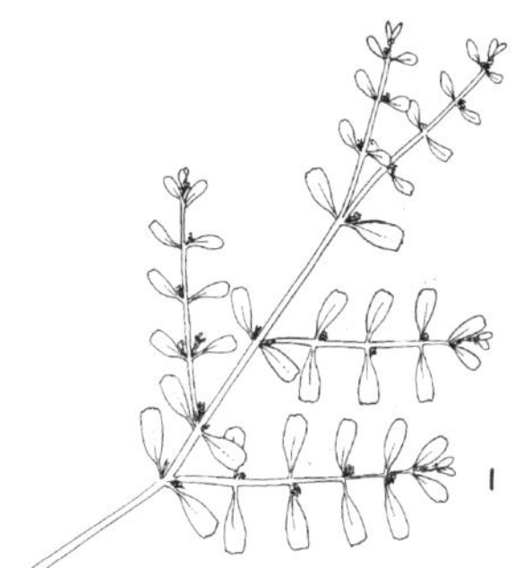

Callitrichaceae WATER STARWORT FAMILY

Water starworts are inconspicuous, aquatic annual or perennial plants found in standing water or on mud. The small leaves grow in opposite pairs on delicate, branched stems, and are borne in tufts at the ends of the branches. The flowers are borne in the leaf axils; they are greenish, tiny and usually number 1 to 3 in an axil, each with a pair of bracts. The fruits are 4-lobed, separating when ripe into 4 single-seeded fruits.

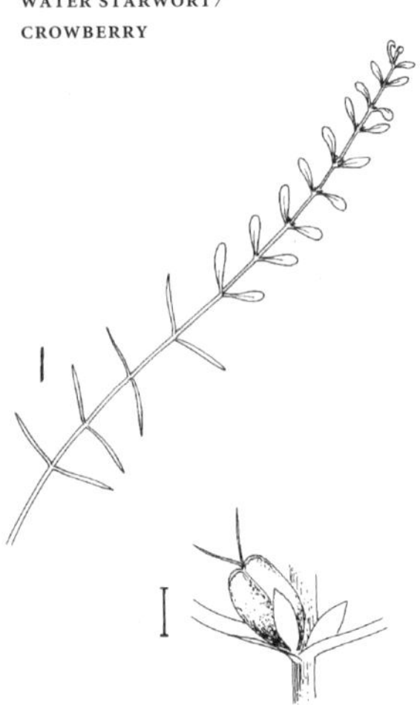

Callitriche verna L.

VERNAL WATER STARWORT

This plant may be submerged or floating; its floating leaves have leaf stalks. The slender stems are 5 to 20 cm long, and the submerged leaves are linear, 5 to 20 mm long. The floating and emergent leaves (above the water) are broadly spoon-shaped, up to 4 mm across. The flowers, in the leaf-axils, have bracts; they are greenish, tiny and unisexual, i.e., with either stamens or an ovary. The staminate flowers usually have only 1 stamen and no petals or sepals; the carpellary (pistillate) flowers usually have only 2 carpels, united, deeply 4-lobed, with 2 styles. After fertilization each develops into a 4-lobed fruit, which separates into 4 single-seeded parts. Vernal water starwort is found in ponds and streams.

NOTE: Vernus *is Latin for "spring."*

Empetraceae CROWBERRY FAMILY

These are small evergreen shrubs, with spreading branches that lie close to the ground and grow up to 4 dm long. The branches divide into slender branchlets that bear tiny glands. The leaves are 4 to 8 mm long and are rolled downwards from the margins; they bear small flowers in their axils. These flowers may be unisexual, lacking either stamens or carpels, or they may possess both; there are 3 sepals, 3 petals (may be absent), 3 stamens with compound pollen grains, and in the centre several carpels (6 to 9) joined together, bearing 6 to 9 stigmas. These form the gynoecium and produce the fruit, a black, berry-like fruit.

Empetrum nigrum L.

CROWBERRY

Crowberry is an evergreen shrub. The spreading branches can grow 4 dm long, with tiny, linear, crowded leaves along their length, with flowers in their axils. These flowers are small, sometimes lacking stamens or carpels, with 3 sepals, 3 petals (sometimes lacking), 3 stamens with compound pollen grains and several carpels joined together in the centre. These produce a black, berry-like drupe, a distinctive feature of the shrub. *Nigrum*, the specific name, means "black" and refers to the fruits. Crowberry is found in muskegs and coniferous forests and on rocky slopes.

NOTES: *The generic name* Empetrum, *comes from 2 words:* em, *meaning "upon" and* petros, *meaning "rock."*

Aceraceae MAPLE FAMILY

Maples are trees and shrubs with opposite leaves; the leaves may be simple or compound. The flowers grow at the tips of short, leafy branches and may be arranged in racemes, umbels or corymbs. The trees may be monoecious (male and female flowers on 1 tree) or dioecious (male and female flowers on different trees). Each flower has 5 sepals and 5 petals, with 4 to 10 stamens in the male (staminate) flower. In the female (carpellary) flower, there are 2 united carpels in the centre. The fruit is a double samara.

Acer glabrum Torr.

MOUNTAIN MAPLE

Mountain maple is a shrub or small tree, up to 10 m tall, with reddish stems. The leaves grow in pairs up to 15 cm long including the slender petiole (leaf stalk); they have the typical maple-leaf shape, usually with 5 sharply pointed main lobes (sometimes only 3). The leaf-margins are sharply toothed. The flowers grow at the tips of short leafy branches ("spurs") in open clusters; they may be male, female or bisexual. There are 5 sepals, 5 petals (sometimes lacking) and 4 to 10 stamens. The fruit is a double samara, 2 to 4 cm long. Samaras are winged achenes and are dispersed by the wind. (The familiar ash "key" is a single samara.) Mountain maple grows on montane slopes in rocky areas, usually in well-protected valley areas; they are also found in ravines.

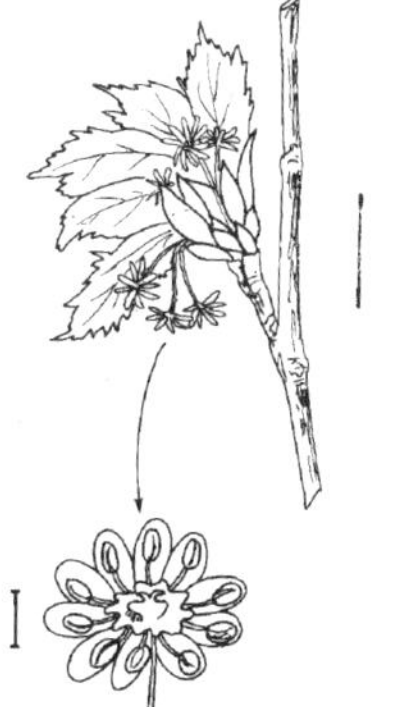

Malvaceae MALLOW FAMILY

Mallows are herbs with characteristic leaves: palmately veined and usually lobed. The flowers are regular (actinomorphic) and arise from the leaf-axils either singly or clustered. There are 5 sepals, 5 petals and numerous stamens. These stamens are united by their filaments into a staminal tube, which surrounds the style. The gynoecium, with 5 or more carpels joined together, becomes the fruit, which splits when ripe into several 1-seeded fruits; these are *mericarps* and the fruit is a *schizocarp*. Hollyhocks belong to this family and so does the exotic hibiscus.

Sphaeralcea coccinea (Pursh) Rydb.

SCARLET MALLOW

Scarlet mallow is an attractive plant with decumbent to ascending stems (1 to 3 dm in length), dainty leaves and flowers of a rich orange-red. It is a perennial plant, spreading by rhizomes to form small colonies, with several stems bearing leaves; the blades are 3 to 5-parted and lobed, palmate, with slender petioles. The whole plant is covered with stellate (star-shaped) hairs. The flowers are arranged in short, terminal racemes with bracts; the pedicels are short. Each flower has a calyx of 5 united sepals and a corolla of 5 petals, more or less joined at the base; the staminal column, formed from numerous stamens with free anthers, is also joined. (This column is typical of the family and is particularly obvious in *Hibiscus.*) The gynoecium in the centre consists of several carpels joined in a flat circle; when ripe, they separate. Each part is a small, single-seeded fruit (*mericarp*) and the whole fruit is called a *schizocarp* or split-fruit. Scarlet mallow is found in prairie grassland and along roadsides.

Violaceae VIOLET FAMILY

Members of this family are small herbaceous plants; some of them have no stems, i.e., they are acaulescent (*caulis* means "stem"). The leaves are either alternate or basal and possess stipules; they are entire. The flowers have a complex structure, adapted to ensure cross-pollination. There are 5 green sepals that have little "ears" or auricles; there are 5 petals, the 2 lateral of which are often bearded on the inside, and the lowest of which has a basal spur. The petals may be blue, white or yellow. There are 5 stamens; the 2 lowermost have appendages bearing nectaries that extend into the spur of the lowest petal. The gynoecium, in the centre, consists of 3 united carpels that turn into a capsule with 3 valves. When ripe, this opens explosively and numerous seeds are ejected. These seeds each have a tiny, white food-body (a caruncle) attached at one end; they are carried off by food-seeking ants to their nests, aiding seed distribution. Violets often produce cleistogamous flowers in late summer. These are enclosed inside a calyx of sepals and lack a corolla. These flowers never open and are self-fertilized.

Viola L. — VIOLET

Violets are low perennials with leafy stems. The leaves are alternate, long-stalked, stipulate, often cordate; the flowers are solitary in leaf axils, of 2 kinds: showy spring flowers and inconspicuous later flowers. The showy flowers are irregular, bearing 5 sepals (auricled) and 5 white or purple petals with vein markings. The 2 lateral petals are often bearded and the lowest has a basal spur. There are 5 stamens, the lowest 2 with appendages bearing basal nectaries that extend into the petal-spur. There are 3 carpels, joined, that form a 3-valved capsule.

1. Flowers purple 2
   Flowers white . . . . *V. canadensis*
2. Slender root stock, early flowering; dry habitats . . . . *V. adunca*
   Thick rhizome, later flowering; wet habitats. . . . . *V. nephrophylla*

Viola adunca M.E. Smith

EARLY BLUE VIOLET

Early blue violet is an attractive spring plant with blue flowers and rounded, oval-shaped leaves; they have toothed stipules and long petioles (leaf stalks). The flower-structure is interesting. There are 5 sepals and 5 petals, including 2 lateral petals and a petal with a long basal spur, either straight or slightly hooked. There are 5 stamens; the 2 lower ones have appendages bearing nectar that extend into the spur. There are 3 joined carpels, which turn into a capsule, splitting into 3 parts to release the seeds. Early blue violet also produces small flowers which never open, called cleistogamous flowers; these are self-fertilized and act as a kind of insurance in case the showy flowers do not produce seed. Early blue violet is found in prairie grassland and open woods.

NOTES: *The specific name* adunca *means "hooked" and probably refers to the prominent spur.*

Viola canadensis L. var. *rugulosa* Greene

WESTERN CANADA VIOLET

Western Canada violet can easily be distinguished from early blue violet because the flowers are usually white, although the petals have yellow bases. (Colour may vary from white to violet.) Western Canada violet is a much larger plant, and its leaves are different: they are definitely heart-shaped and often end in a sharp point; they do not have toothed stipules. Western Canada violet also produces cleistogamous flowers. It is found in damp meadows and moist woods.

Viola nephrophylla Greene

BOG VIOLET

Bog violet grows from a stout, creeping rootstock. The plant is acaulescent (stemless), and the leaves grow from the rootstock. It has no stolons. The leaves are heart-shaped and very similar to those of *V. adunca,* but are larger. The flowers are large, up to 3 cm, and each flower has the usual 5 sepals, 5 petals and 5 stamens. The petals are purple and irregular; the lowest 3 are bearded at the base and the lowest one is spurred. The fruit is a 3-valved capsule. Bog violet also has cleistogamous flowers. It is found in moist meadows, on stream banks and moist open woods.

NOTES: *The species name* nephrophylla *is a misnomer: it means "kidney-leaf" but the leaves are more heart-shaped than kidney-shaped.*

Elaeagnaceae OLEASTER FAMILY

These are shrubs or small trees, with branches and twigs covered with silvery or brown scales; the leaves are either alternate or paired, and they are often covered with scales. The flowers arise in the leaf-axils, and there are 3 types; there are staminate flowers, with a cup-shaped hypanthium bearing 4 sepals, no petals, 4 or 8 stamens, and no ovary. There are carpellary flowers, with a tubular hypanthium, constricted above the ovary, bearing 4 sepals, no petals, no stamens and an ovary. The third type of flower is perfect, with a tubular hypanthium, bearing 4 sepals, no petals, 4 to 8 stamens and an ovary in the centre. The ovary is simple; it consists of 1 carpel with a single ovule, and develops into a 1-seeded, drupe-like berry.

Leaves alternate, silver-coloured .... *Elaeagnus*
Leaves opposite, green.. *Shepherdia*

Elaeagnus commutata Bernh. ex. Rydberg

**WOLF WILLOW,
SILVER-BERRY**

Wolf willow is a common, well-known prairie shrub. It grows up to 4 m in height and often forms large colonies. The leaves are characteristic of the shrub: silvery in colour, with a wavy outline, 2 to 10 cm long. They grow alternately up the twigs and bear pairs of flowers in their axils. These flowers have no petals. The sepals are yellow inside and silvery outside, tubular in shape, 4-parted; they give off a heavy, spicy scent. There are 8 stamens, and the ovary in the centre produces a silvery fruit. Wolf willow is found in valleys and on shores, as well as on prairie grassland.

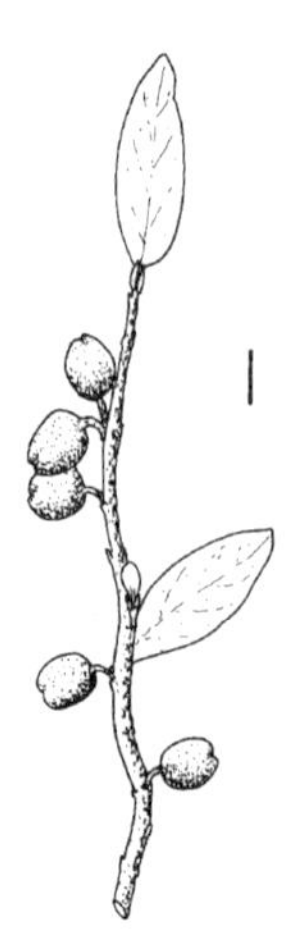

NOTES: *The common name "wolf willow" is a misnomer: the plant is definitely* not *a willow and appears to have no connection with wolves. "Silver-berry" is more appropriate because the dry, mealy, drupe-like fruit is silver in colour.* ~ *Inside the fruit is a stony "seed" with 8 ridges. Native women used to dry these "seeds" and make them into necklaces and as decoration on antelope skin dresses.*

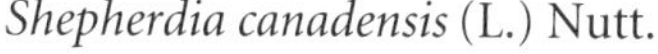

Shepherdia canadensis (L.) Nutt.

~ **CANADIAN BUFFALOBERRY**

Canadian buffaloberry is another well-known shrub. It is a bushy shrub with scurfy branches, 1 to 3 m tall; the leaves are usually broadly elliptic and have the characteristic brown scurfy scales on the undersurface. The flowers are small (4 mm wide) and may be either male or female, that is, lacking either stamens or carpels. The shrubs are dioecious: they either bear all male or all female flowers. These flowers appear in early spring before the leaves; the female flowers have 4 sepals and a single ovary (with conspicuous stigma) which turns into a red or yellow berry. The male flowers have 4 sepals and 8 stamens. The male and female flowers grow on twigs of the previous season. Canadian buffaloberry is found on prairie grassland, open woods and shores.

NOTES: *Canadian buffaloberry berries are a favorite food of black bears.*

Onagraceae EVENING PRIMROSE FAMILY

These are herbaceous plants, annual or perennial, with simple leaves that are either alternate or paired. The flower parts are arranged in 4s (except in *Circaea*, where they are arranged in 2s); this is unusual in dicotyledonous flowers. There are 4 sepals, 4 petals and usually 8 stamens in each flower. The gynoecium, of 4 united carpels, is enclosed in the hypanthium, so it is inferior. The other flower-parts are inserted on the hypanthium above the carpels (the gynoecium), so the flower is epigynous. The fruit is usually a capsule; in *Gaura*, it is 1-seeded and nut-like.

Flowers yellow . . . . *Oenothera*
Flowers pink, purple or white . . . . *Epilobium*

Epilobium L. **WILLOWHERB**

Willowherbs are perennials. The leaves are simple, ovate, lanceolate or linear; flowers grow in racemes or solitary in leaf axils. There are 4 sepals, 4 purple, white or pink petals, 8 stamens and 4 joined carpels (gynoecium is markedly inferior); the fruit is a long, 4-valved capsule with numerous tufted seeds.

1. Leaves alternate 2
   Leaves opposite or whorled 3
2. Flowers 1 to 7; petals 13 to 32 mm long .... *E. latifolium*
   Flowers numerous; petals 10 to 20 mm long .... *E. angustifolium*
3. Pods 2 to 4 cm long; alpine slopes .... *E. clavatum*
   Pods 3 to 9 cm long; bogs and marshes .... *E. palustre*

Epilobium angustifolium L.

GREAT WILLOWHERB, COMMON FIREWEED

Fireweed is a tall plant (up to 2 m high). It grows from a creeping rootstock and forms large colonies. The leaves are 1.5 to 20 cm long and lanceolate. The flowers grow in a tall raceme; they vary in colour from deep pink to purple, and the flower parts are in 4s. Fireweed flowers have 4 sepals, 4 petals, 8 stamens and 4 carpels; there are 4 styles, joined, with a 4-cleft stigma. The 4 carpels turn into a long, thin purple capsule, 4 to 10 cm long, which splits into 4 parts and releases numerous seeds. Fireweed often grows in areas devastated by fires—it is one of the first plants to colonize the ground.

NOTES: *The seeds have a tuft of silky hairs at the upper end, called a* coma, *which helps in wind dispersal. Many fruits are plumed, but plumed seeds are not so common.* *The species name* angustifolium *means "narrow-leaved."* *The common name "willowherb" is appropriate because the leaves resemble the leaves of the willow.*

Epilobium clavatum Trel.

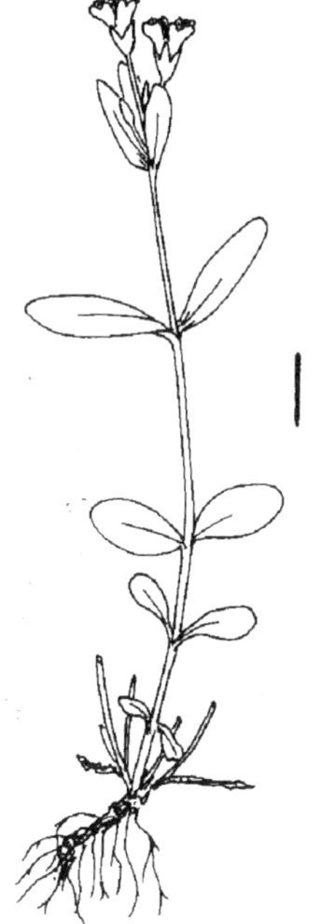

WILLOWHERB

(E. alpinum L.*)*

This small, perennial willowherb is only 0.5 to 2 dm high and spreads by suckers, sometimes called *soboles*; the plants are often matted. The leaves are usually in pairs (until the upper part of the stem is reached) and are lance-shaped to nearly elliptical; the blades are 1.2 to 2.8 cm long. The flowers grow in the axils of the upper leaves; each flower has 4 sepals, 4 petals, varying from white to pink to lilac, and 8 stamens. There are 4 united carpels in the centre of the flower; these are inferior, long and narrow and are sometimes mistaken for the flower stalk. The carpels develop into a long capsule which splits into 4 parts. As these 4 parts unroll, they release numerous plumed seeds which are dispersed by the wind. The fruit is 2 to 7 cm long, narrow or club-shaped (clavate), giving rise to the specific name, *clavatum*. Willowherb is found in moist to boggy places on alpine slopes, up to 2300 m.

NOTES: *Willowherb can be distinguished from marsh willowherb* (E. palustre) *because the former spreads by suckers, not by stolons and turions.*

Epilobium latifolium L.

RIVER BEAUTY

River beauty can be distinguished from fireweed because it is a smaller plant (0.4 to 7.0 dm tall) and its leaves are smaller and broader; the species name *latifolium* means "wide-leaved." The flower arrangement is also different: the raceme is much shorter, with fewer flowers (1 to 7, but occasionally 15). The common name river beauty is fitting because the plant has striking rose-purple flowers and often grows on gravelly river-bars. The plant is found on stream banks, gravel-bars and scree slopes to alpine elevations.

NOTES: *This species has been found at timberline in Marmot Basin, at 7300 ft (2225 m) at the Highwood Pass and elsewhere.* ~ *River beauty can be easily distinguished from marsh willowherb* (E. palustre): *it lacks stolons and turions.*

Epilobium palustre L.

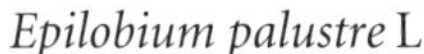

MARSH WILLOWHERB

Marsh willowherb is a small perennial, 1 to 4 dm high, which produces, from its base, thread-like stolons bearing *turions.* Turions are small, dark, bulb-like offsets (like scaly winter buds) used for asexual reproduction. The plant may be finely strigillose (covered with appressed hairs) throughout or only sparsely. Each flower is long-stalked and has 4 sepals, 4 white, occasionally pink, petals and 8 stamens. There are 4 united carpels in the centre forming a 4-valved fruit that bursts open to release numerous fluffy seeds. Marsh willowherb is found in bogs and marshy ground.

NOTES: *The specific name* palustris *means "of the marshes."* ~ *Marsh willowherb can be easily distinguished from fireweed* (E. angustifolium), *because it is a smaller plant, it does not produce racemes of deep pink flowers and its leaves are smaller and narrower. Fireweed does not have filiform stolons and turions.*

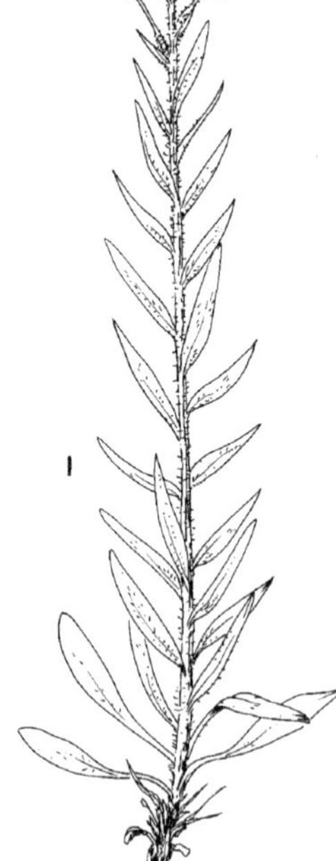

Oenothera biennis L.

YELLOW EVENING PRIMROSE

Yellow evening primrose is a typical biennial plant, as its specific name, *biennis,* implies. During its first year, a tight rosette of leaves, often pinkish, grows from a stout taproot. These leaves overwinter, and the following summer tall, leafy flowering-stalks appear, 5 to 15 dm high, hairy and often pinkish, sometimes with a few basal branches. The leaves are lanceolate, and the lower leaves are up to 15 cm by 3 cm, including the leaf-stalk. The flowering stalks are crowned with showy yellow flowers and the flower-buds often open in the evening; these flowers are up to 6 cm long. Each flower grows in the axil of a bract and has all its parts in multiples of 4—there are 4 bent-back sepals, 4 petals, 8 stamens and 4 carpels. The carpels are below the sepals, petals and stamens (inferior) and turn into a large (3.5 cm) 4-valved capsule, which has the remains of the flower-parts at its tip and opens to release numerous seeds. Yellow evening primrose is found in disturbed places, such as roadsides, and in dry, open areas.

Haloragaceae WATER-MILFOIL FAMILY

Members of this family are aquatic perennials with delicate, feathery, submerged leaves on slender stems, in whorls of 3 or 4. The flowers are sessile (lacking stalks) in the axils of tiny bracts. They are whorled, in an interrupted spike which grows just above the leaves on the leafy stem. These flowers may be perfect or unisexual: there are 4 sepals, 4 petals (falling early), 8 stamens and an inferior ovary. The fruit is like a nut and splits into 4 pieces, each with 1 seed.

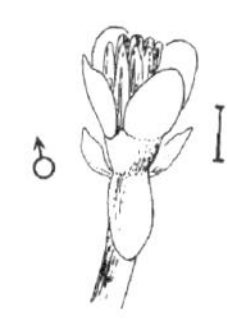

Myriophyllum exalbescens Fern.

SPIKED WATER-MILFOIL

Water-milfoil is an aquatic plant with several slender stems, usually submerged. The leaves grow in whorls of 4 and are finely dissected into hair-like filaments, giving a feathery appearance. The main flower-stalks appear above the water-level, and the flowers grow in an interrupted spike, in the axils of bracts; these flowers may be perfect or uni-sexual. If they are uni-sexual, the male (staminate) ones grow above the female (carpellary) ones. Each perfect flower has 4 sepals and 4 petals, 2.5 mm long; the petals are deciduous, falling quickly. There are 8 stamens. The ovary in the centre has 2 to 4 feathery stigmas; it develops into a nut-like fruit, splitting into 4 single-seeded parts. Water-milfoil is found in quiet waters of sloughs and lakes, and also in streams.

NOTES: *The generic name* Myriophyllum *comes from two Greek words:* myrius, *"countless" and* phyllos, *"a leaf," referring to the feathery leaves.*

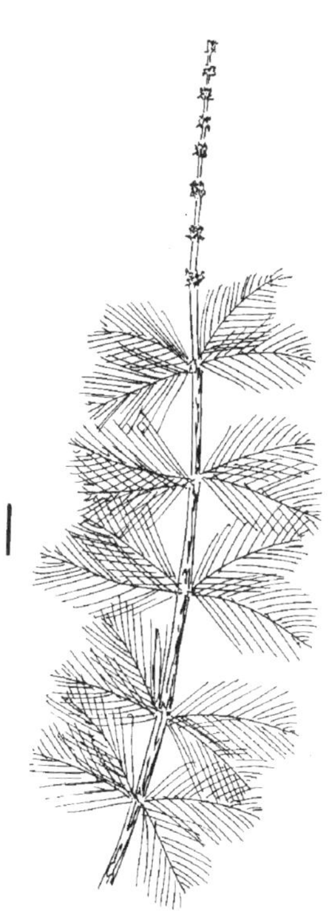

Hippuridaceae MARE'S-TAIL FAMILY

Mare's-tails are aquatic perennials, with linear, whorled leaves, 1 to 3 cm long. The unbranched stems are up to 6 dm long and arise from a creeping rhizome. The flowers are green and sessile, borne in the axils of the leaves. Each flower has a calyx but no petals; there is 1 stamen with a large, 2-loculed anther and an inferior ovary which develops into a nut-like fruit.

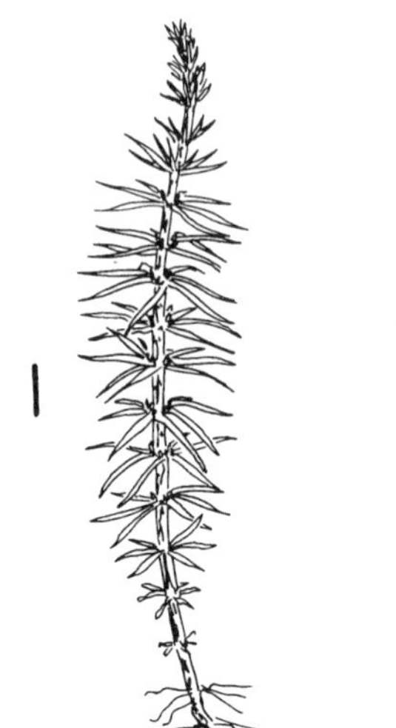

Hippuris vulgaris L.

COMMON MARE'S-TAIL

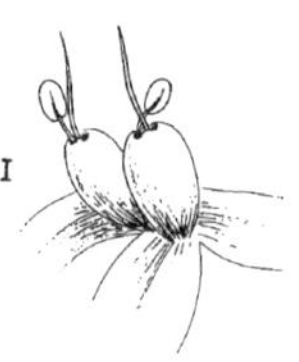

Mare's-tail is an aquatic perennial plant with a creeping rhizome that produces several unbranched stems up to 6 dm long with leaves in whorls. These leaves are quite different from those of water-milfoil, with which mare's-tail grows, because they are linear and definitely not feathery; they are usually 6 to 12 per whorl and vary from 10 to 50 mm long and 1 to 2 mm wide. The tiny flowers are found in the axils of the upper leaves, above water-level; they are greenish. Each flower has an entire calyx (not cleft), usually no petals and only 1 stamen; the gynoecium in the centre consists of 1 carpel containing 1 ovule. The carpel turns into an indehiscent (not opening) nut-like fruit, with 1 seed. Mare's-tail is found in shallow pools, on muddy shores and in lakes and streams.

Araliaceae GINSENG FAMILY

Members of the ginseng family may be shrubs, woody perennials or herbaceous plants, with simple or compound leaves and bisexual or unisexual flowers, arranged in an umbellate inflorescence. (This family is closely related to the carrot family, formerly Umbelliferae.) Each flower has 5 petals (the 5 sepals are usually minute or absent), 5 stamens and an inferior ovary, which develops into a berry-like drupe.

Aralia nudicaulis L.

WILD SARSAPARILLA

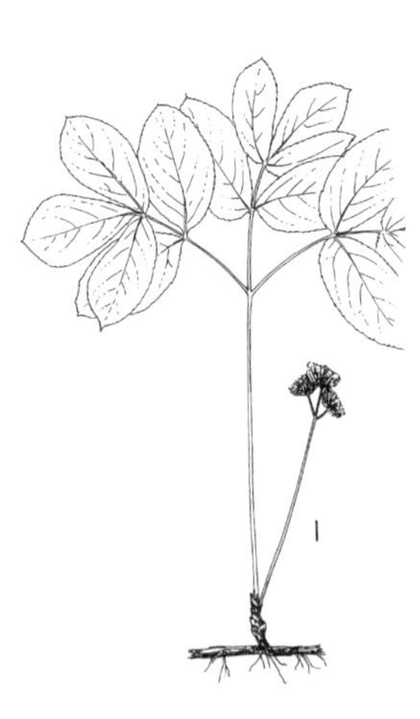

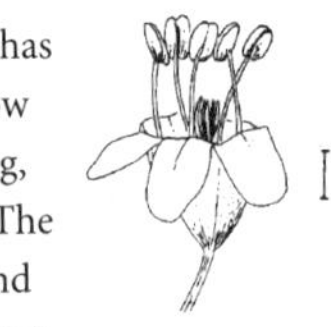

Wild sarsaparilla is an easy plant to identify. It has one large leaf, 6 to 8 dm long, with a stalk; below the leaf, a smaller flowering stalk, 2 to 3 cm long, grows, usually with 3 umbels of small flowers. The leaf is biternate, that is, split into 3 divisions, and each division is further split, pinnately, into 3 to 5 leaflets. The leaf arises from a long, creeping rhizome and a short caudex. The small, greenish flowers of the umbels each have a calyx of minute sepals, often absent, 5 petals, 5 stamens and 5 united carpels—these are inferior and covered by an epigynous disc, from which the petals and stamens arise. The fruit is a globose, fleshy, berry-like drupe, which is green but turns purplish black. These fruits are very similar to those of ivy, which belongs to the same family, Araliaceae. Wild sarsaparilla is found in moist woodland.

NOTES: *Wild sarsaparilla was once used as a substitute for sarsaparilla, which is extracted from the roots of the tropical climber* Smilax, *and is popular as a cooling summer drink.* ~ *Fernald states that the generic name* Aralia *is the Latinized form of* Aralie, *the Quebec habitant word for this plant.* ~ *The specific name* nudicaulis *means "naked stem" and refers to the flowering stalk.*

Apiaceae (Umbelliferae) CARROT FAMILY

The former name of this family, Umbelliferae, was appropriate because it means "bearing umbels"; in an umbel, the flowers are arranged in a flower-cluster with all the flower-stalks arising from the same point, like an umbrella. The umbels may be simple or compound. There is usually a ring of bracts underneath the umbel, called an *involucel.* The flowers are usually small, and the calyx-teeth are small or absent; there are 5 petals (the tips are often bent inwards) and 5 stamens. The gynoecium is inferior, that is, enclosed by the hypanthium, which forms a disc; the stamens are inserted on this disc. The gynoecium consists of 2 united carpels with 2 styles and forms a *schizocarp* or split-fruit (another important characteristic of this family). The schizocarp splits into 2 single-seeded fruits called *mericarps.*

1. Basal leaves simple, upper leaves compound . . . . *Zizia*
   All leaves compound 2
2. Leaves palmate . . . . *Sanicula*
   Leaves pinnate. 3
3. Plants lacking stems 4
   Plants with well-developed stems 5
4. Fruits more or less round in cross-section . . . . *Musineon*
   Fruits flattened in cross-section . . . . *Lomatium*
5. Leaves trifolioate . . . . *Heracleum*
   Leaves not as above 6
6. Umbels with 8 to 20 rays; flowers white to pink . . . . *Perideridia*
   Umbels with 2 to 5 rays; flowers white to purple . . . . *Osmorhiza*

Heracleum lanatum Michx.

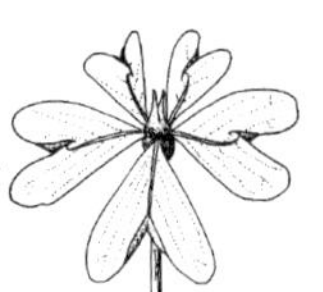

COW PARSNIP

Cow parsnip is a leafy plant 1 to 2 m tall and is found in damp places, such as moist woods, stream banks and ditches. The large leaves are characteristic: the leaflets are 1 to 3 dm broad and the petioles (leaf stalks) have dilated sheaths (characteristic of Apiaceae). The flowers are arranged in an umbel, and the bracts below are lance-shaped (lanceolate). The umbel is 1 to 2 dm wide, and the 5 petals of the flowers are white; the calyx is usually lacking, and there are 5 stamens. The gynoecium is inferior, that is, below the petals and stamens; it is covered by a disc on which the stamens are inserted. The fruit is a schizocarp, a fruit that splits into 2 fruits (mericarps) when ripe. Each mericarp contains a seed and has narrow ribs and dark oil-tubes.

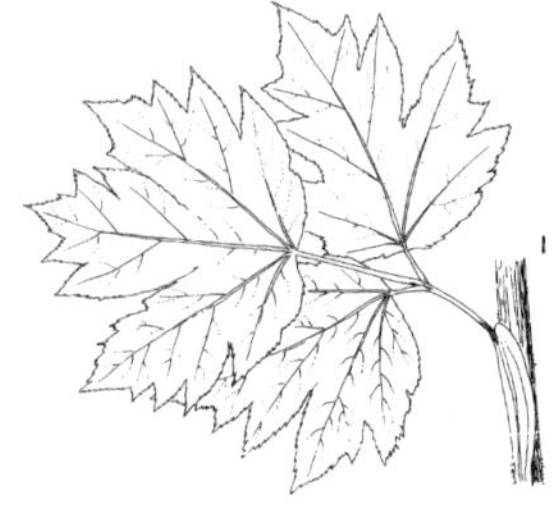

NOTES: *Cow parsnip is named for the god Hercules (Herackles was the Greek name). The species name* lanatum *means "hairy"; the plant is conspicuously hairy when young. The spring stalks of cow parsnip were eaten by the Blackfoot after being roasted on hot coals—they were peeled first. The Blackfoot also used a cow parsnip stalk in the Sundance ceremony. The roots were used by several tribes to treat arthritis and rheumatism; they were also made into a poultice and applied to boils and bruises.*

Lomatium Raf. PRAIRIE PARSLEY

This genus features caulescent perennials, some with tall stems. The leaves are usually ternately compound; the flowers grow in compound umbels and may be yellow or purplish. The fruit is a schizocarp, splitting into 2 mericarps.

Leaves divided into numerous segments; fruit 10 to 12 mm long .... *L. dissectum*
Leaves divided into a few segments; fruit 3 to 6 mm long ..... *L. triternatum*

Lomatium dissectum
(Nutt.) Mathias and Constance

PRAIRIE PARSLEY, MOUNTAIN WILD PARSLEY

Prairie parsley is a tall plant, 3 to 12 dm in height; its large leaves, 1 to 3 dm broad, are deeply dissected, hence the species name. There are usually 3 main leaflets, each with several divisions, all toothed. The leaf-stalks (petioles) are sheathing at the base. The compound umbel has yellow or purple flowers which give rise to the schizocarp fruits. The plant grows on rocky slopes and open montane woods.

NOTES: *In Kananaskis Country this species has been found in the Sheep River Wildlife Sanctuary, with purple flowers, at Marmot Creek in a meadow, at 6850' (2090 m), also with purple flowers, and elsewhere. The roots of this plant were used by the Blackfoot to make a hot drink that was taken as a tonic.*

Lomatium triternatum
(Pursh) Coulter and Rose

PRAIRIE PARSLEY, WESTERN WILD PARSLEY

The prairie parsleys are easy to distinguish, because the leaves of *L.triternatum* are entirely different from those of *L. dissectum*: they are divided into leaflets which are divided again, then again, but the leaflets are all narrow (linear) and entire, not toothed. The leafstalks are sheathing, an important characteristic of this family. The flowers of *L. triternatum* are yellow and the umbels are only 2 to 5 cm in diameter. The fruits have yellow marginal wings and brown stripes. The plant grows in prairie grassland and has been found in Kananaskis Country near the Highwood River on a grassy slope, elevation 1500 m, and in other places.

NOTES: *The roots of this plants were eaten by the Blackfoot, either raw or roasted. The species name* triternatum *refers to the triternate leaves.*

Musineon divaricatum (Pursh) Nutt.

PRAIRIE PARSLEY, LEAFY MUSINEON

Musineon is yet another prairie parsley and can be separated from the others because its growth-habit is different. It is often acaulescent, close to the ground, with the leaves and flowering stems arising from a thick tap-root. The leaves are also different because they have 7 to 11 leaflets which are not divided again. The plant is usually only 1 to 2 dm high. The flowers are yellow and the fruits are flattened and ribbed; they lack the flat marginal wings of the *Lomatium* species. The plant grows on dry prairies and craggy hillsides.

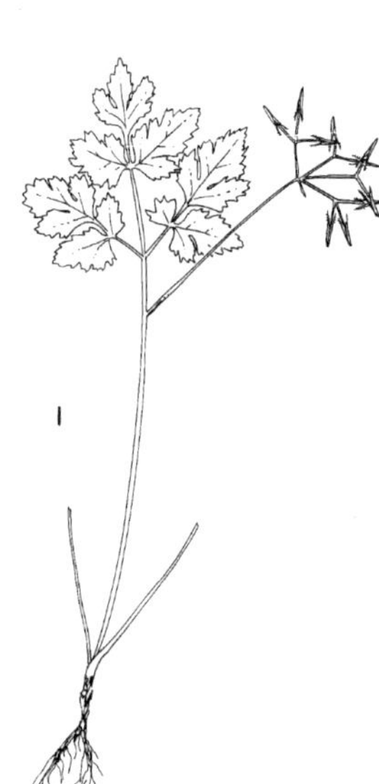

Osmorhiza depauperata Philippi

SPREADING SWEET CICELY

Spreading sweet cicely is a perennial, aromatic plant, 2 to 6 dm high, growing from a thick root-stock. The stems of sweet cicely bear large, hairy leaves at the base, with long leaf-stalks; these leaves are biternate, that is, they are divided into 3 leaflets, and each leaflet is divided into 3 again. The leaflets are toothed. The small flowers are greenish white and arranged in compound umbels, with 3 to 5 rays which spread widely in the fruiting stage. Each flower has 5 minute sepals, often absent, 5 petals, 5 stamens and 2 carpels joined together. These carpels are inferior (below the petals and stamens) and bear 2 styles with expanded bases. The fruits of sweet cicely are very characteristic and help to identify the genus: they are thin, linear-oblong, 10 to 20 mm long, clavate (club-shaped) and covered with bristles. These fruits are schizocarps. Sweet cicely is found in moist woods.

NOTES: *When this rootstock is crushed it gives off a sweet, licorice-like odour, hence the name "sweet cicely." The word* cicely *is a corruption of* Seseli, *an ancient Greek name for an aromatic plant.* ~ *The generic name,* Osmorhiza *comes from 2 Greek words:* osme *(scent) and* rhiza *(root).*

Perideridia gairdneri (Hook. & Arn.) Mathias

SQUAWROOT, YAMPA

Squawroot is a perennial plant with tuberous roots and delicate stems up to 6 dm tall. There are few leaves, but the leaves are quite characteristic because they have 3 to 7 pinnate leaflets; these are very slender, strap-shaped, up to 6 cm long. The upper leaves are much reduced. The small, creamy-white flowers are arranged in a compound umbel; the terminal umbel is the largest, up to 6 cm wide. The flowers are very small, and each has the typical family pattern, with 5 calyx-teeth, 5 petals, 5 stamens and 2 united carpels. The carpels are inferior and split into 2 separate fruits when ripe. Squawroot is found in meadows and in woodlands.

NOTES: *The native name for this plant is yampa. The common name "squawroot" was given because native women gathered the tuberous roots.* ~ *The roots can be eaten fresh, cooked or dried; they were also ground into flour. Lewis and Clark mention the roots frequently in their journals.*

Sanicula marilandica L.

SNAKEROOT

Snakeroot is a glabrous (hairless) perennial plant, with stems up to 6 dm high, growing from a cluster of strong fibrous roots. The basal leaves have long stalks, and the blades are palmately cut, with 5 to 7 divisions, broadly lance-shaped and toothed. The stem (cauline) leaves are few, reduced upwards. The small, yellowish green flowers are arranged in compound umbels. At first the umbellets are about 1 cm wide and are often hidden by the upper leaves. But in the fruiting stage the main stalk and the rays lengthen, and the umbels are wider and well above the leaves. Each flower has 5 narrow sepals (persistent), 5 petals, 5 stamens and 2 united carpels, which form the fruit, a schizocarp. The fruits are 5 mm long and densely covered with stout, hooked bristles. Snakeroot is found in moist meadow and woodland.

NOTES: *The species name* marilandica *indicates this plant was first found in Maryland.*

Zizia aptera (A. Gray) Fern.

HEART-LEAVED ALEXANDERS, MEADOW PARSNIP

Heart-leaved alexanders is an erect plant usually about 4 dm high. The common name "heart-leaved alexanders" refers to the heart-shaped basal leaves, although the stem leaves are often divided into 3 leaflets. All the lower leaves have leaf-stalks clasping the stem, but higher up the stem the leaves are sessile. The flowers are bright yellow and the umbels are 3 to 3.5 cm wide at flowering time. The fruits are 2 to 4 mm long, greenish brown and ribbed. The plant grows in moist meadows and on dry, exposed slopes.

Cornaceae DOGWOOD FAMILY

Members of this family are either shrubs or herbaceous perennials with entire, ovate leaves. The leaves are usually paired, but in bunchberry they appear to grow in a whorl at the top of the stem. The flowers are small and grow in a cymose arrangement, as in red osier dogwood, or in a dense cluster surrounded by an involucre of petalloid bracts, as in bunchberry. The flowers are epigynous, that is, the carpels (gynoecium) are enclosed by the hypanthium, so that the sepals, petals and stamens are borne above them. There are 4 tiny sepals, 4 petals, 4 stamens and usually 2 carpels, forming a 2-loculed ovary; this turns into a drupe with a seed enclosed in a stony endocarp, like a plum.

Shrub with red bark; leaves opposite .... *C. stolonifera*
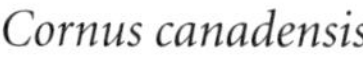
Herb; leaves appearing whorled .... *C. canadensis*

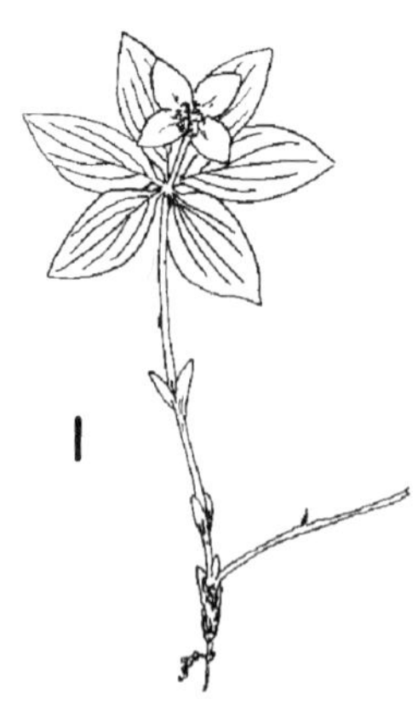

Cornus canadensis L.

BUNCHBERRY, DWARF DOGWOOD

This plant is only 8 to 10 cm high. The upper leaves grow in a group (whorl) of 4 to 6, below the flower-head. The tiny true flowers in the centre are surrounded by 4 large white bracts, often mistaken for petals. The true flowers have 4 minute sepals, 4 small, white petals, 4 stamens and an inferior ovary. The ovaries develop into bright red drupes. Bunchberry is found in moist woods and often spreads like a carpet from creeping rootstocks.

NOTES: *The common name "bunchberry" arises from the distinctive red fruits that appear in the fall.*

Cornus stolonifera Michx.

RED OSIER DOGWOOD, KINNIKINNICK

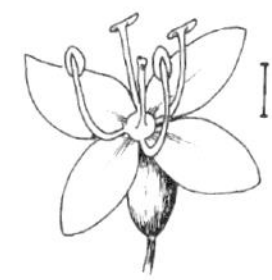

There is never any difficulty in distinguishing between the *Cornus* species. Red osier dogwood is a shrub 1 to 3 m high. Its flowers are not subtended by a ring of white bracts, like those of bunchberry, but grow in a flat-topped cyme. The fruits are white drupes with a bluish tinge. Red osier dogwood grows in moist woods and on river banks.

NOTES: *The species name* stolonifera—*"bearing stolons"—is a misnomer because the shrub does not possess stolons. But the lower branches are prostrate and often produce roots where they touch the ground, so they look rather like stolons.* ~ *This shrub has various common names. It is called "dogwood" because its wood was used to make "dags," skewers or wedges. The name "red osier" arises because the bark is red, and its tough, smooth canes, like those of osier willows, were used for making baskets.* ~ *The other common name,* kinnikinnick, *is an Algonquian word meaning "that which is mixed." The bark was shredded and mixed with tobacco.* ~ *The inner bark was made into a tea and taken for various complaints.* ~ *Common bearberry* (Arctostaphylos uva-ursi) *has been given the same name, kinnikinnick.*

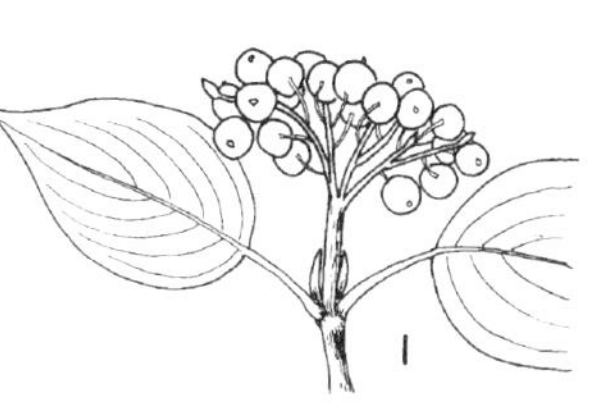

Pyrolaceae WINTERGREEN FAMILY

Members of this family are evergreen perennials, hence the family name. The dark green, shiny leaves often grow in a basal cluster, with a flower-stalk arising from the centre. The flowers may be arranged in a raceme or corymb, or they may be solitary. Each flower has a 4 or 5-parted calyx, 4 or 5 petals, and 8 or 10 stamens. The stamens are interesting, because the filaments are somewhat swollen at the base, and the anthers are inverted at flowering time; the pollen escapes through tubular basal pores. The gynoecium is superior, formed from 4 or 5 united carpels. The styles and stigmas are quite conspicuous; the style is often curved, and the stigmas are more or less 5-lobed. The fruit is a globose loculicidal capsule which bursts open to release numerous tiny seeds.

1. Stems leafy . . . . *Chimaphila*
   Stems not leafy 2
2. Flowers 1 per stem . . . . *Moneses*
   Flowers 2 or more per stem 3
3. Flowers borne on 1 side of the stem . . . . *Orthilia*
   Flowers borne around the stem . . . . *Pyrola*

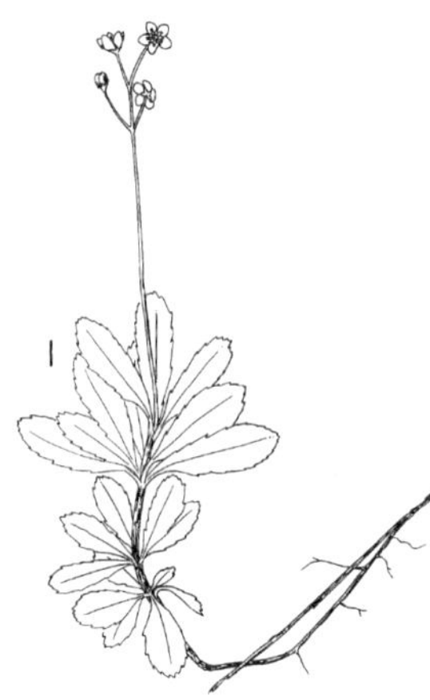

Chimaphila umbellata (L.) Bart.

PRINCE'S PINE, PIPSISSEWA

Prince's pine is an evergreen with woody stems and few branches. It grows 1 to 3 dm high, and the leaves are crowded in clusters up the stem. They are dark and shiny on the upper surface and are finely toothed (serrulate). The flowers are pinkish or rose-pink and are sometimes fragrant. The calyx has 4 or 5 sepals joined together; there are 5 petals and 8 or 10 stamens; the anthers open by basal pores. The fruit is a capsule, splitting from the top to release the seeds. Prince's pine is found in dry woods, often in coniferous woods, at middle elevations.

NOTES: *The genus name* Chimaphila *comes from 2 Greek words:* cheima, *meaning winter, and* philem, *to love; that is, "winter-loving," referring to the evergreen leaves.* *The species name* umbellata, *suggesting flowers arranged in an umbel, is not strictly accurate: the flowers are sometimes arranged in a corymb.* *The leaves were mixed with tobacco leaves by the Blackfoot, who also made a tea used to reduce fever from the plant.*

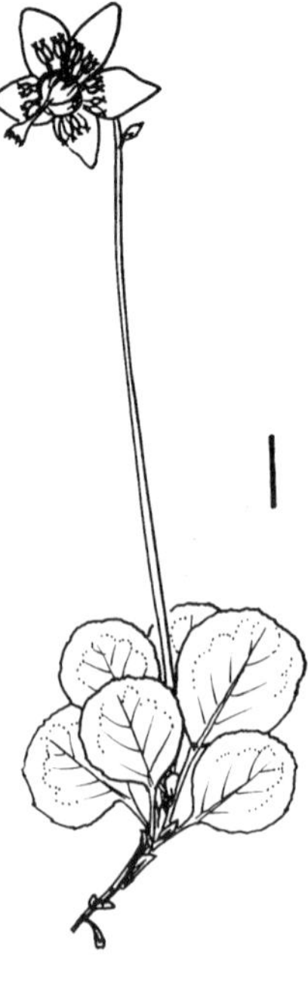

Moneses uniflora (L.) A. Gray

ONE-FLOWERED WINTERGREEN, SINGLE DELIGHT

Single delight is only 3 to 10 cm tall, and its rounded leaves are mostly basal. The slender flowering-stalk, bearing a single flower, arises from the centre of the leaf-cluster; the flower itself is nodding. There is a 5-parted calyx, and the 5 petals are waxy-white, fragrant and spreading; there are 8 to 10 stamens. The ovary has 5 locules and a distinct, large stigma which is peltate, with 5 lobes. The fruit is a capsule. The plant is found in moist woods.

NOTES: *This species has been found in Kananaskis Country east of the Spray River Dam, in pine* (Pinus contorta) *woods at 1600 m, near the Kananaskis Field Stations at 1500 m and in other places.*

Orthilia secunda (L.) House

ONE-SIDED WINTERGREEN

(Pyrola secunda L.*)*

One-sided wintergreen often forms dense colonies in coniferous woods because the plants spread with long, creeping rootstocks. The leaves are mostly basal, evergreen and broadly lance-shaped or ovate. The flowering-stalk arises from the centre of the leaves and bears numerous flowers, in a raceme, all 1-sided. Each flower has 5 sepals, 5 greenish white petals and 10 stamens. There is a disc at the base of the gynoecium with 10 small glands. The ovary has a long, straight style; the fruit is a capsule. One-sided wintergreen has been found south of Barrier Lake (Reservoir), on Kananaskis Highway 40, in moist forests dominated by pine (*Pinus contorta)* and spruce (*Picea glauca).* It has also been found beside Spray Lake, east of the dam at 1600 m, in pine forest, and near Cataract Creek, 1600 m high, in pine forest, and elsewhere.

Pyrola L. — WINTERGREEN

Wintergreen is a glabrous perennial with slender, creeping rhizomes and shiny, petiolate, basal leaves, with a slender peduncle arising from the centre, bearing a terminal, several-flowered raceme of pink-rose or greenish white flowers. The flowers have 5 sepals, partly united, 5 petals, distinct, 10 stamens, opening by 2 pores, and 5 carpels, united, with conspicuous stigmas, often 5-lobed. The fruit is a capsule.

Flowers pink . . . . *P. asarifolia*
Flowers greenish white . . . . *P. chlorantha*

Pyrola asarifolia Michx.

COMMON PINK WINTERGREEN

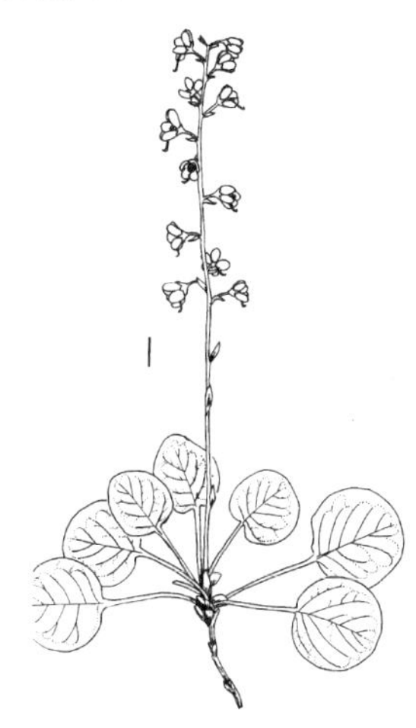

Common pink wintergreen is an attractive plant found, like so many other members of the family, in damp woods. It can be easily distinguished from the greenish-flowered wintergreen by its pink flowers and taller flowering stalk (scape), 15 to 25 cm tall. The flowers are arranged in a raceme and have 5 sepal-lobes, 5 petals, 10 stamens and an ovary with a curved style. The flowers are not 1-sided so the plant can be distinguished from one-sided wintergreen.

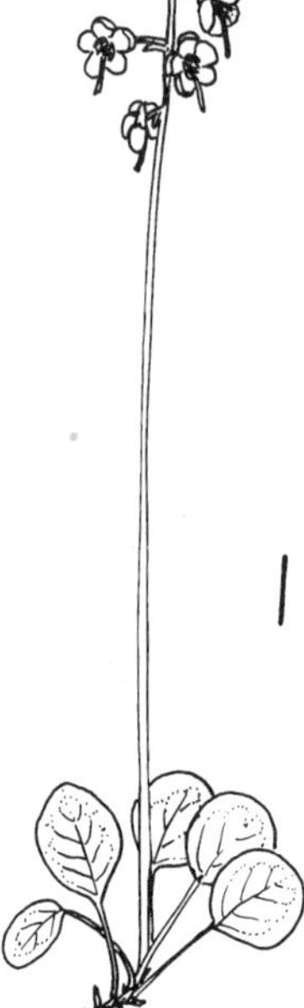

Pyrola chlorantha Sw.

GREENISH-FLOWERED WINTERGREEN

Greenish-flowered wintergreen has rounded, basal leaves in a cluster and a tall leafless flowering-stalk (8 to 20 cm) arising from the centre of the cluster. There are only a few flowers, arranged in a raceme; each flower has 4 to 5 calyx-lobes (sepals), 5 greenish petals and 10 stamens. The ovary has a prominent style and stigma, characteristic of Pyrolaceae; the fruit is a capsule. The plant is found in damp woods.

NOTES: *The species name* chlorantha *means green-flowered, from* chlorus *(green) and* anthos *(flower).*

Monotropaceae INDIAN-PIPE FAMILY

This is one of the most interesting plant families represented in Kananaskis Country because its members are saprophytes. They possess no chlorophyll and do not perform photosynthesis, but rather obtain their food from organic matter in the soil. The word *saprophyte* comes from the Greek words *sapros*, meaning "rotten," and *phytos*, meaning "plant"—they feed on dead plant material.

These plants may be brownish red or white or yellowish, but never green. The fleshy white stems arise from a mass of matted roots. The roots are mycorrhizal, that is, they are invested with a fungus, with which they have a symbiotic relationship. The leaves (with no function) are reduced to mere scales along the stem. The flowers are unusual—they are waxy-white on a waxy-white stem, 1 to 2 dm tall. Each flower has 2 to 5 sepals (occasionally absent), 4 to 5 petals, 4 to 10 stamens and 5 united carpels, which develop into a capsule.

Monotropa uniflora L.

INDIAN PIPE

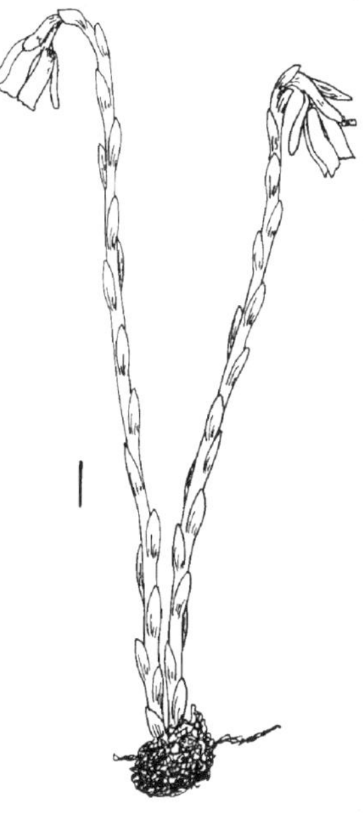

Indian pipes have a curious appearance: the whole plant is waxy-white, sometimes pinkish or yellowish, and grows 1 to 3 dm tall, often in clumps. When found in deep shade in a coniferous forest, they are quite striking. The leaves, having no photosynthetic function, have been reduced to overlapping scales that cover the stem. The flowers are solitary and nodding, 1 at the top of each stem, hence the specific name *uniflora.* Each flower has a calyx of 2 to 4 sepals, rather like bracts, 5 erect petals and 4 to 8 stamens; the anthers release pollen through 2 transverse clefts. The gynoecium in the centre consists of 4 united carpels; these turn into a capsule, which splits from the top down to release numerous minute seeds. The plant is found in deeply shaded, mossy coniferous woods.

NOTES: *Indian pipe gets its common name from its waxy-white flower which grows at right angles to the stem and is shaped like the bowl of a pipe. When dried, Indian pipe often turns black.* ~ *Indian pipes obtain their food from the humus litter of coniferous woods and have a symbiotic relationship with fungi.* ~ *This plant has a long history, and specimens have been found in rocks of the Tertiary age, when early mammals were evolving; it might almost be called "a living fossil." It has long been used in folk medicine for the treatment of nervous troubles and for sore eyes.*

Ericaceae HEATH FAMILY

Members of this family are shrubs with simple leaves, often leathery and evergreen. The flowers are attractive, each with 4 or 5 small sepals forming the calyx, sometimes united, and a corolla with 4 or 5 lobes. The corolla is often urn-shaped, as in bearberry; the flower-shape is characteristic of several members of the family. There are either 4 or 5 stamens (the same number as the corolla lobes) and the anthers face inwards; they open by terminal pores and sometimes have awns. The gynoecium in the centre consists of 2 to 5 carpels joined together, usually superior. The fruit is usually a drupe or a berry, sometimes a capsule, as in Labrador tea *(Ledum groenlandicum)* and white-flowered rhododendron *(Rhododendron albiflorum).*

NOTE: *This family includes red heather, yellow heather and white heather, but these are* not *true heathers; true heathers are species of* Erica *or* Calluna *and do not grow in Alberta.*

1. Fruit fleshy and berry-like 2
   Fruit dry 4
2. Ovary superior . . . . *Arctostaphylos*
   Ovary inferior 3
3. Corolla deeply 4-lobed, pink and curved backwards . . . . *Oxycoccus*
   Corolla cup-shaped, 4 to 5-lobed, white or pink . . . . *Vaccinium*
4. Leaves deciduous 5
   Leaves evergreen 6
5. Flowers white . . . . *Rhododendron*
   Flowers greenish yellow tinged with bronze . . . . *Menziesia*
6. Petals not united; leaf underside rust-coloured and woolly . . . . *Ledum*
   Petals united; leaf underside not rust-coloured 7
7. Leaves alternate . . . . *Phyllodoce*
   Leaves opposite 8
8. Flowers white; leaves 2 to 4 mm long . . . . *Cassiope*
   Flowers pink; leaves 10 to 15 mm long . . . . *Kalmia*

Arctostaphylos Adans. — BEARBERRY

Bearberry is a trailing, densely branched shrub, with alternate, evergreen or deciduous leaves. Its flowers grow in terminal racemes or clusters, bearing 4 to 5 sepals, a corolla of 5 white, often pink-tipped, urn-shaped (urceolate) petals and 10 stamens. The anthers bear awns and open by pores. The fruit is red, globose and berry-like.

Leaf-edges toothed; leaves turning red in autumn . . . . *A. rubra*
Leaf-edges smooth; leaves remaining green in autumn . . . . *A. uva-ursi*

Arctostaphylos rubra (Rehder & Wils.) Fern.

ALPINE BEARBERRY

Alpine bearberry is a prostrate, creeping shrub with many branches, 0.5 to 3 dm tall. Many of the branches are ascending and bear spatulate (spoon-shaped) or oval leaves with tiny teeth. These leaves turn reddish purple in the fall and then wither; the species name *rubra* (red) refers to the colour. The flowers are white and urn-shaped; they are arranged in a terminal cluster of 2 to 4 flowers. They each have a calyx of 4 to 5 sepals, joined together and deeply lobed, 5 petals, joined together, and 10 stamens. There are 2 to 5 united carpels, and when ripe, they form a juicy, red, berry-like fruit. Alpine bearberry grows on moist, montane slopes.

Arctostaphylos uva-ursi (L.) Spreng.

~ COMMON BEARBERRY, KINNIKINNICK, SANDBERRY

Bearberry is a trailing, evergreen shrub. It forms mats with prostrate, rooting branches and is valuable as ground cover for checking erosion on watersheds. The branches are covered with reddish, shreddy bark and bear dark green oval leaves. The flowers are small but attractive; they are white or pink, urn-shaped, and grow in clusters at the ends of branches. Each flower has a deeply lobed calyx, a 4 or 5-parted corolla and 10 stamens that are included, that is, they do not protrude beyond the corolla tube. The anthers of the stamens open by pores and have 2 reflexed awns. The fruit is globose, mealy and bright red, enclosing 5 nutlets. Bearberry grows on dry, sandy soils, gravel terraces and woodlands.

NOTES: *The generic name* Arctostaphylos *comes from 2 Greek words:* arctos, *a bear and* staphylos, *a berry or bunch of grapes, literally translated as "bear-berry." The specific name* uva-ursi *comes from the Latin* uva *(berry) and* ursus *(bear). So although the scientific name looks complex, it simply means "bear-berry, berry-bear."* ~ *The common name* kinnikinnick *is an Algonquian word meaning "a plant whose leaves can be mixed with tobacco leaves and smoked." The Blackfoot and other peoples used the leaves in religious smoking ceremonies.* ~ *Bearberry extract has a high content of gallic and tannic acids, which make it an astringent. It also contains arbutin, which acts as a germicide in the urinary tract (the arbutus, the strawberry tree, belongs to the same family).* ~ *The Thompson Indians of British Columbia used bearberry for kidney diseases, and the Blackfoot made its leaves into a salve for skin diseases. They also made an infusion of the leaves for sore gums.*

Cassiope tetragona (L.) D. Don

WHITE MOUNTAIN HEATHER

The leafy stem of white mountain heather is distinctive because the tiny leaves are arranged in 4 ranks, hence the species name *tetragona*, which means "four-angled." The rows of overlapping leaves are unmistakable, and the leaves have deep grooves on the undersurface. The shrub grows 1 to 3 dm high and has white, sometimes pinkish, flowers, bell-shaped and solitary, growing at intervals up the stem. There are 4 or 5 sepals, 4 or 5 petals, joined, with the tips bent back, and 8 or 10 stamens with awned anthers. The fruit is a capsule with 4 or 5 valves, which splits open to release the seeds. White mountain heather is an alpine shrub.

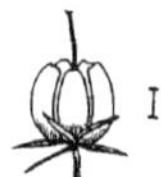

NOTES: *This plant has been found in the alpine zone on Pasque Mountain, on Plateau Mountain, near the summit of Moose Mountain and in other places.*

Kalmia microphylla (Hook.) Keller

SWAMP LAUREL, MOUNTAIN LAUREL

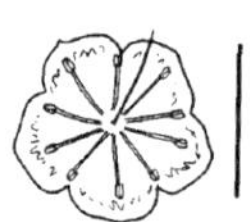

Mountain laurel is an attractive shrub 5 to 20 cm tall, with creeping, rooting stems. The evergreen leaves are 10 to 15 mm long, dark green and lance-shaped with no teeth. The large, attractive flowers are found at the tips of the branches in clusters, with reddish stalks. There are 5 green and pink sepals, 5 bright reddish pink petals, 10 stamens, which lie in small pouches on the petals, and 5 carpels, which form a capsule. The stamens have arched filaments which are under tension. When touched by an insect, they snap inwards, releasing the anthers. If the anthers are ripe, the insect gets dusted with pollen. Swamp laurel is found in bogs and moist mountain meadows.

NOTES: *The genus name* Kalmia *commemorates Peter Kalm, an enthusiastic botanical collector and pupil of Linnaeus. Kalm travelled widely in North America from 1748-1751.* ~ *The species name* microphylla *is not appropriate because it means "small-leaved."*

Ledum groenlandicum Oeder

COMMON LABRADOR TEA

Common Labrador tea is a well-known shrub, 4 to 8 dm high. Its twigs are covered with rusty hairs; the lance-shaped leaves also have a dense covering of rusty hairs on the undersurface and are strongly revolute (rolled inwards). The white flowers are arranged in showy bunches (corymbs) at the tips of the branches. Each flower has a tiny calyx of sepals with 5 white petals and 5 to 7 stamens; the fruit is a capsule. The shrub is found in cold bogs and swamps and in damp coniferous forest.

NOTES: *The leaves of this plant were used by the natives and early fur-trappers to make "tea," hence the common name. The "tea" was used by the Cree for chest colds and fevers, and by the Blackfoot as a diuretic.* *All* Ledum *species contain ledol in their leaves. In high concentrations, this substance can induce cramps and paralysis, but in weak concentrations, such as tea, it may act like caffeine and have restorative powers.*

Menziesia ferruginea J.E. Smith

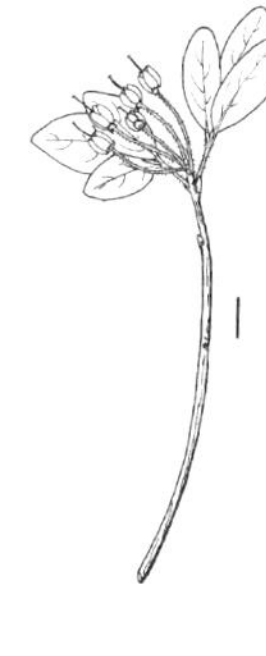

FALSE AZALEA

This shrub is found in montane woods. The leaves are deciduous and oval, and the flowers are in terminal clusters. The calyx has 4 sepals, and the petals are fused together in a bell-shape, with 4 lobes; they are greenish yellow. There are 8 stamens which lie inside the bell-shaped flower. The fruit is a capsule.

NOTES: *The specific name* ferruginea *(from* ferrus, *meaning "iron") refers to the rusty-red glandular hairs that cover the young twigs. The generic name* Menziesia *honours Archibald Menzies (1754-1842), who took the post of physician-botanist with Captain George Vancouver in his explorations along the coast of western North America at the end of the eighteenth century, and who took the original specimen of this plant back to England.*

Oxycoccus microcarpus Turcz.

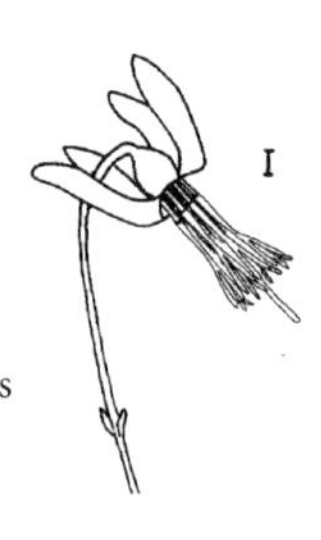

SMALL BOG CRANBERRY

Small bog cranberry is a tiny, creeping plant, rooting at the nodes, found in *Sphagnum* moss bogs. The stems are 1 to 5 dm long with alternate leaves. There are 1 to 3 attractive, terminal flowers that look like tiny shooting star flowers because the petals are reflexed and the stamens protrude; they are exserted. There are 4 sepals, 4 petals, 8 stamens and 4 carpels, joined, which turn into a reddish, globose berry.

NOTES: *The specific name* microcarpus *means a small fruit.*

Phyllodoce Salisb. MOUNTAIN HEATHER

Phyllodoce species are low, densely branched shrubs. Their densely crowded evergreen leaves have down-curled margins. The flowers bloom in terminal clusters with a calyx of 5 united sepals, a corolla of 5 bell-shaped or urn-shaped, rose or yellow petals, and an androecium of 10 stamens. The anthers open by pores; the fruit is a 5-valved globose capsule.

Flowers purple. . . . . *P. empetriformis*
Flowers yellow. . . . . *P. glanduliflora*

Phyllodoce empetriformis (Smith) D. Don

RED HEATHER, PURPLE HEATHER

Red heather is a small, heath-like shrub, only 1 to 3 dm high, with crowded leaves which are alternate and evergreen. The flowers grow in a group at the tips of the stems (terminal) and are urn-shaped, like those of *Menziesia.* There are 5 sepals, 5 petals (joined), 10 stamens and 5 carpels. These turn into a capsule with 5 valves, which open to release the seeds. Red heather is found in subalpine and alpine meadows.

NOTES: *The generic name* Phyllodoce *was the name of a sea-nymph in early Greek mythology; the specific name* empetriformis *was given because the leaves resemble those of* Empetrum nigrum, *crowberry.*

Phyllodoce glanduliflora (Hook.) Coville

YELLOW HEATHER

It is fairly easy to distinguish the red and yellow heathers. Yellow heather has honey-coloured flowers and its flower is slender, bell-shaped and constricted at the mouth. Yellow heather is glandular-hairy, hence the specific name *glanduliflora*, while the urn-shaped flower of red heather lacks glands. Yellow heather grows in sub-alpine and alpine meadows, the same habitat as red heather.

Rhododendron albiflorum Hook.

WHITE-FLOWERED RHODODENDRON

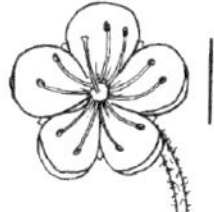

White-flowered rhododendron is a deciduous shrub, 5 to 12 dm tall; the young branches are covered with reddish hairs. The leaves turn orange and crimson in the fall. The flowers are usually axillary and saucer-shaped. The calyx is 5-parted, and there are 5 white petals, hence the species name *albiflorum* ("white-flowered") There are 10 stamens, with filaments that are all hairy at the base. The fruit is a woody capsule. White-flowered rhododendron is found in damp montane and subalpine forests.

NOTES: *The generic name* Rhododendron *comes from 2 Greek words:* rhodon *means "rose" and* dendron *means "tree."* *This plant rarely grows in areas that have suffered forest fires.*

Vaccinium L. — BLUEBERRY

Blueberry is a profusely branched, small, deciduous shrub with numerous simple leaves. The flowers are often nodding, either solitary in leaf axils or in clusters; the calyx has 5 united sepals and 5 united petals forming an urn or a bell. The androecium has 8 to 10 stamens, sometimes awned, and the fruit is a red, blue or black berry.

1. Branches reddish brown; fruit light blue. . . . . *V. caespitosum*
   Branches green; fruit red or black. 2
2. Plant broom-like with numerous branches; fruit red. . . . . *V. scoparium*
   Plant not broom-like; fruit black . . . . *V. myrtillus*

Vaccinium caespitosum Michx.

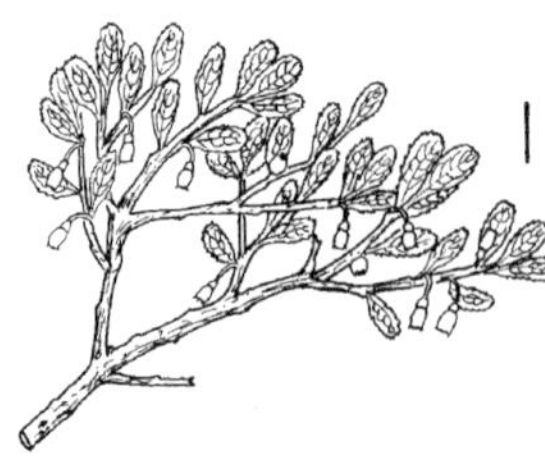

DWARF BILBERRY

Dwarf bilberry is a tiny shrub, 0.5 to 2.5 dm high, densely branched. The leaves are 1 to 3 cm long, with reticulate (branched) veins; the flowers are axillary (growing in the leaf-axils). They each have 5 joined sepals and 5 petals, usually pink, joined to form a tiny bell. There are 8 to 10 stamens. The fruit is a light blue, many-seeded berry, characteristic of the genus *Vaccinium.* Dwarf bilberry is found in woods and on open slopes and high rock-ridges.

NOTES: *The specific name* caespitosum *means "tufted." Dwarf bilberry berries are edible and were popular among the tribes of interior British Columbia. They were eaten fresh or made into cakes.* *The common names given to members of the genus* Vaccinium *are very confusing. They are called bilberries, cranberries, blueberries, and one of them,* V. membranaceum, *is even called a huckleberry.*

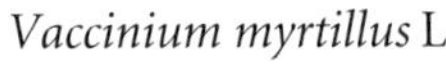

Vaccinium myrtillus L.

LOW BILBERRY

Low bilberry is only 1 to 3 dm tall with wide-spreading branches; its leaves are 1 to 2 cm long, usually oval, with tiny teeth. The flowers are axillary. There are 5 minute sepals, 5 petals joined to form a bell-shape, with 5 teeth. There are 8 to 10 stamens, and the 5 carpels are inferior, that is, below the calyx, corolla and stamens. The carpels form a edible black berry. Low bilberry is found on mountain slopes and subalpine woods.

NOTES: *This species grows in moist white spruce forest areas in Peter Lougheed Provincial Park and is found in other places in Kananaskis Country.* *It is not easy to distinguish low bilberry from dwarf bilberry because they have the same growth-habit. But there are 3 small differences. Low bilberry has leaves that are definitely serrulate (i.e., with tiny teeth), while those of dwarf bilberry are only serrulate in the upper half of the leaf. The young branches of low bilberry are reddish or brownish, while those of dwarf bilberry are greenish. The fruit of low bilberry is usually black, while that of dwarf blueberry is light blue.*

Vaccinium scoparium Leiberg

GROUSEBERRY

Grouseberry can be distinguished from the other species because its growth-habit is different: it is broom-like with erect branches, giving it a rather "straggly" appearance. The leaves are well-spaced apart; they are not so crowded as those of the other two, and the berries are red, not black or light blue. Grouseberry is found on open mountain slopes and in montane woods.

Primulaceae PRIMROSE FAMILY

Members of this family are herbaceous annuals or perennials with simple, usually entire leaves and perfect flowers. Each flower has a calyx of 4 or 5 sepals, more or less united (*gamosepalous*) and a corolla of usually 5 petals (sometimes 4 to 9). The petals are united, forming a corolla-tube which is lobed or deeply cleft; there are as many stamens as there are corolla lobes, and these are usually inserted inside the corolla tube; the anthers open inwards. In the centre of the flower is the superior gynoecium with 1 style, which turns into the fruit, a capsule opening by 2 to 6 valves or by splitting horizontally, to release many tiny seeds. These seeds arise from a "knob" in the centre of the ovary—this arrangement is called *free-central placentation* and the "knob" is a *placenta.*

1. Flowers white or whitish yellow . . . . *Androsace*
   Flowers pink or purple 2
2. Corolla-lobes reflexed . . . . *Dodecatheon*
   Corolla-lobes not reflexed . . . . *Primula*

Androsace L.

These are tufted or stoloniferous herbaceous plants bearing leaves in rosettes. The flowering stems are leafless (scapes), each terminating in a few-flowered umbel. There is a calyx of 5 united sepals, a corolla of 5 united funnel-shaped or salver-shaped petals, and an androecium of 5 stamens; the gynoecium is superior. The fruit is a capsule.

Leaf-edges hairy; petals cream-coloured . . . . *A. chamaejasme*
Leaf-edges not hairy; petals white . . . . *A. septentrionalis*

Androsace chamaejasme Host

SWEET-FLOWERED ANDROSACE, ROCK JASMINE

Sweet-flowered androsace is a welcome sight on mountain slopes in June. It is a creeping perennial only 2 to 10 cm high. The plant forms small, loose mats, and the flowering stems rise from leaf-rosettes. The flowering stems (scapes), 2 to 10 cm tall, are leafless and bear clusters of flowers at the tips, surrounded by an involucre of bracts. The flowers are white or cream-coloured, with a yellow eye; they are fragrant. The whole plant is dainty and attractive. Each flower has a 5-parted calyx (sepals), a 5-lobed corolla of petals, 5 stamens, and the fruit is a capsule. The plant grows at lower elevations, as well as on alpine slopes.

NOTES: *The specific name* chamae *means "dwarf".*

Androsace septentrionalis L.

NORTHERN FAIRY CANDELABRA

Fairy candelabra is another dainty plant but can be distinguished from sweet-flowered androsace because the flower stalks are much longer and are arranged in a conspicuous umbel, hence the candelabra of the common name. It is also taller than androsace, up to 25 cm. The flowering stems rise from a rosette of leaves, and the tiny flowers are white or pinkish. Fairy candelabra usually grows in dry meadows, on prairies and even as a weed in gardens; it is also occasionally found in higher elevations up to 2400 m.

Dodecatheon L. — SHOOTING STAR

Shooting stars are perennials, rising from a short caudex with fibrous roots. The leaves are entire and grow in a rosette. Flowering stalks are leafless, 2 to 20 cm high; there are 1 or more flowers, with small bracts beneath. Flowers are nodding when open. There is a calyx of 5 deeply cleft, united sepals, a tubular corolla of 5 united petals, reflexed from the base, exposing the 5 stamens inserted on the corolla tube, with prominent connectives, smooth or wrinkled. The fruit is a capsule.

Leaves hairy . . . . *D. conjugens*
Leaves not hairy . . . . *D. pulchellum*

Dodecatheon conjugens Greene

MOUNTAIN SHOOTING STAR

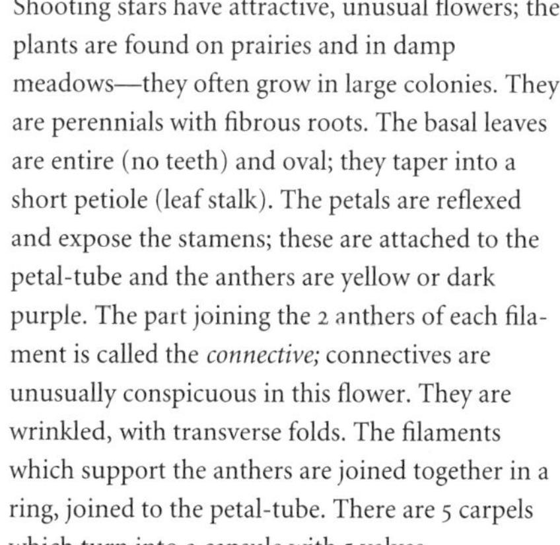

Shooting stars have attractive, unusual flowers; the plants are found on prairies and in damp meadows—they often grow in large colonies. They are perennials with fibrous roots. The basal leaves are entire (no teeth) and oval; they taper into a short petiole (leaf stalk). The petals are reflexed and expose the stamens; these are attached to the petal-tube and the anthers are yellow or dark purple. The part joining the 2 anthers of each filament is called the *connective;* connectives are unusually conspicuous in this flower. They are wrinkled, with transverse folds. The filaments which support the anthers are joined together in a ring, joined to the petal-tube. There are 5 carpels which turn into a capsule with 5 valves.

NOTES: *The generic name* Dodecatheon *comes from 2 Greek words:* dodeka, *which means "twelve," and* theos, *which means "gods." Why Linnaeus gave the genus this name is not clear.*

Dodecatheon pulchellum (Raf.) Merr.

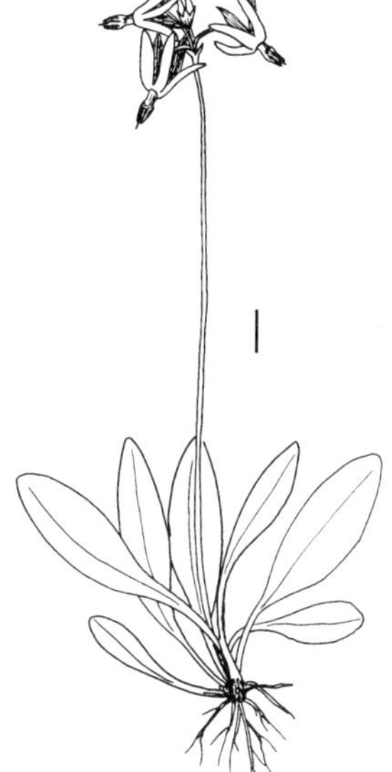

SALINE SHOOTING STAR

D. pulchellum is similar to *D. conjugens*—in fact, it is often difficult to tell them apart. There are 3 differences: *D. conjugens* has glandular-hairy leaves, while those of *D. pulchellum* are glabrous; *D. conjugens* has wrinkled connectives with transverse folds between the anthers, while *D. pulchellum* has smooth connectives which, when dry, are usually wrinkled longitudinally; the fruit capsule of *D. conjugens* has 5 teeth that are square-shaped at the tips, while *D. pulchellum* has pointed capsule teeth. The petals of both species are pink to purple, very rarely white.

Primula mistassinica Michx.

DWARF CANADIAN PRIMROSE

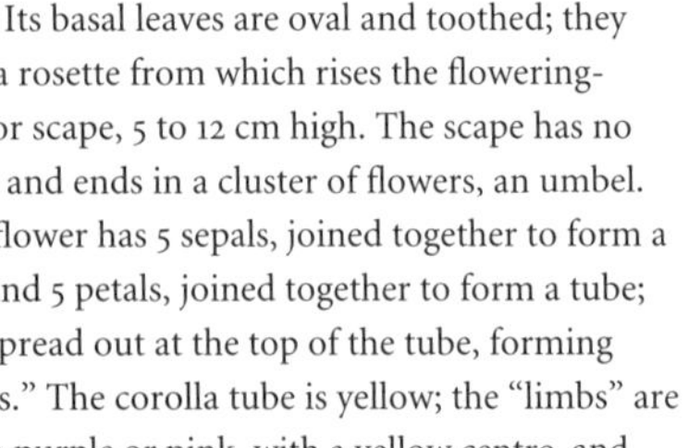

Dwarf Canadian primrose is a dainty, attractive plant. Its basal leaves are oval and toothed; they form a rosette from which rises the flowering-stem or scape, 5 to 12 cm high. The scape has no leaves and ends in a cluster of flowers, an umbel. Each flower has 5 sepals, joined together to form a cup, and 5 petals, joined together to form a tube; they spread out at the top of the tube, forming "limbs." The corolla tube is yellow; the "limbs" are bluish purple or pink, with a yellow centre, and notched. The fruit is a capsule with 5 valves which open to release the seeds. The plant is found on marshy ground and along shorelines.

NOTES: *André Michaux (1770-1855), a French botanist, discovered this primrose on the shores of Lake Mistassinica in northern Quebec, then New France. He was sent there by King Louis XVI to find plants suitable for the King's gardens at Versailles. Michaux called the plant "fairy primrose" because it is only 5 to 12 cm tall.*

Gentianaceae GENTIAN FAMILY

The members of this family are herbaceous plants (annuals, biennials or perennials) with simple, opposite leaves with no teeth and with perfect flowers, often showy. Each flower has a calyx of 4 or 5 sepals, united (gamosepalous), and 4 or 5 petals, united to form a tubular or funnelform corolla (wheel-shaped in marsh felwort, *Lomatogonium*). There are 4 or 5 stamens (as many as the corolla lobes); these alternate with the lobes and are partly joined to the inside of the corolla tube. The gynoecium in the centre of the flower consists of 2 carpels joined together; these are superior, above the other flower-parts, so the flower is hypogynous. The fruit is a capsule, splitting into 2 valves to release the seeds.

Gentians are reputedly named after King Gentius of Illyria, who is credited by Pliny the Elder, in his *Historia Naturalis* (A.D. 77), with the discovery that the yellow gentian (not found in Alberta) had several medicinal qualities. A tonic, bitters, is still made from this plant by Europeans.

1. Flowers with a spur . . . . *Halenia*
   Flowers without a spur 2
2. Corolla lobes with 3 veins . . . . *Gentiana*
   Corolla lobes with 5 to 9 veins . . . . *Gentianella*

Gentiana prostrata Haenke

MOSS GENTIAN

Moss gentian is a biennial, only 1 to 10 cm high, erect or prostrate (hence the specific name *prostrata*). It has numerous appressed leaves. The single flowers are found at the tips of the branches, and the colour varies from light blue to greenish purple. Each flower has 4 to 5 sepals joined together and 5 petals joined to form a tubular corolla; there are 5 stamens and 2 carpels. The carpels form the fruit, a capsule, usually with 2 valves. Moss gentian is found in alpine meadows and on subalpine slopes.

Gentianella Moench — FELWORT

These plants are annuals or biennials. The leaves are usually opposite, simple and sessile; the flowers are regular. There is a calyx of 4 to 5 sepals, partly united, and a corolla of 4 to 5 petals, tubular, lobed and lacking folds between the lobes. There are 4 to 5 stamens and a gynoecium of 2 carpels, united. The fruit is a capsule.

Plants of alpine slopes . . . . *G. propinqua*
Plants of lower elevations. . . . . *G. amarella*

Gentianella amarella (L.) Börner

FELWORT

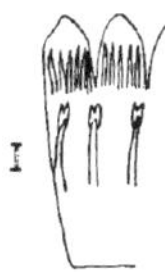

Felwort is an erect annual or biennial which may be branched or unbranched and grows 0.5 to 3 dm high. The leaves vary in shape from ovate to narrowly lanceolate, and are opposite, in pairs. The flowers grow from the leaf-axils, often in clusters; each flower has the typical structure found in this family: 5 sepals, joined, 5 petals forming a slender tubular corolla, 5 stamens and 2 carpels which turn into a capsule. The petals are fringed and light violet or pale blue. Felwort can be distinguished from moss gentian because of its growth-habit; it is an erect plant with clusters of flowers growing in the leaf-axils—quite different from the prostrate habit of moss gentain. It is found in meadows, moist woods, and in rocky habitats.

NOTES: *The specific name* amarella *comes from the Latin word* amarus, *meaning "bitter," because of the bitter alkaloids in the juice of this plant.*

Gentianella propinqua (Richards.) J.M. Gillett

FOUR-PARTED GENTIAN

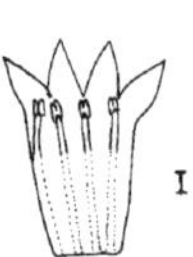

This plant is similar to felwort, but does not have a fringed corolla and its corolla lobes end in a point. Four-parted gentian, as its name implies, has 4 sepals and 4 petals. It has a short stem, 0.3 to 2.5 dm high; the flowers are large, often purple. This plant is found on alpine slopes, up to 2200 m.

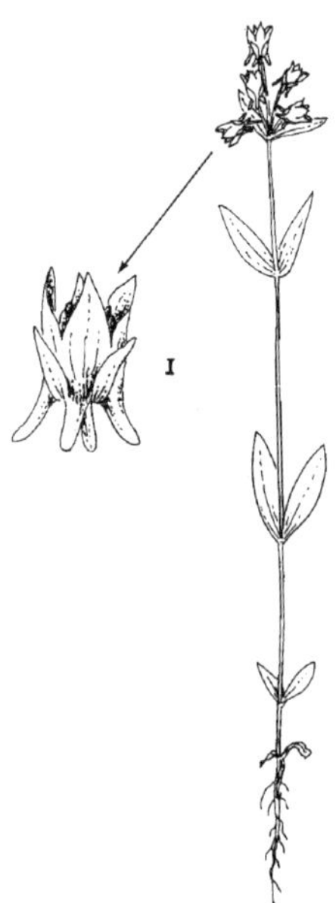

Halenia deflexa (Sm.) Grisebach

SPURRED GENTIAN

Spurred gentian is a winter annual, 0.3 to 1.9 dm tall, with a simple stem, branched above, with leaves in pairs. The basal leaves are oblong-spatulate, while the stem-leaves are oval and rather pointed. The flowers are arranged in a terminal or axillary cyme; each flower has a united calyx deeply 4-cleft and a tubular corolla of 4 united petals, with 4 lobes. This corolla is purplish green or bronze, and the lobes have slender spurs at the base, hence the common name "spurred gentian." There are 4 stamens, arising from the corolla-tube, and 2 carpels in the centre, joined, with 2 stigmas; this develops into a 2-valved capsule when fertilized. Spurred gentian is found in cool, damp woods.

Menyanthaceae BUCKBEAN FAMILY

Members of the buckbean family are perennial bog plants with stout rhizomes, which are covered with persistent leaf-bases and have fleshy rootlets. The leaves are basal, rather crowded, and have very long petioles and a large basal sheath; there are 3 distinct leaflets, oval or elliptic in shape, entire, 3 to 8 cm long. The flowers are arranged in a dense raceme on a tall flowering-stalk 1 to 3 dm high. Each flower has a 5-parted calyx and a 5-parted corolla, extending into a tube, white or pinkish, with densely hairy upper surfaces on the petals. There are 5 stamens, inserted inside the corolla tube, with the ovary in the centre with a persistent style and a 2-lobed stigma. The ovary develops into an ovoid capsule which opens to release large shiny seeds.

Menyanthes trifoliata L.

BUCKBEAN

The leaves of this plant rise from a stout rhizome (which is eaten by muskrats); they have long leaf-stalks, sheathing the stem, and each leaf is split into 3 leaflets, hence the specific name, *trifoliata.* These leaflets are oval, 3 to 8 cm long, and have no teeth; they are entire. The tall, leafless flowering-stalk, up to 3 dm high, bears 10 or more stalked flowers in a crowded raceme. Each flower has a 5-parted calyx and a 5-parted corolla. The petals are white with a purplish tinge, and are covered with thick, hair-like appendages; bearded petals are unusual, so this helps to identify the buckbean. There are 5 stamens, arising from the corolla tube, which is funneliform. The ovary has a persistent style with a 2-lobed stigma and develops into a capsule, which opens to release shiny seeds. Buckbean is a bog plant; the common name is probably a corruption of "bog-bean."

Apocynaceae DOGBANE FAMILY

Members of this family are perennial plants containing milky juice (latex). The forked stems bear simple, opposite, entire leaves. The flowers are arranged in cymes at the tips of the branches. Each flower has a 5-lobed calyx and a 5-lobed corolla. The corolla is often bell-shaped, but may be tubular; the petals are united and form a corolla-tube; the 5 stamens are joined to the inside of this tube, at the base. The tube also bears 5 tiny appendages. The gynoecium consists of 2 separate carpels, united above to form a large stigma. The fruit is characteristic: 2 long, slender follicles.

Apocynum androsaemifolium L.

INDIAN HEMP, SPREADING DOGBANE

Spreading dogbane is a perennial plant, 2 to 10 dm high, with an erect stem, well-branched above (often forked), and with leaves in pairs on the stems. These leaves are oval to broadly lanceolate, often with a hairy undersurface. The flowers are in cymes at the tips of the branches; individual flowers are often drooping. Each flower has a calyx of 5 small, united sepals and 5 petals, united, forming a bell-shaped corolla-tube with spreading petal-lobes, pink with darker veins. There are 5 stamens rising from the base of the corolla tube; these alternate with 5 small appendages. There are 2 carpels, joined above by a 2-lobed stigma, which develop into a characteristic fruit: a pair of slender follicles, 8 to 10 cm long, which split open (along 1 side) to release numbers of tufted seeds. These follicles are sometimes red, and their shape is so characteristic that they help to identify the plant. Dogbane is found in dry situations, in sandy areas and dry woods.

Asclepiadaceae MILKWEED FAMILY

The milkweeds have their name because of the milky sap (latex) in their stems (*Asclepias tuberosa* is an exception). The plants are herbaceous, with opposite leaves, and have a complex flower structure. The calyx of 5 sepals is persistent, and the sepals are bent back (reflexed). They are free or sometimes joined at the base. The 5 petals forming the corolla are united into a corolla-tube, deeply 5-cleft. There are 5 stamens attached to the base of the corolla tube, united into a staminal tube. A staminal tube is an unusual feature of flowers. Attached to the staminal tube is a curious structure, a 5-hooded corona; each hood has an appendage called a "horn," which is inwardly curved. The hoods alternate with the petals. The stamens are also unusual because their pollen is in the form of pollinia: single suspended masses of pollen. The gynoecium in the centre consists of 2 carpels, which are free and ripen into 2 long, narrow follicles, characteristic of this family. They open along 1 side to release many plumed seeds.

Asclepias speciosa Torr.

SHOWY MILKWEED

Showy milkweed has spreading rhizomes. It is a large, coarse, leafy plant up to 2 m high. The leaves are large, lanceolate to oval, growing in pairs. The flowers are arranged in numerous umbels. In this species, the sepals are greenish, tinged with red; the petals are pink to reddish purple; the hoods are pink and longer than the petals. The fruits are white-hairy, narrowly ovoid; the tufted seeds are flattened. Showy milkweed is found in moist sites, in grassland and along roadsides.

NOTES: *The genus name* Asclepias *was given by Linnaeus in honour of Asclepias, the Greek doctor known as the father of medicine. Statues show him carrying a caduceus, a staff with 2 snakes twined round it; after 2000 years, this remains doctors' emblem.*

Polemoniaceae PHLOX FAMILY

Members of this family are annual or perennial herbs, with leaves in pairs or alternate; these may be simple or compound (as in *Polemonium*). Each flower has 5 sepals, joined to form a 5-parted calyx, and 5 petals that are coiled in the bud. These petals are united to form a corolla-tube with 5 spreading free lobes; there are 5 stamens inserted on the corolla tube (epipetalous). There are 3 carpels, united, in the centre of the flower, with a 3-lobed style. The fruit is a loculicidal capsule opening with 3 slits to release the seeds.

1. Leaves compound . . . . *Polemonium*
   Leaves simple 2
2. Leaves opposite . . . . *Phlox*
   Leaves alternate . . . . *Collomia*

Collomia linearis Nutt.

NARROW-LEAVED COLLOMIA

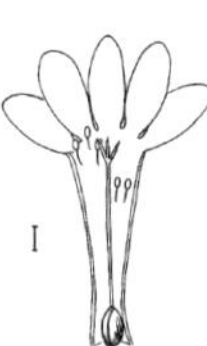

Collomia is an erect annual, usually unbranched, which grows up to 4 dm high. The leaves near the base are narrowly lance-shaped, but further up the stem they become broader, up to 5 cm long. The flowers grow in a dense cluster at the tops of the leafy stems; they are pink and inconspicuous. Each flower has a 5-lobed calyx; the 5 petals form a tubular corolla, with 5 spreading "limbs," and 5 stamens arise from inside the tube. There are 3 carpels joined together in the centre of the flower, with a long style and 3 stigmas; these form a capsule with 3 valves. Collomia is found on roadsides and in dry or moist meadows.

NOTES: *The species name* linearis *refers to the lower leaves, which are often linear in shape.*

Phlox L. — PHLOX

Phloxes are cushion-like perennials growing from a branched woody base, with opposite linear or needle-shaped leaves. The flowers grow 2 to 3 or solitary. Each has a calyx of 5 sepals, united, lobed; a corolla of 5 petals, united, tubular, lobed; 5 stamens, epipetalous on the inside of the corolla-tube; and a gynoecium of 3 carpels, united. The fruit is a loculicidal capsule.

Petals pink, white or pale purple, 15 to 18 mm long; leaves more than 2 mm wide . . . . *P. alyssifolia*
Petals white, 8 to 10 mm long; leaves less than 2 mm wide . . . . *P. hoodii*

Phlox alyssifolia Greene

BLUE PHLOX

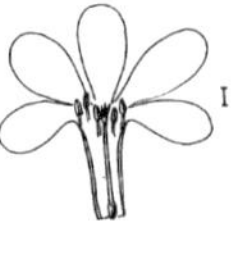

Blue phlox is a low perennial with nearly prostrate branches and lance-shaped leaves; these are fringed with hairs (ciliate). The flowers are usually solitary and have 5 sepals, joined at the base, 5 petals, united, forming a corolla-tube, with 5 spreading "limbs" at the top, which are pink, white or pale purple. There are 5 stamens arising from inside the corolla-tube; the 3 united carpels, which produce a 3-valved capsule, are in the centre of the flower. Blue phlox is found on open, rocky crests and dry, gravelly slopes.

NOTES: *The specific name* alyssifolia *suggests that the leaves resemble those of* Alyssum, *in the mustard family.*

Phlox hoodii Richards.

MOSS PHLOX

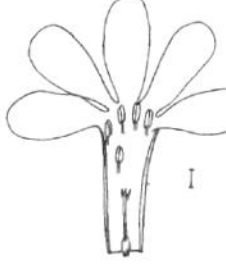

Moss phlox is found on prairie grassland and is densely tufted; both phloxes have a rather woody base. Moss phlox forms "cushions" 3 or 4 cm high, covered with flowers, usually white but sometimes tinged with pink or purple—a welcome sight in early spring. The flowers are similar to those of blue phlox, but the plants can be distinguished because the leaves of moss phlox are awl-shaped and have cobwebby hairs, while those of blue phlox are linear to oblong with ciliate margins (fringed) and few cobwebby hairs. The plants are found in different habitats: moss phlox is a typical prairie plant.

NOTES: *The specific name* hoodii *has an interesting history. Robert Hood and Dr. Richardson accompanied Sir John Franklin on his expedition to the Coppermine River and the Arctic coast. Hood was killed by an Iroquois. Richardson found a new species of phlox on his wanderings and named it after his dead friend, Hood.*

Polemonium L.

Polemonium species are perennials growing from a branched crown. The leaves are alternate and pinnate; the flowers grow in cymes and are bright blue with a yellow eye. The calyx has 5 sepals, cup-shaped; the corolla has 5 petals, funneliform, with rounded lobes. The androecium has 5 stamens, epipetalous; the gynoecium, 3 carpels, united. The fruit is a capsule.

Plants sticky, with a skunky odour .... *P. viscosum*
Plants not sticky .... *P. pulcherrimum*

Polemonium pulcherrimum Hook.

SHOWY JACOB'S-LADDER

Showy Jacob's-ladder is an attractive, dainty plant which grows 1 to 3 dm tall with terminal clusters of showy blue flowers. The common name "Jacob's-ladder" derives from the plant's leaves, which have 11 to 23 leaflets arranged in pairs up the midrib, ladder-like. Each flower has a cup-shaped calyx, enclosing 5 petals which are fused together at the base, forming a short yellowish corolla-tube from which 5 stamens arise. The spreading petal lobes are blue. The fruit is a 3-valved capsule. This plant is found on rocky, well-drained sites and on alpine slopes.

NOTES: *This species has been found in the Bragg Creek area, near the Marmot Creek Experimental Station, in the alpine zone on Mount Allan at 2100 m and other places.* *The specific name* pulcherrimum *means "very beautiful."*

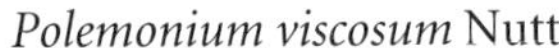

Polemonium viscosum Nutt.

SKUNKWEED, SKY PILOT

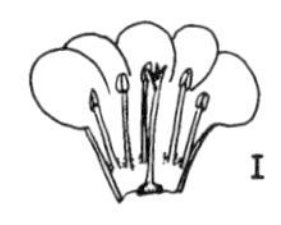

Skunkweed is densely glandular-hairy (hence the specific name *viscosum*) and sticky. The flowers are a brilliant blue with a yellow eye; they have 5 sepals, 5 petals and 5 stamens, and the fruit is a 3-valved capsule. Skunkweed can be distinguished from Jacob's-ladder because it is strongly glandular-hairy and sticky to touch. The leaves of the 2 plants are quite different, those of skunkweed growing in tiny bunches up the stem, appearing whorled; this arrangement is never found in Jacob's-ladder. Skunkweed grows on exposed scree slopes.

NOTES: *This species has been found on alpine slopes on Pasque Mountain at 2400 m, and elsewhere in Kananaskis Country.* *It is unfortunate that such an attractive plant should have the unpleasant name skunkweed, but its leaves do have a strong skunk-like smell.*

Hydrophyllaceae WATERLEAF FAMILY

The members of this family are herbaceous plants, either annuals or perennials, with alternate leaves and perfect flowers. Each flower has a calyx with 5 sepals, more or less united, and 5 petals, united into a cup-shape or a funnel-shape, often with appendages. There are 5 stamens inserted near the base of the corolla-tube; the gynoecium in the centre consists of 2 carpels, united. The fruit is a capsule, opening by 2 valves.

Plants hairy; stamens longer than petals .... *Phacelia*
Plants glabrous; stamens not longer than petals .... *Romanzoffia*

Phacelia Juss. SCORPION WEED

These are hairy annuals or perennials, with taproots. The leaves are entire or pinnately divided, and the flowers are borne in helicoid cymes (a sympodial cyme with the apparent main axis curved in a helix;the successive lateral branches all arise on the same side). The flowers are crowded and often showy, bearing a calyx of 5 strap-shaped sepals deeply 5-parted, a corolla of 5 petals (purple or blue, sometimes white), united to form a bell-shaped or funnel-shaped corolla-tube. 5 stamens are inserted inside the base of the corolla-tube, exserted (extending beyond the tube). The fruit is a loculicidal capsule.

Stamens longer than the petals .... *P. sericea*
Stamens the same length as the petals .... *P. franklinii*

Phacelia franklinii (R.Br.) A. Gray

FRANKLIN'S SCORPION WEED

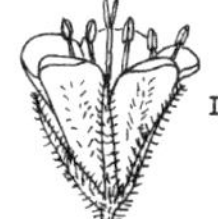

Franklin's scorpion weed is an erect plant 2 to 5 dm tall; the lower leaves have petioles (leaf-stalks) and are pinnate or bi-pinnate, while the upper leaves are usually sessile and are less divided. The flowers are borne in clusters, with stalks, in an arrangement rather like a raceme, well above the leaves. The flowers are pale blue. They have a 5-parted calyx, and the corolla-tube has 5 lobes. There are 5 stamens arising from the corolla; the gynoecium consists of 2 united carpels, and the fruit is a capsule. Franklin's scorpion weed is found in dry, open areas and on disturbed ground.

Phacelia sericea (Graham) A. Gray

MOUNTAIN PHACELIA, SILKY SCORPION WEED

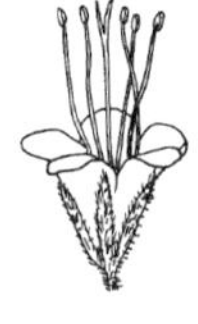

Mountain phacelia has stamens with violet filaments and orange anthers. The flower spike (technically a thyrse or a scorpioid cyme) is up to 2 dm tall. The flower-structure is the same as that of Franklin's scorpion weed, but the flower-spike is much larger and more striking. The stamens protrude well above the level of the petals. The leaves of are pinnate or bi-pinnate, and the lower leaves have long petioles. Mountain phacelia is found on mountain rock-ledges and in open woods and meadows.

NOTES: *This species has been found near Ribbon Creek bridge in rocky soil, beside the Three Isle Lake Trail, on a dry rock scree north of Cataract Creek beside Highway 40 and elsewhere.* ~ *Mountain phacelia is one of the most striking flowers in Kananaskis Country: the tall, dense flower-spike of deep purple flowers with orange anthers rises well above the greenish grey leaves, making an attractive foil for the flowers.* ~ *The specific name* sericea, *meaning "silky," refers to the silky hairs covering the leaves.*

Romanzoffia sitchensis Bong.

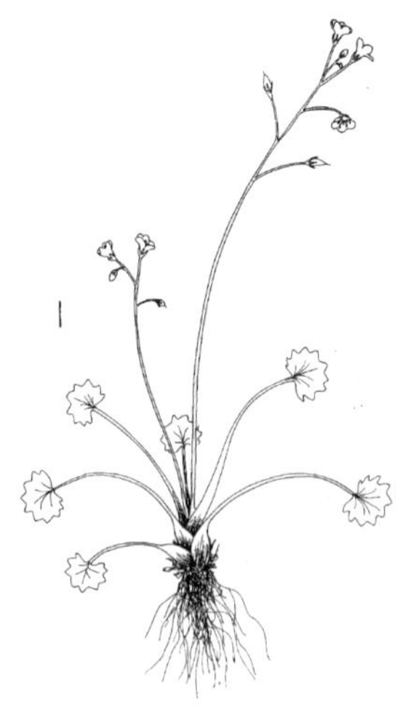

MIST MAIDEN, SITKA ROMANZOFFIA, CLIFF ROMANZOFFIA

Mist maiden is an attractive little plant with dainty leaves and flowers, found on moist cliffs and rock ledges, and on slopes near the melting snows—it is often found far above timberline. The plant has fibrous roots, from which grow a rosette of leaves. Each leaf is kidney-shaped in outline, with large, rounded marginal lobes and long petioles with sheathing bases; these are scalloped and, with the petiole, 7 cm long. The dainty leaves help to identify the plant. The flowers are in a loose raceme at the tips of the flowering stalks; each flower has 5 sepals, partly united, and 5 white petals, united, forming a funnel-shaped corolla with rounded lobes. There are 5 stamens, inserted near the base of the corolla-tube. There are 2 carpels in the centre, united, superior; these develop into a capsule with 2 valves which open to release many seeds.

NOTES: *The genus name* Romanzoffia *was given by the botanist Chamisso to honour his patron, Count Romanzoff, a Russian Minister of State who was enthusiastic about sending botanical excursions to other countries; Eschscholtz found this plant on an expedition financed by Count Romanzoff.* ~ *The species name* sitchensis, *given by Bongard, means "of Sitka," a town on the coast of Alaska where this plant was found.* ~ *Mist maiden is often mistaken for a saxifrage, but the flower has 5 stamens whereas saxifrage flowers have 8 to 10 stamens.*

Boraginaceae BORAGE FAMILY

Members of the borage family are herbaceous plants, usually covered with hairs, with simple, alternate leaves, occasionally paired below. The inflorescence is unusual because the flowers are usually arranged in circinate or scorpioid cymes; that is, they are arranged in series and uncoil gradually as they grow, rather like fern "fiddleheads." This characteristic, together with the hairiness, helps to identify this family. The flowers each have a calyx of 5 united sepals and 5 petals fused into a cylindrical corolla-tube, with free petal-lobes, often with appendages at the mouth of the tube. There are 5 stamens, attached to the inside of the corolla-tube. The gynoecium in the centre consists of 4 united carpels; the fruit is usually 4 nutlets, each with a single seed inside. These nutlets, as in stickseed (*Hackelia*), often have prickles that aid in fruit dispersal.

1. Flowers yellow or greenish white 2
   Flowers blue 3
2. Flowers yellow .... *Lithospermum*
   Flowers greenish white .... *Onosmodium*
3. Corolla salverform, or somewhat tube-shaped; fruits with hooked prickles 4
   Corolla tube-shaped; fruits without hooked prickles 5
4. Flowering-stalk with leafy bracts; fruits with rim, bearing prickles.......... .... *Hackelia*
   Flowering-stalks without leafy bracts; margin of fruits bearing 2 rows of prickles.. *Lappula*
5. Tall perennial; leaves ovate-lanceolate, soft-hairy; inflorescence drooping; flowers large, corolla-tube without appendages .... *Mertensia*
   Low perennial; inflorescence not drooping; corolla-tube short, throat with appendages (crested).. *Myosotis*

Hackelia Opiz **STICKSEED**

Members of this genus are perennials or biennials, 5 to 10 dm tall, with hairy stems and leaves. Leaves are linear-lanceolate; flowers grow in scorpioid cymes. Each flower bears a calyx of 5 sepals, united and 5-parted, and a corolla of 5 petals, united, lobed, There are 5 stamens, epipetalous, and a gynoecium of 4 carpels. The fruit is 4 nutlets; the back of each nutlet has a rim, bearing barbed prickles.

Fruit with a semi-circle of very flat prickles . . . . *H. floribunda*
Fruit with a few short spines . . . . *H. jessicae*

Hackelia floribunda (Lehm.) I.M. Johnston

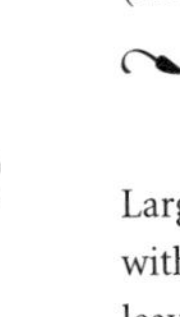

LARGE-FLOWERED STICKSEED

Large-flowered stickseed grows 5 to 10 dm tall, with stalked, narrowly lanceolate leaves; the upper leaves are sessile and pointed at both ends. The inflorescence is freely branched with numerous flowers, either blue or white. These are arranged in a scorpioid raceme: the flowers are coiled-up in bud and uncoil later. Each flower has 5 sepals, united, and 5 petals, also united; they form a corolla-tube with 5 lobes at the top and 5 stamens arising from the tube. The fruit consists of 4 nutlets. In stickseed these nutlets are attached to a pyramidal base, and the back of each nutlet has a rim with barbed prickles, hence the common name. Intramarginal prickles are usually absent. Stickseed is found in moist woods and meadows.

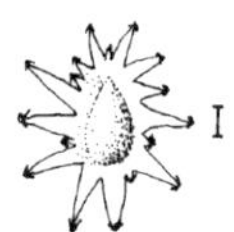

Hackelia jessicae (McGreg.) Brand

JESSICA'S STICKSEED

This plant is similar to *H. floribunda* but can be distinguished by several characteristics. *H. jessicae* has several stems, while *H. floribunda* is single-stemmed. In *H. jessicae* the fruit consists of 4 nutlets. The back of each nutlet has a rim with barbed prickles; a few intramarginal prickles can be seen. In *H. jessicae* the style is about as long as the sepals, while in *H. floribunda* the style of the ovary is much shorter than the sepals. *H. jessicae* is found in open montane woods and meadows.

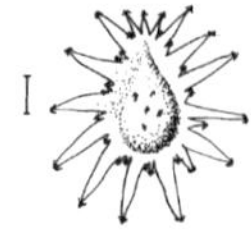

Lappula squarrosa (Retzius) Dumort.

BLUEBUR

(L. echinata Gilibert*)*

Bluebur may be single-stemmed or profusely branched, with narrow leaves up to 6 cm long. The flowers are borne in racemes; each flower, 3 to 4 mm broad, is subtended by a bract and has a 5-parted calyx and a 5-lobed corolla of blue petals. (The flower is sometimes mistaken for a forget-me-not.) There are 5 stamens, inserted on the corolla, and a 4-lobed gynoecium of 4 carpels. These carpels develop into 4 nutlets, and each has 2 rows of slender marginal prickles on the back, hence the "bur" of the common name. Bluebur is an introduced species, an annual or winter annual, found on waste ground, on roadsides and in fields and gardens.

Lithospermum L. — PUCCOON

Members of this genus are perennials with thick roots. They have hairy, sessile leaves and linear-lanceolate, yellow flowers arranged in spike-like cymes. The flowers bear 5 sepals, united and 5-parted, and 5 salverform petals, united and lobed. There are 5 stamens, epipetalous, and 4 carpels. The carpels form 4 nutlets, white and "stony." *L. incisum* forms cleistogamous flowers which produce most of the nutlets.

Corolla less than 10 mm long . . . . *L. ruderale*
Corolla more than 10 mm long . . . . *L. incisum*

Lithospermum incisum Lehm.

INCISED PUCCOON, NARROW-LEAVED PUCCOON

Incised puccoon has stems 1 to 5 dm high and is found on dry hills and plains, often on prairie grassland. The flowers are bright yellow and follow the "borage pattern" with a 5-lobed calyx, a corolla-tube of 5 lobes with 5 stamens arising from the inside of the tube. The fruits are ovoid nutlets, white, shiny and smooth— rather like tiny stones.

NOTES: *The fruits give rise to the generic name* Lithospermum, *which literally means "stone-seed." The common name incised puccoon refers to the yellow petals, which are notched.* *The Blackfoot obtained a violet dye from the roots.*

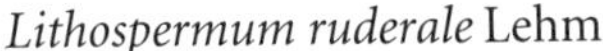

Lithospermum ruderale Lehm.

PUCCOON, WOOLLY GROMWELL

Puccoon is a stout perennial plant, 2 to 5 dm high, which often forms a large "cushion" with many stems arising from the base. The roots are reddish brown and yield a purple dye, which can often be seen as discolourations on a herbarium sheet. The flower-structure is similar to that of incised puccoon, with a 5-lobed calyx and a corolla-tube with stamens attached inside the tube. The nutlets are also similar. An important difference between the 2 species is that the flowers of puccoon are pale yellow and often nearly hidden by its leafy stems. Puccoon is found in dry slopes and grassland.

NOTES: *The Assiniboines, the Shoshones and the Navaho used an extract of this plant as an oral birth control. The plant contains natural estrogens that suppress the formation of certain hormones without which ovulation cannot occur.* ~ *A tea made from the leaves was used as a laxative.*

Mertensia paniculata (Ait.) G. Don

TALL MERTENSIA, TALL LUNGWORT

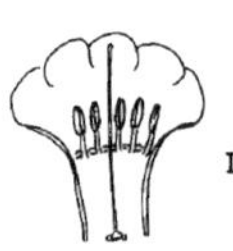

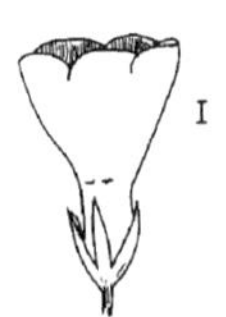

Tall mertensia is an erect plant, 2 to 8 dm tall, growing from a woody rootstock. The basal leaves have petioles and are hairy, while those higher up the stem have no petioles. The flowers are in clusters at the top of the stem; at first, these are congested, but later they separate into spreading panicles (compound racemes): hence the specific name, *paniculata*. Each flower has a 5-parted calyx, a bright blue corolla-tube with 5 lobes, and 5 stamens inside the corolla-tube. The fruit consists of 4 nutlets, each with 1 seed inside. Tall mertensia is found in moist woods and moist open slopes.

NOTES: *This plant is sometimes called bluebells, because the flower is somewhat bell-shaped and is deep blue.* ~ *The generic name,* Mertensia, *honours an early German botanist, Mertens (1764-1831).*

Myosotis alpestris Schmidt

ALPINE FORGET-ME-NOT

Alpine forget-me-not is a fibrous-rooted perennial up to 20 cm tall, with hairy, spatulate basal leaves. The flowers are clustered at the top of the stem in 1-sided racemes; later, when they are fruiting, they separate. Each flower has a hairy, sometimes glandular, 5-parted calyx, a corolla-tube with 5 well-marked lobes, and is bright blue with a yellow eye. It bears a group of 5 appendages, like a tiny "corona." The 5 stamens are in the throat of the corolla tube, and the 4 nutlets of the fruit are nearly black. Alpine forget-me-not is found on stable, open alpine slopes and meadows.

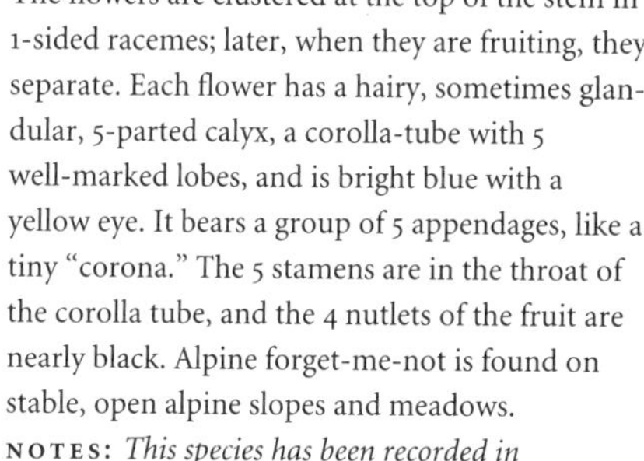

NOTES: *This species has been recorded in Kananaskis Country from Raspberry Ridge, from near the summit of Moose Mountain, from the alpine zone on Pasque Mountain, from the alpine zone on Plateau Mountain in an alpine meadow (elevation 2300 m), also from Hailstone Butte and in other places.*

Onosmodium molle Michx. var. *occidentale* (Mack.) Johnston

WESTERN FALSE GROMWELL

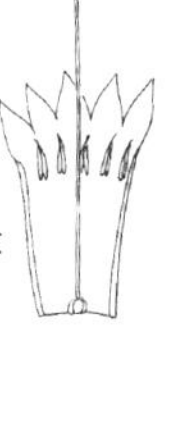

False gromwell is a coarse perennial, finely greyish-pubescent, growing 4 to 6 (sometimes 8) dm high with oval or lance-shaped leaves, 4 to 6 cm long. The midribs and the veins (usually 5 to 7) are clearly marked. The whole plant is covered with hairs, hence the specific name *molle*, meaning "soft." The flowers are arranged in scorpioid racemes in the axils of leafy bracts. Each flower has a silky, hairy calyx, 7 to 12 mm long, of united sepals, with 5 erect, linear lobes. The tubular corolla is greenish white or yellowish with 5 triangular lobes; the stamens arise from inside the corolla-tube, in the throat, but do not extend beyond it: they are intruded. The fruits, nutlets, are about 4 mm long and shiny. False gromwell is found in dry situations, in dry open woods and on gravel banks.

NOTES: *The varietal name,* occidentale, *means "western" and is sometimes seen on old maps.*

Lamiaceae (Labiatae) MINT FAMILY

Members of the mint family are generally aromatic herbs, either annuals or perennials, with 4-angled stems and opposite leaves. These leaves are often dotted with small glands containing volatile oils. The flowers are zygomorphic; that is, they are bi-laterally symmetrical and arranged in cymose clusters in spikes or racemes. In spikes the flowers are sessile; in racemes the flowers have stalks. The flower has 5 sepals, partly fused, and 5 petals, united, usually forming a 2-lipped corolla. The former family name, Labiatae, was a descriptive one, because the Latin *labium* means "lip." There are 2 or 4 stamens inserted in the corolla-tube (epipetalous), and the gynoecium, 4-lobed, consists of 2 fused carpels with a 2-cleft style. The fruit consists of 4 small, 1-seeded nutlets.

1. Corolla rose-purple 2
   Corolla blue or purple 3
2. Stamens 4; plants densely hairy . . . . *Stachys*
   Stamens 2; plants not densely hairy . . . . *Monarda*
3. Stamens longer than petals. . . . . *Mentha*
   Stamens not longer than petals. 4
4. Calyx 5-lobed, the upper wider than the lower 4 . . . . *Dracocephalum*
   Calyx 2-lobed, the upper with 3 awns. . . . . *Prunella*

Dracocephalum parviflorum Nutt.

AMERICAN DRAGONHEAD

(Moldavica parviflora (Nutt.) Britt.*)*

American dragonhead is a rather coarse annual or biennial that grows up to 6 dm high. The stems may be simple or branched and bear lanceolate leaves with leaf-stalks. These leaves have prominent coarse marginal teeth and are in pairs, often with shoots in the axils. The flowers are arranged in a dense terminal spike. The leafy bracts below the flowers on the spike are numerous, crowded and toothed; these teeth have a stiff spine. The individual flowers each have a tubular calyx with 5 lobes; the upper lobe is longer and wider than the other 4. There is a tubular corolla of 5 bluish violet petals; 2 of these form a hooded upper lip, and 3 form a lobed lower lip—the middle lobe is larger than the other 2. There are 4 stamens inserted on the corolla-tube, and there are 2 carpels, united, in the centre; these develop into the fruit, 4 nutlets, all 1-seeded. American dragonhead is found in grasslands, open woods and waste ground; it is rather a weedy species.

NOTES: Dracocephalum *comes from 2 Greek words:* draco *(dragon) and* cephalos *(head). The specific name* parviflorum *means "small-flowered."*

Mentha arvensis L.

WILD MINT

Wild mint is a perennial, 2 to 5 dm tall, with aromatic leaves growing in pairs up the stem. Small flowers are found in the axils of the leaves, which are toothed and dotted with glands. Each flower has 5 sepals, fused together; these are often purplish, have 5 lobes and enclose the corolla-tube, of 4 or 5 petals, light pink to purple. 4 exserted stamens arise from inside the corolla-tube. The fruit consists of 4 small nutlets, enclosed in the calyx of persistent sepals, and each nutlet has a single seed. Wild mint is found in boggy places and sloughs.

NOTES: *The species name* arvensis, *meaning "of the fields," is not appropriate in Alberta, but Linnaeus named the plant in Europe where it is often found in fields. The Blackfoot used the leaves to flavour meat or pemmican. The dried leaves were used to treat heart ailments and chest pains.*

Monarda fistulosa L.

WILD BERGAMOT, HORSE MINT

Wild bergamot, with its large, rose-coloured flowers, is not easily overlooked. The flowers grow in a dense terminal cluster surrounded by bracts. Each flower has a tubular calyx with 5 lobes, enclosing the large (2 to 5 cm), 2-lipped corolla. The upper lip has a fringe of hairs, and the lower lip has 3 lobes. There are 2 stamens, arising from the upper lip, exserted. The long style, arising from the ovary, can be seen above the lower lip. The fruit consists of 2 nutlets. Wild bergamot is found in prairie grassland, open woods and roadsides. The plants often grow in large colonies.

NOTES: *The generic name* Monarda *honours Nicholas Monardes (b. 1493), a Spanish physician interested in New World plants. The Blackfoot made an eyewash from the blossoms, and a tea made from the leaves and blossoms was used to cure stomach pains.*

Prunella vulgaris L.

SELF-HEAL, HEAL-ALL

Heal-all is a small perennial with an erect or decumbent stem, with stalked leaves in pairs up the stem, which is only 1 to 3 dm high. The dense flower-clusters are terminal spikes, subtended by 2 or more leaves. The individual flowers are interesting because the calyx is 2-lipped: the upper lip is squarish (truncate) and bears 3 small awns, and the lower lip has 2 lobes. The corolla is also 2-lipped. The corolla can be purple, pink or white, and the upper lip is a large hood, enclosing the stamens and style. The lower lip is 3-lobed, with the middle lobe larger than the other 2. There are 4 stamens, and the fruits are nutlets. Heal-all is found in moist places in fields and open woods.

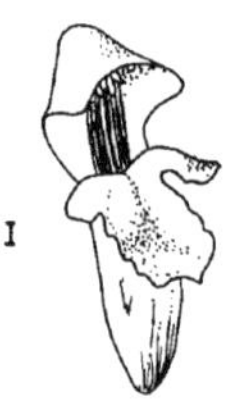

NOTES: Prunella *has been known as a healing herb since the time of the Greeks; Gerard's* Herball, *1597, describes it as a "wound herbe."*

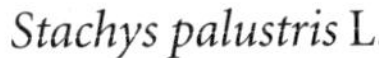

Stachys palustris L.

MARSH HEDGE NETTLE

Marsh hedge nettle is an unbranched, soft-hairy perennial, with hairy leaves in opposite pairs up the stem, which is 3 to 8 dm tall. The flowers grow in an interrupted terminal spike, with conspicuous bracts, and each flower has 5 sepals, united into a cup with 5 lobes, enclosing the corolla tube. The corolla has 2 lips; the upper one is hooded, and the lower one (which is mottled) is divided into 3 unequal lobes. Both lips are bright lilac. There are 4 stamens hidden inside the upper lip; the fruits are nutlets. Hedge nettle is found in damp places such as wet meadows and along stream banks.

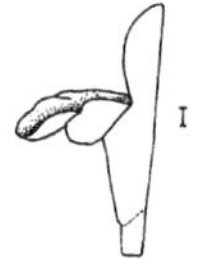

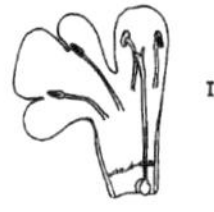

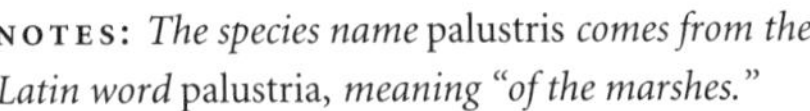

NOTES: *The species name* palustris *comes from the Latin word* palustria, *meaning "of the marshes."*

Scrophulariaceae FIGWORT FAMILY

Members of the figwort family are annuals or biennials, usually herbaceous but occasionally woody. The leaf-shape is varied: many plants have entire leaves, like toad-flax (*Linaria*), but others have serrated leaves, and the paintbrushes (*Castilleja*) often have 3-cleft leaves. The leaves of some of the louseworts (*Pedicularis*) are fern-like. The flowers grow in a spike-like or raceme-like arrangement and are bi-laterally symmetrical (zygomorphic). Each flower has 2 to 5 sepals, often more or less united, and 5 petals joined to form a 2-lipped corolla; the upper lip is usually 2-lobed, and the lower lip 3-lobed. There are 2 or 4 stamens—in beard-tongues (*Penstemon*) there is a fifth stamen, but this is sterile and usually bearded. The gynoecium in the centre consists of 2 fused carpels, and the fruit is a 2-valved capsule. Several members of this family are semi-parasites: the paintbrushes (*Castilleja*), owl clover (*Orthocarpus*), lousewort (*Pedicularis*) and yellow rattle (*Rhinanthus*).

1. Leaves mostly basal . . . . *Besseya*
   Leaves alternate or opposite 2
2. Leaves alternate. 3
   Leaves opposite. 7
3. Petals absent; stamens 2. . . . . *Besseya*
   Petals present; stamens 4 4
4. Bracts coloured, petal-like, calyx 4-lobed. . . . . *Castilleja*
   Bracts usually green; calyx deeply 5-lobed 5
5. Corolla spurred . . . . *Linaria*
   Corolla not spurred 6
6. Stem-leaves entire . . . . *Orthocarpus*
   Stem-leaves lobed. . . . . *Pedicularis*
7. Stamens 2 . . . . *Veronica*
   Stamens 4 8
8. Corolla yellow, hooded, persistent calyx constricted at neck, inflated in fruit . . . . *Rhinanthus*
   Corolla purple, rarely yellow, (corolla-tube cleft) calyx not inflated in fruit. 9
9. Delicate annual, with small blue-white flowers, corolla deeply 2-lipped, lower lip blue violet, reflexed . . . . *Collinsia*
   Perennial herbs, corolla tubular, purple or yellow, 2-lipped, upper lip 2-lobed, lower lip not reflexed . . . . *Penstemon*

Besseya wyomingensis (A. Nels.) Rydb.

KITTENTAILS

(B. cinerea (Raf.) Pennell)

Kittentails is a small perennial plant, 1.5 to 3 dm high, grey-hairy on all the green parts; the fruit, a capsule, is hairy also. There are several basal leaves with long petioles (leaf-stalks); the blades are toothed and broadly lance-shaped. The stem leaves (cauline) are smaller, have no petioles and grow alternately up the stem. The flowers are arranged in a dense terminal cluster, a spike, on the flowering-stalk; they are subtended by bracts and have no individual stalks (pedicels). Each flower has 2 sepals, partly fused, no petals and 2 stamens. The stamens help to identify the flower: they are long, deep purple and extend beyond the sepals (extruded). There are 2 carpels, joined, and these develop into a capsule. The plant is found on grasslands and alpine slopes.

NOTES: *The species name* wyomingensis *refers to the state of Wyoming where, presumably, the "type-specimen" was found. The former name,* Besseya cinerea, *was descriptive because* cinerea *means "ash-like" and refers to the grey hairs that cover the plant. The common name "kittentails" is supposed to be derived from the halo-like cluster of purple stamens, but this is a fanciful resemblance.*

Castilleja Mutis ex L.f. — INDIAN PAINTBRUSH

These are leafy perennials, hairy or smooth. The leaves are linear-caudate, and the flowers grow in leafy, bracted, terminal spikes. Large bracts are brightly coloured, but the flowers are smaller than the bracts and not so conspicuous; there is usually 1 flower to a bract. Each flower has a calyx of 4 sepals, partly united and 4-cleft, and 5 united petals, bilabiate (3 in upper lip, lobed, and 2 in lower lip). The androecium consists of 4 stamens inserted on the inside of the corolla tube; the fruit is a loculicidal capsule.

Paintbrushes are semi-parasites: they make food from normal leaves but are dependent on other plants as well. They attach themselves to the roots of nearby plants by means of small suckers (*haustoria*) and are probably never completely independent.

The flowers of this genus are unusual because the colour of the so-called "flower" comes not from the petals but from the showy bracts that enclose the true blooms. These blooms are also unusual because they have a long, narrow 2-lipped corolla of 5 petals joined to form a tube, and the upper lip, the *galea*, is like a beak, practically enclosing the 4 stamens. The lower lip is much reduced in size.

1. Bracts of flowers red or pink . . . . *C. miniata*
   Bracts of flowers yellow 2
2. Bracts with 1 to 2 lobes . . . . *C. occidentalis*
   Bracts with 3 to 7 lobes . . . . *C. lutescens*

Castilleja lutescens (Greenm.) Rydb.

STIFF YELLOW PAINTBRUSH

Yellow paintbrush is 3 to 6 dm tall and is usually branched above. The short spike of flowers, enveloped in yellow bracts, is found at the top of the stem. The flower has a 4-lobed calyx, enclosing the tubular 2-lipped corolla. The *galea* (upper lip) is less than 1/2 the length of the corolla tube, and the lower lip is 1/5 to 1/2 the length of the galea. There are 4 stamens, in pairs, enclosed by the galea. The fruit is a capsule, with many seeds. Yellow paintbrush is found in grassy meadows at low elevations.

NOTES: *The generic name* Castilleja *honours an 18th-century Spanish botanist, Don Domingo Castillejo.*

Castilleja miniata Dougl. ex Hook.

COMMON RED PAINTBRUSH

Common red paintbrush can be distinguished from yellow paintbrush by the colour of the bracts (although the red bracts can sometimes show yellowish) and by the calyx which in yellow paintbrush has acute lobes, but in common red paintbrush is sub-equally cleft above and below. In common red paintbrush the galea is more than half the length of the corolla tube. Common red paintbrush is a familiar sight in Kananaskis Country, in open woods and meadows and along roadsides. It has been recorded from Cat Creek, Highwood River, Spray Lake area, near Gibraltar Mountain, Elbow Falls, Ribbon Creek, Bragg Creek area and other places.

Castilleja occidentalis Torr.

ALPINE YELLOW PAINTBRUSH

C. occidentalis is a smaller plant than *C. lutescens*: it grows only up to 2 dm high, while *C. lutescens* is 3 to 6 dm tall. The bracts in *C. occidentalis* are often greenish yellow. The leaves are also different in the 2 yellow species: in *C. lutescens* the upper leaves are lobed, while in *C. occidentalis* they are usually entire. Alpine yellow paintbrush is often found on alpine slopes and stable scree areas. NOTES: *This plant has been found near the summit of Plateau Mountain and at 2000 m at Marmot Creek, but is sometimes found at lower elevations in Kananaskis Country.*

Collinsia parviflora Lindl.

BLUE-EYED MARY

Blue-eyed mary is a dainty little annual with slender stems, 1 to 3 dm high; there are minute white hairs on the stems, the lower midribs of the leaves and the sepals. The leaves are lance-shaped, either in pairs or whorled, and grow 1 to 2 cm long. The flowers grow in the axils of the leaves, on very slender stalks up to 15 mm; they may be solitary or whorled. Each flower has 5 sepals, narrowly pointed and joined at the base, and 5 petals joined to form a 2-lipped corolla tube; the upper lip of 2 petals is blue, and the lower lip is blue-violet, hence the common name. 4 stamens arise from inside the corolla-tube. There are 2 carpels, joined, which develop into the fruit, a straw-coloured capsule, which splits into 4 when ripe to release the seeds. Blue-eyed mary is found in rocky open places and moist open woods.

Linaria vulgaris Hill

TOADFLAX, BUTTER-AND-EGGS

Toadflax has attractive flowers, like small snapdragons. It grows 2 to 8 dm tall, with numerous linear leaves, and the flowers are borne in a tall terminal raceme. Each flower has a 5-cleft calyx, and the yellow corolla-tube is 2-lipped. The upper lip is 2-lobed, and the lower one is 3-lobed, with an orange patch in the throat. This orange patch and the yellow corolla give rise to the common name "butter-and-eggs." The corolla is spurred at the base, and there are 4 stamens. The fruit is a capsule which opens by valves at the tip. Toad-flax is a common introduced weed, found on waste ground, roadsides and fields; as it spreads easily by rootstocks, it forms large colonies and merits the description "noxious weed."

Orthocarpus luteus Nutt.

OWL-CLOVER

Owl-clover is quite common in dry prairie grassland, but is often overlooked because the flowers are not conspicuous. The plant is only 1 to 4 dm tall with numerous leaves, the lower ones usually entire and the upper deeply lobed. The plant is often branched. The small flowers are in terminal spikes, subtended by 3-cleft bracts. The calyx is 4-cleft. The corolla is bright yellow and 2-lipped, enclosing 4 stamens. The fruit is a capsule containing ridged seeds.

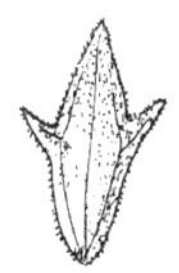

NOTES: *Owl-clover is a semi-parasite.* ~ *Owl-clover foliage was used by the Blackfoot to dye small skins and feathers a reddish tan.*

Pedicularis L. **LOUSEWORT**

These species are perennials, often with pinnately lobed leaves (fern-like in *P. bracteosa*). The flowers grow in a tall raceme (conspicuously bracted in *P. bracteosa*), bearing a calyx of 5 sepals, united, with 2 to 5 lobes, and 5 petals, united and bi-labiate. There are 4 epipetalous stamens; the fruit is a capsule.

1. Plants less than 10 cm tall, reddish purple; alpine slopes. . . . . *P. flammea*
   Plants taller than 10 cm 2
2. Corolla reddish purple . . . . *P. groenlandica*
   Corolla yellow, often tinged with purple . . . . *P. bracteosa*

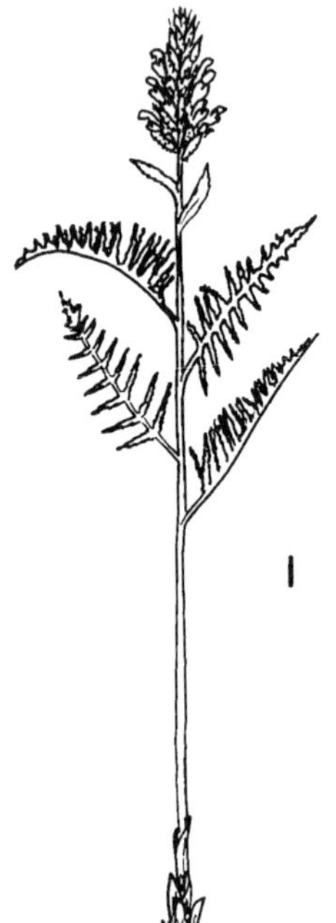

Pedicularis bracteosa Benth.

BRACTED LOUSEWORT

These plants, like the paintbrushes, are semi-parasites. *P. bracteosa* is a tall plant, up to 1 m high, with fern-like leaves, pinnately divided. The lower leaves have petioles, but the upper leaves are sessile. The flowers are arranged in a tall spike, 5 to 15 cm, with many leafy bracts; each flower has 5 sepals, fused into a short tube, and 5 petals, yellow to brownish purple, fused at the base and separated into 2 lips. The upper lip of 2 petals forms a hood, while the lower lip has 3 lobes. There are 4 stamens, and the fruit is a flattened capsule which splits along 2 seams. Bracted lousewort is a woodland species, and is also found on open montane slopes.

NOTES: *It is unfortunate that these attractive plants are called louseworts.* Pediculus *is the Latin word for louse, and it was an early European belief that cattle feeding on* Pedicularis *plants would become covered with lice.*

Pedicularis flammea L.

❧ FLAME-COLOURED LOUSEWORT

This is a much smaller plant than *P. bracteosa*: it is only 6 to 10 cm high, and the leaves, although pinnately lobed, are not really fern-like—the lobes are too short. The flowers have a deep crimson galea or hood (this is the upper lip), distinctly different from *P. bracteosa* (although it is sometimes yellow and is then called *forma flavescens)*. The spike-like raceme of flowers is only 2 to 5 cm long.

NOTES: *This plant is found on alpine slopes and has been recorded from Plateau Mountain at 7800 ft (2380 m), Marmot Basin at 7000 ft (2140 m) and other places.* ❧ *Flame-coloured lousewort is* calciphilous, *that is, it grows on calcareous rocks.* ❧ *The specific name* flammea *comes from the flame-like colour.*

Pedicularis groenlandica Retz.

❧ ELEPHANT'S HEAD, LITTLE RED ELEPHANT

The common names suggest this is an unusual flower, and this is certainly true. The petals of each flower are a light reddish purple, and together they resemble an elephant's head, with the upper lip modified into the head and upcurved trunk and the 2 lateral lobes of the lower lip representing the elephant's ears. The whole flower is 10 to 15 mm long. Little red elephant is found in wet meadows and marshes, but it can also be found on alpine slopes.

NOTES: *In Kananaskis Country this species has been found at Ribbon Creek Bridge, elevation 1400 m, also on Plateau Mountain, at 3400 m, and elsewhere.* ❧ *The curious flower makes this plant easy to distinguish from either* P. bracteosa *or* P. flammea.

Penstemon Mitch. BEARDTONGUE

Members of this genus are perennials, sometimes woody at the base. The leaves are paired and reduced in size upwards. The flowers grow in terminal racemes or in clusters; the calyx is deeply 5-parted, the corolla is bi-labiate and tubular, and the fruit is a septicidal capsule. The beardtongues are distinguished from other members of the figwort family because of their curious "bearded" stamens. Each flower has 5 stamens (hence the generic name *Penstemon*); 4 of these are fertile stamens, with anthers bearing pollen, but the fifth is sterile and is called a staminode. It is "bearded," that is, it is densely hairy at the tip, and as it lies in the throat of the flower with the bearded part visible the common name "beard-tongue" is quite appropriate.

1. Corolla densely bearded . . . . . *P. eriantherus*
   Corolla not densely bearded 2
2. Corolla yellow . . . . . *P. confertus*
   Corolla purplish blue or lilac. 3
3. Corolla lilac, 3.5 to 4.5 cm long . . . . . *P. fruticosus*
   Corolla purplish blue. 4
4. Corolla 6 to 10 mm long . . . . . *P. procerus*
   Corolla 15 to 18 mm long . . . . . *P. nitidus*

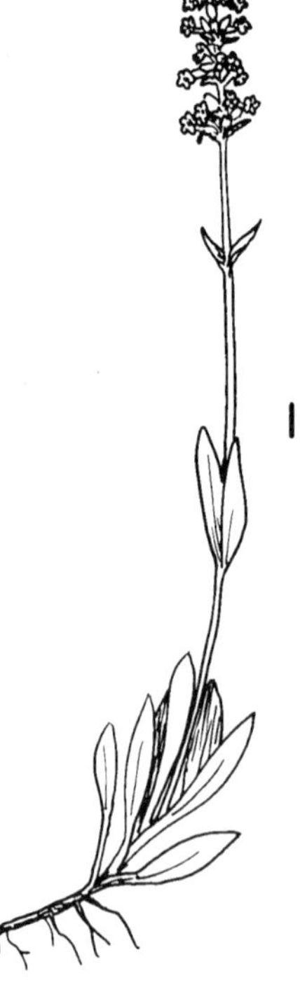

Penstemon confertus Dougl.

YELLOW BEARDTONGUE

Yellow beardtongue has a slender stem, 1 to 5 dm tall, and the lanceolate leaves grow in pairs up the stem. The flowers are borne in the axils of the upper leaves and grow in clusters. Each flower has 5 sepals, united at the base, with jagged margins, 5 petals, yellow, joined in a corolla-tube with 2 lips, and 5 stamens. The fruit is a capsule. Yellow beardtongue is found on dry, open slopes of mountains, in meadows and in open woods.

NOTES: *This species has been found on Plateau Mountain at 2300 m, in the Highwood River Valley and in other places in Kananaskis Country.*

Penstemon eriantherus Pursh

CRESTED BEARDTONGUE

Crested beardtongue is 1 to 3 dm tall with a stout, leafy stem, with oblong-lanceolate, rather pointed leaves. The flowers are borne in a terminal raceme, and each flower has 5 united sepals, deeply 5-parted, enclosing a blue-purple 2-lipped corolla of 5 petals, 2 cm long. The corolla-tube has an enlarged hairy throat and bears 4 stamens and the staminode. The staminode is densely hairy with yellow hairs up to 2 cm long. The fruit is a capsule. Crested beardtongue is found on dry buffs and slopes, up to 2000 m. It can be distinguished from yellow beardtongue because in yellow beardtongue the flowers are borne in the axils of the upper leaves and are in clusters; in crested beardtongue, the flowers are arranged in a raceme.

Penstemon fruticosus (Pursh) Greene

SHRUBBY BEARDTONGUE

Shrubby beardtongue can easily be distinguished because its flowers are relatively large—up to 5 cm long and blue-lavender in colour. Shrubby beardtongue is a woody, shrub-like plant, growing 1 to 4 dm high. It is found at alpine and sub-alpine elevations, usually on dry, rocky slopes.

NOTES: *The specific name* fruticosus *means "shrub-like."*

Penstemon nitidus Dougl. ex Benth.

SMOOTH BLUE BEARDTONGUE

Smooth blue beardtongue is an attractive plant with brilliant blue flowers and distinctive, bluish green leaves. It is a typical prairie plant, 2 to 3 dm high, and often grows in colonies, giving a bright blue splash of colour. The flowers grow in bunches in the axils of the upper leaves, and each flower has a 5-parted calyx, a tubular corolla with an upper and a lower lip, and 5 stamens. The fruit is a capsule.

NOTES: *Smooth blue beardtongue has been found at Lusk Creek in Kananaskis Country, at 1300 m, and elsewhere.*

Penstemon procerus Dougl. ex Graham

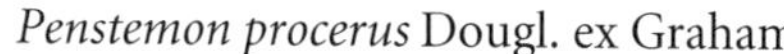

SLENDER BLUE BEARDTONGUE

Slender blue beardtongue is an attractive plant found in meadows and open slopes; in some areas, it reaches timberline. The slender stems are 1.5 to 4 dm high and sometimes decumbent at the base; the basal leaves are lanceolate and stalked, but the stem leaves are smaller, narrower and sessile. The flower arrangement is characteristic: the flowers are in dense clusters at the nodes of the upper stem and look as if they were whorled. Each flower has a calyx of 5 sepals, each 5 mm long, fused at the base; there are 5 petals, united to form a 2-lipped, tubular corolla, deep purplish blue, 6 to 10 mm long. The upper lip is 2-lobed, and the lower lip is 3-lobed. 4 fertile stamens are inserted into the corolla tube, with the staminode. The 2 carpels join together to develop into the fruit, a capsule with 2 valves.

Rhinanthus minor L.

YELLOW RATTLE

(Rhinanthus crista-galli L.*)*

Yellow rattle gets its rather curious name from its yellow flowers and rattle-like fruit. It is semi-parasitic on grasses and other neighbouring plants. The plant grows 3 to 6 dm tall and has pointed leaves growing in pairs up the stem. The flowers grow in the axils of the upper leaves; each has a yellow corolla-tube, 2-lipped, enclosing the 5 stamens. The fruit is a flat, rounded capsule with a few flat seeds inside, which rattle in the wind. The capsule is hidden in the inflated calyx-tube of 5 sepals. Yellow rattle is found in prairies and meadows, sometimes in open woodlands.

Veronica L. SPEEDWELL

These plants are perennials or annuals, often found in damp locations. The leaves are lanceolate-ovate and usually opposite; the flowers grow the in leaf axils, or in spike-like racemes. The calyx is deeply 4-cleft, and the corolla is united and 4-lobed, but not bi-labiate. The 2 stamens produce the fruit, a loculicidal capsule. The genus *Veronica* is unique in the figwort family because the corolla is 4-lobed instead of 5-lobed, and is only slightly irregular, whereas in other members of the family the corolla is 2-lipped and definitely irregular.

1. Flowers in terminal clusters; alpine and subalpine habitats 2
   Flowers in leaf axils; streambanks and marshy ground . . . . *V. americana*
2. Leaves stalked; flowers 3 to 4 mm across . . . . *V. serpyllifolia*
   Leaves not stalked; flowers 4 to 5 mm across . . . . *V. alpina*

Veronica alpina L.

ALPINE SPEEDWELL

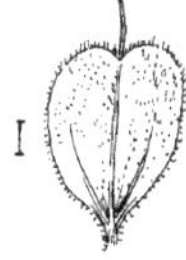

Alpine speedwell is an attractive plant of the alpine meadows, with a cluster of deep blue flowers at the top of the stem, 0.5 to 3 dm high. The stem is covered with fine hairs and has pointed leaves, growing in pairs—the top 2 leaves subtend the flower cluster. Each flower has a 4-cleft calyx, a 4-cleft corolla and only 2 stamens, which helps to distinguish the speedwells from the beardtongues, which have 4 fertile stamens and 1 infertile stamen. The fruit, a capsule, is heart-shaped and glandular-hairy. Alpine speedwell is found in alpine and subalpine zones and has been found on the summit of Plateau Mountain and elsewhere.

Veronica americana (Raf.) Schw.

AMERICAN BROOKLIME

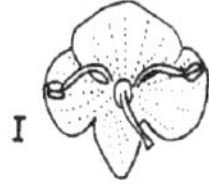

American brooklime grows 2 to 6 dm in length and has decumbent stems which root at the base. The flowers grow in short racemes in the axils of the upper leaves, and the petal colour varies from pale blue to violet. American brooklime can be easily distinguished from alpine speedwell because its flowers grow in definite racemes and are pale blue to violet, while those of alpine speedwell grow in a dense little cluster and are deep blue. As its name implies, American brooklime is found along stream banks and in damp, marshy ground.

Veronica serpyllifolia L.

THYME-LEAVED SPEEDWELL

This speedwell is a creeping perennial, 1 to 2 dm in length, and its glandular-hairy stems root at the base. The blue flowers are borne in a tall, unbranched raceme. The fruit, a capsule, has a distinct notch at the top from which the prominent persistent style arises. Thyme-leaved speedwell is similar to American brooklime, but can be distinguished from the latter by the elongated raceme which grows at the top of the unbranched stem. The racemes of the American brooklime arise from the axils of the upper leaves and give a branched appearance to the plant. Thyme-leaved speedwell is found on moist, montane slopes.

Orobanchaceae BROOMRAPE FAMILY

The plants of the broomrape family are parasitic on the roots of other plants. The family name, broomrape, indicates that some of these plants attack broom, *Sarothamnus scoparius* (*Cytisus scoparius*), in the rose family. The plants possess no chlorophyll, so they are true parasites, and obtain food from the roots of other plants by means of root-suckers called *haustoria*. The leaves, having no photosynthetic function, are reduced to scales. The stems are erect and may be brown or yellow, sometimes purplish, and the flowers may be solitary or arranged in spikes or racemes. Each flower has a 4 to 5-lobed calyx, a 5-lobed corolla of petals, usually 2-lipped like a snapdragon, and 4 stamens. The gynoecium is superior and develops into a capsule with numerous, tiny, dust-like seeds.

Stem with a single flower . . . . *O. uniflora*
Stem with 3 to 10 flowers. . . . . *O. fasciculata*

Orobanche fasciculata Nutt.

CLUSTERED BROOMRAPE, CANCER-ROOT

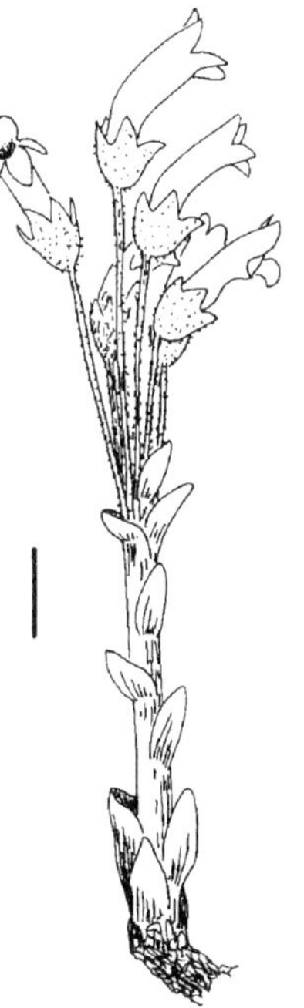

Clustered broomrape is a root-parasite, parasitic on the members of the Asteraceae family (composites), especially *Artemisia* (sagewort). It has fleshy stems covered with scales; these stems arise from the woody caudex, a thick underground stem-base, and they are 3 to 10 cm high, bearing 3 to 10 curved purplish flowers on long (6 cm) flower-stalks. Each flower has a 5-parted calyx and a corolla of 5 petals, slightly 2-lipped with 2 petals forming the upper lip and 3 petals the lower lip. There are 4 stamens, and the gynoecium is superior, developing into a capsule. Clustered broomrape is found on prairie grassland.

NOTES: *The generic name* Orobanche *comes from 2 Greek words:* orobus *(vetch) and* anchien *(to strangle).* ~ *The species name* fasciculata *refers to the fact that the flowers cluster together: the word* fascicle *means "a cluster."*

Orobanche uniflora L.

ONE-FLOWERED CANCER-ROOT, ONE-FLOWERED BROOMRAPE

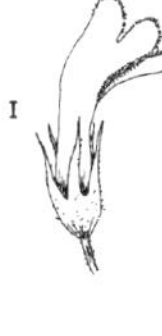

One-flowered broomrape is an inconspicuous, flesh-coloured plant. The short stem is covered with fleshy scales from which arise the tall, leafless flower-stalks, 5.5 to 5.9 cm long, each bearing a single whitish mauve flower, curved and slightly 2-lipped. One-flowered broomrape has a soft caudex, and its stems, which are much shorter than those of clustered broomrape, have only 1 or 2 flowers, hence the specific name *uniflora.* Clustered broomrape has pointed petals and triangular sepals, while one-flowered broomrape has rounded petals, narrow, slender sepals and a taller flower-stalk (9 cm). One-flowered broomrape is found in moist woods.

NOTES: *Specimens of one-flowered broomrape were found in Bow Valley Provincial Park, in the northern part of Kananaskis Country, by Kathleen Wilkinson on June 21, 1990. This was not only a new species for Kananaskis Country but probably the most northerly record for the plant in Alberta; it was reported by Joan Williams.*

Lentibulariaceae BLADDERWORT FAMILY

Members of this family are perennial herbaceous plants, growing in wet places, sometimes submerged in water. They can also be found on damp ground. The plants are particularly interesting because they are insectivorous: they can trap insects and other small organisms by means of specially adapted leaves, either sticky, as in butterwort *(Pinguicula)*, or transformed into bladders, as in bladderwort *(Utricularia)*. The flowers are irregular; there are 2 to 5 sepals, united, and 5 petals forming a 2-lipped corolla, with a long spur. There are 2 stamens, joined to the base of the corolla tube. The fruit is a capsule with 2 valves. These plants are not only adapted for trapping insects but produce enzymes so they can digest their prey. The soils where they grow are low in essential nutrients, especially nitrogen, and their diet of insects helps to remedy this shortfall.

1. Leaves basal; flowers purple; plants terrestrial . . . . *Pinguicula*
   Leaves alternate or opposite; flowers yellow; plants aquatic . . . . . . *Utricularia*

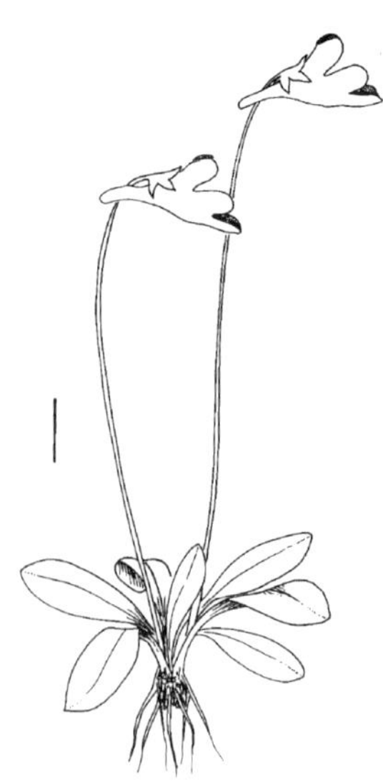

Pinguicula vulgaris L.

COMMON BUTTERWORT

The butterwort flower grows at the top of a tall, leafless flower-stalk, 4 to 20 cm long, and is quite striking, violet-purple in colour. Each flower has a 5-parted calyx, and the corolla consists of 5 petals united to form a bi-lobed, cylindrical corolla-tube. 2 petals form the upper lip, and 3 petals the lower lip; the corolla-tube terminates in a spur. There are 2 stamens attached to the lower part of the corolla-tube, inside it; the stamen filaments are thick and curved, each capped by a spherical anther. The spherical gynoecium has a short style with 2 stigma-lobes; 1 lobe is very short, the other much longer. When fertilization has taken place, the gynoecium develops into a 2-valved capsule. The plant is found in acid bogs, on mossy stream-sides and on wet mossy banks.

NOTES: *In Kananaskis Country, butterwort is often found in calcareous springs around the base of Mount Lorette and in other places.* ~ *Common butterwort traps and digests insects by means of its glandular, stalkless, incurled leaves, which grow in a rosette. These leaves are yellowish green, broadly lanceolate and pointed; their upper surfaces are covered with microscopic glands which produce a greasy exudate. The leaves trap insects on their greasy surfaces, then the glands produce digestive juice to digest their prey.* ~ *The Latin word* pinguis *means "fat," from which derives the generic name* Pinguicula.

Utricularia vulgaris L.

COMMON BLADDERWORT

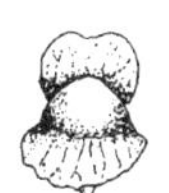

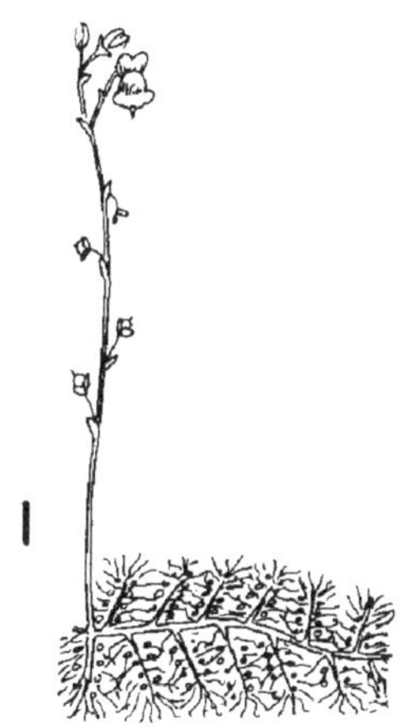

Common bladderwort is a floating or submerged aquatic with its flower stalks above water. It has long, sparsely branched shoots with alternate leaves; these are dissected into feathery filaments bearing small, flat, numerous bladders called utricles, 1 to 3 mm wide and 3 to 5 mm long. These bladders trap tiny water insects and produce digestive enzymes, so that nutrients can be absorbed by the plant. The yellow flowers are borne at the top of a flowering-stalk, 6 to 20 cm tall, raised well above the water. The flower-structure is very similar to that of butterwort, with a 2-cleft calyx, a spurred 2-lipped corolla and 2 stamens, attached to the base of the corolla-tube; the stamens each have a stout filament and 1 anther. The ovary has 2 stigmas, 1 short and 1 long, and the fruit is a capsule which splits to release numerous wrinkled seeds. Vegetative reproduction is carried out with winter buds. Bladderwort is found in shallow standing water (beaver ponds) and lakes at low elevations.

NOTES: *The generic name* Utricularia *refers to the numerous utricles; the specific name* vulgaris *means "common."* *A detailed, illustrated description of the bladderwort trap appears on pp.* *XLV–XLVI.*

Plantaginaceae PLANTAIN FAMILY

Members of the plantain family are coarse plants, either annuals or perennials, and lack leafy stems. The leaves grow in a basal cluster with well-marked veins. There are usually several flowering stalks (peduncles); these are tall and leafless, with cylindrical flower-spikes at the tips. The individual flowers are small, greenish white and inconspicuous; they are usually perfect, except *Plantago elongata*, which is imperfect. Each flower has 4 sepals, united, 4 petals, united, and 2 or 4 stamens. The gynoecium in the centre consists of 2 carpels; the fruit is a *pyxis*.

Plantago major L.

COMMON PLANTAIN, WHITEMAN'S-FOOT

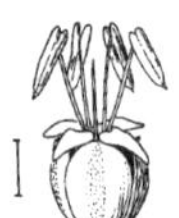

Common plantain is a coarse, weedy perennial with a mass of fibrous roots; its leaves grow in a basal cluster, with ovate leaf-blades up to 8 cm wide, long petioles and well-marked veins. The small flowers grow at the top of a tall flowering-stalk and are densely crowded into a narrow spike. Each flower has 4 sepals and 4 petals fused into a corolla-tube with 4 lobes. There are 4 stamens attached to the inside of the corolla-tube, and the fruit is a capsule, splitting around the centre. This type of capsule is called a *pyxidium* or *pyxis* and is unusual. Common plantain is found in disturbed areas and roadsides and is often a weed in lawns. NOTES: *The upper half of the pyxis is supposed to resemble a pixie's cap. The common name "whiteman's-foot" is supposedly derived from the fact that plantain was brought to North America by Europeans and was often found around their dwellings.*

Rubiaceae MADDER FAMILY

Members of this family are herbaceous plants, either annuals or perennials, with slender stems that are usually 4-angled. The stems bear entire leaves in whorls of 3 to 8; this is a distinctive characteristic of bedstraw (*Galium*) and woodruff (*Asperula*), but is not found in bluets (*Houstonia*), where the leaves are opposite. The small, regular flowers are in branched clusters, arising from the leaf-whorls (if these are present). Each flower has no sepals, 3 or 4 petals joined together (gamopetalous) with 3 or 4 lobes, and 3 or 4 stamens arising from inside the corolla-tube. The gynoecium of 2 carpels is inferior, so the flower is epigynous. The fruit is either a capsule or 2 globose, 1-seeded nutlets.

Galium L. — BEDSTRAW

These species are slender perennial herbs with 4-angled stems. The leaves are simple and sessile, in whorls of 4 to 6. The small flowers are terminal, axillary or in a 3-forked cluster. The calyx is absent, and the corolla has 4 petals, united and lobed. The androecium has 4 stamens, the gynoecium has 2 carpels and the fruit forms as 2 1-seeded nutlets.

1. Leaves in whorls of 4 2
   Leaves in whorls of 5 or 6, fruits covered with hooked bristles . . . *G. triflorum*
2. Leaves 1-nerved, fruits smooth . . . . *G. trifidum*
   Leaves 3-nerved, fruits densely hairy . . . . *G. boreale*

Galium boreale L.

NORTHERN BEDSTRAW

Northern bedstraw is a perennial with erect square stems, 3 to 6 dm high, and often grows in large colonies. The leaves are arranged in whorls of 4 up the stem and are linear-lanceolate. The flowers grow in dense clusters at the tops of the stems or in the axils of the leaf-whorls, and each flower has 4 white petals in the shape of a cross (no sepals). There are 4 stamens alternating with the petals, and the fruit is a densely hairy schizocarp. Although the individual flowers are small, they grow in dense clusters, and the tall plants grow in colonies, so when all the plants are in bloom, the effect is quite striking. Northern bedstraw is found on prairies, roadsides and moist valleys.

NOTES: *The specific name* boreale, *which means "northern," is derived from Boreas, the Greek god of the north wind.* *The common name "bedstraw" comes from the use of a related species, sweet-scented bedstraw* (G. triflorum), *to "sweeten" straw mattresses: the leaves are sweet-scented when dried.* *The Blackfoot boiled the roots of northern bedstraw to obtain a red dye.*

Galium trifidum L.

SMALL BEDSTRAW

Small bedstraw grows only 1 to 3 dm high, and the slender, weak stems are many-branched and often matted—a different growth-habit from that of northern bedstraw, which makes it easy to distinguish the species. The corolla is 3-lobed, not 4-lobed, and the stem-leaves are smaller. The fruits are also different: in northern bedstraw they are densely hairy; in small bedstraw they are smooth. Small bedstraw grows in bogs and marshes and along stream banks.

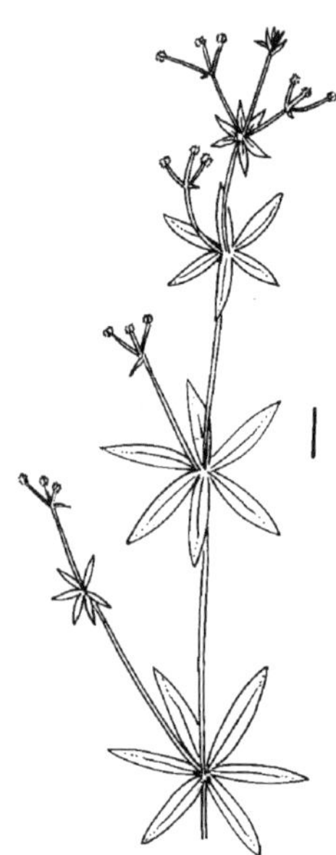

Galium triflorum Michx.

SWEET-SCENTED BEDSTRAW

Sweet-scented bedstraw is a perennial with weak stems, often spreading. It is sparsely branched and bears 5 or 6 leaflets in whorls up the stem, a distinguishing trait. Sweet-scented bedstraw is similar to small bedstraw in growth-habit, although much larger, but their fruits are quite different: the fruits of sweet-scented bedstraw are covered with hooked bristles. Sweet-scented bedstraw grows in moist woods.

NOTES: *Blackfoot women dried sweet-scented bedstraw flowers and used them as perfume.*

Caprifoliaceae HONEYSUCKLE FAMILY

Members of the honeysuckle family are shrubs, erect or twining; the twin-flower, *Linnaea*, is exceptional because it is a perennial, semi-woody, trailing herb. The leaves are in pairs, and the flowers are perfect; they are arranged in terminal clusters or in lateral pairs with a common stalk. Each flower has a 5-lobed calyx, sometimes absent, and 5 petals, fused, with 5 lobes or sometimes 2-lipped, as in twining honeysuckle (*Lonicera dioica*). There are 5 stamens (4 in twinflower), which are joined to the corolla; the gynoecium is inferior, that is, it is enclosed by the hypanthium, with the sepals, petals and stamens above it. The fruit is either a berry-like drupe, as in elderberry, or dry and 1-seeded, as in twinflower (*Linnaea borealis*).

1. Leaves compound . . . . *Sambucus*
   Leaves simple 2
2. Prostrate shrub; leaves with 2 pairs of notches near the tip . . . . *Linnaea*
   Erect shrubs and woody vines 3
3. Flowers orange or yellow, woody vine . . . . *Lonicera*
   Flowers white 4
4. Leaves lobed or toothed . . . . *Viburnum*
   Leaves entire . . . . *Symphoricarpos*

Linnaea borealis L.

TWINFLOWER

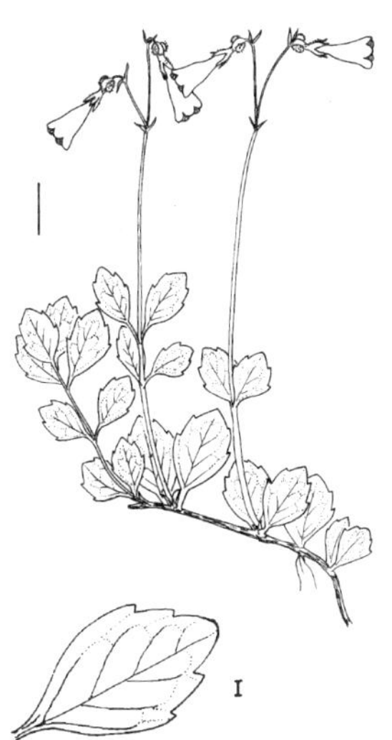

Twinflower is a dainty plant with long runners that creep over mossy forest floors and rotting logs. These runners produce little stems, 5 to 10 cm tall, which branch, bearing the "twin-flowers"—small, trumpet-shaped, rose-coloured flowers in pairs with a common stalk. Each flower has 4 calyx-lobes, a corolla-tube with 5 lobes, and 4 stamens attached to the inside of the corolla-tube; the gynoecium has 3 carpels. The fruit is dry, 1-seeded and enclosed by bracts, which are glandular-pubescent. This delicate, attractive flower is found in damp woodlands.

NOTES: *The genus,* Linnaea, *was named in honour of Carolus Linnaeus (1707-1778) by his friend, Gronovius. Linnaeus himself named this species* borealis. *He was fond of the twinflower which grows in the forests of Sweden and Lapland; fortunately for us, it is found in western Canada in British Columbia and Alberta. It is a circumpolar plant: it grows in Alaska, Greenland, Russia and China.* *The specific name* borealis *comes from Boreas, the Greek god of the north wind.*

Lonicera L. — HONEYSUCKLE

Honeysuckles are erect or twining shrubs. The leaves are entire and opposite; the flowers grow in pairs or in terminal spikes. The sepals are fused into a lobed rim above the inferior ovary. Each flower bears a corolla of 5 petals, united, with 5 lobes. An androecium of 5 stamens rises from the corolla-tube. The fruit is a red or purple-black berry.

Stems twining; flowers orange to red . . . . *L. dioica*
Stems erect; flowers yellow, borne in pairs . . . . *L. involucrata*

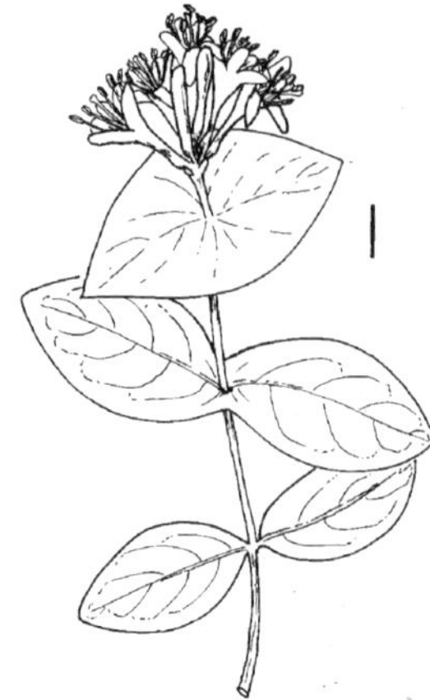

Lonicera dioica L.

TWINING HONEYSUCKLE

Twining honeysuckle, as its common name implies, is a climbing shrub that twines around low shrubs and tree-trunks in open woods and on rocky slopes. The leaves grow in pairs up the stem, and the pair near the top of the stem fuses together. The flowers grow at the top of the stem, in a cup formed by the top 2 joined leaves. These flowers are yellow when they first open, then change to orange-red. The calyx-teeth are short, the corolla-tube is 2-lipped, there are 5 stamens and the fruit is a red berry.

NOTES: *The generic name* Lonicera *honours Adam Lonitzer, a German botanist and physician.*

Lonicera involucrata (Richards.) Banks

BLACK TWINBERRY, BRACTED HONEYSUCKLE

Bracted honeysuckle is a bushy shrub, 1 to 2 m tall, with bright green leaves growing in pairs on the twigs. The flowers grow in the axils of the leaves in pairs; they are subtended and partly enclosed by 4 bracts, which are greenish purple and distinctive. These bracts form an involucre around the flower, hence the specific name *involucrata*. The calyx-teeth of the flower are short, and the tubular corolla is yellow, dilated at the base, 5-lobed. There are 5 stamens, and the fruit is a purplish black berry, with the persistent bracts attached to it. Bracted honeysuckle is found in moist woodlands.

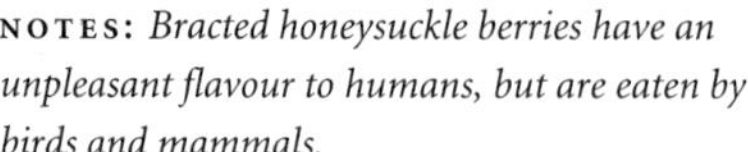

NOTES: *Bracted honeysuckle berries have an unpleasant flavour to humans, but are eaten by birds and mammals.*

Sambucus racemosa L.

ELDERBERRY, RED-FRUITED ELDER

Elderberry is a large shrub, 1 to 3 m tall, that often forms colonies; the twigs are filled with pith and bear leaves in pairs. These leaves are pinnate, with 5 to 7 serrate (toothed) leaflets. The flowers grow in a large terminal panicle (a raceme of racemes), which yields the specific name *racemosa*. The panicle is quite striking with numerous, creamy yellow flowers. Each flower has a 5-parted corolla-tube (no sepals) with 5 stamens, and the fruit is either red (variety *pubens*) or blackish purple (variety *melanocarpa*). Elderberry is found in moist woods, usually at low altitudes, but it also occurs at Highwood Pass just below timberline.

Symphoricarpos Duhamel SNOWBERRY, BUCKBRUSH

These species are low, bushy shrubs that form large colonies from underground rootstocks. The leaves are ovate to elliptic, and the flowers are terminal or axillary, in short clusters. Each flower has a calyx of 5 sepals, united, with short calyx teeth; a corolla of 4 to 5 petals, united; and an androecium of 4 to 5 stamens, epipetalous. The fruit is soft, white or greenish, 2-seeded and berry-like.

Stamens shorter than the corolla; fruit white .... *S. albus*
Stamens longer than the corolla; fruit green to purple .... *S. occidentalis*

Symphoricarpos albus (L.) Blake

SNOWBERRY

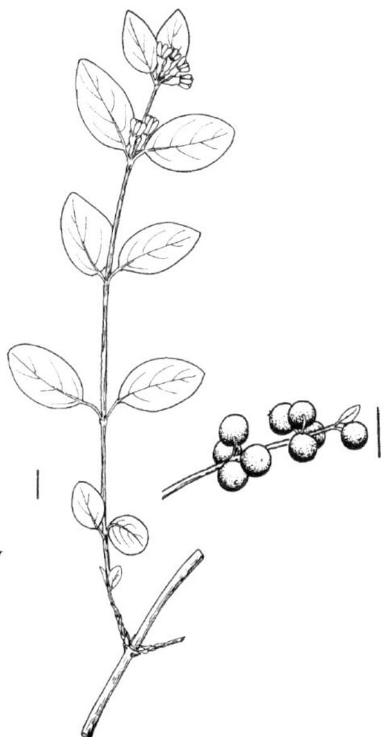

Snowberry is a low, branched, erect shrub up to 1 m high, with oval leaves growing in pairs on the branches. It forms large colonies from underground rootstocks. The flowers grow in a dense cluster in the axils of the upper leaves; they are white or pink. Each flower has a calyx with short teeth, a bell-shaped corolla of 5 petals, and 4 to 5 stamens arising from inside the corolla tube, which is hairy within. These stamens are usually "included," that is, they do not extend beyond the corolla tube. The fruit is white and berry-like, hence the common name "snowberry." Snowberry is found on valley slopes and in open woodland, usually on exposed sites.

NOTES: *The specific name* albus *is Latin for "white."*

Symphoricarpos occidentalis Hook.

WOLFBERRY, BUCKBRUSH

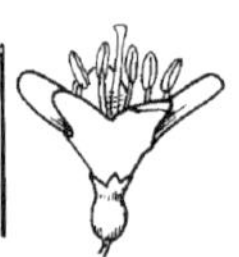

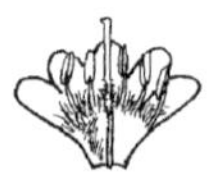

Buckbrush is similar to snowberry and it can be difficult to tell them apart. There are 3 characteristics that help to identify them. Snowberry has stamens that are included, while the stamens in the flower of buckbrush are exserted (extend beyond the corolla tube). The style, extending from the ovary, is much longer in buckbrush and can also be seen beyond the corolla tube. The fruits of snowberry, when ripe, are pure white, but the fruits of buckbrush are smaller and are greenish white, later purplish. Buckbrush is found on prairie grassland and in open woodland.

NOTES: *Both snowberry and buckbrush were used by the Blackfoot. The leaves were steeped to make an infusion for sore eyes, and the slender twigs were used to make arrow shafts. The fruits were used as famine food.*

Viburnum edule (Michx.) Raf.

LOW-BUSH CRANBERRY, MOOSEBERRY

Low-bush cranberry is a shrub, 1 to 2 m high, with large leaves, 6 to 10 cm long, growing in pairs up the twigs. The white flowers grow in cymes (flat-topped clusters); these clusters have long stalks (peduncles) which arise in the leaf axils. The outer flowers of the cluster are neutral (no stamens or ovaries) but the true flowers each have a 5-toothed calyx and a white corolla-tube with 5 lobes. There are 5 stamens, and the fruit is 1-seeded, light red and edible, hence the specific name *edule*. Low-bush cranberry is found in moist woods.

NOTES: *Unlike low-bush cranberry, "bog cranberries" belong to the heather family,* Ericaceae.

Valerianaceae VALERIAN FAMILY

Valerians are herbaceous perennials with erect, leafy, unbranched stems arising from scented rootstocks. The leaves grow in pairs and are either entire or pinnately divided. The flowers grow in dense clusters at the tops of the stems; later these clusters become elongated and open. The flowers are epigynous; the sepals form a ring of bristles, and there are 5 petals, joined to form a corolla-tube, with 5 rounded lobes. There are 3 stamens arising inside the corolla-tube, but protruding from it. When the fruit is ripe, it is crowned by the calyx (bristles), which develops into a feathery pappus, similar to the pappus found on the fruit of the dandelion. A pappus is useful for fruit dispersal.

Stem-leaves with 3 to 7 lobes; corolla 5 to 8 mm long. . . . . *V. sitchensis*
Stem-leaves with 9 to 15 lobes; corolla 1 to 3 mm long. . . . . *V. dioica*

Valeriana dioica L.

NORTHERN VALERIAN

(*V. septentrionalis* Rydb.)

Northern valerian is 3 to 7 dm tall, and the basal leaves are 2 to 5 cm long. These leaves are lanceolate or spatulate and entire; they have long leaf-stalks. The cauline (stem) leaves, 2 to 4 pairs, are pinnately lobed, with 3 to 7 lobes. The flowers grow together in clustered cymes at the top of a tall stalk (peduncle). Each flower has a calyx made up of several bristles, a white, tubular, 5-lobed corolla, 3 stamens arising from inside the corolla-tube and an inferior gynoecium, with 1 ovule. The gynoecium ripens into a dry, 1-seeded fruit, crowned by the calyx bristles, which become feathery. Northern valerian is found in wet mountain meadows and bogs.

NOTES: *The Blackfoot made a hot drink from the root of northern valerian and used it to treat stomach trouble.*

Valeriana sitchensis Bong.

SITKA VALERIAN, MOUNTAIN VALERIAN

Mountain valerian is 3 to 10 dm tall. It is similar to the other species of valerian, but can be distinguished by several features. The basal leaves are usually 3 to 5-lobed, the stem leaves have 3 to 5 lobes and the corolla is 5 to 8 mm long. Mountain valerian is found in mountain woods and subalpine slopes. It has been recorded from near the Highwood Pass, near the summit of Moose Mountain, on the Elbow Lake Trail and in other areas.

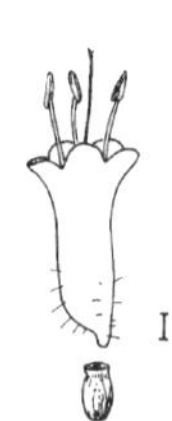

NOTES: *The species name* sitchensis *records the fact that the botanist Mertens first collected this plant at Sitka, Alaska, shortly before 1833.* *Both the valerians are malodorous.*

Dipsacaceae TEASEL FAMILY

Members of the teasel family are perennial, herbaceous plants with erect stems and opposite leaves. The leaves are pinnately divided into lanceolate leaflets, and the small flowers are arranged in dense heads, similar to the *capitula* of Asteraceae (Compositae). The tall flowering-stems bear these dense heads, surrounded by involucral bracts. Each flower has a cup-shaped calyx with 8 awns, a pink or purplish corolla 9 to 12 mm long, bearing 4 stamens. The gynoecium is inferior and wrapped in an involucel; it develops into an achene, a dry, 1-seeded fruit.

Knautia arvensis (L.) Duby

BLUE BUTTONS

Blue buttons is a perennial plant growing from a taproot, 4 to 9 dm tall, with a branched stem. The lower leaves are usually coarsely toothed, but the other leaves are compound, pinnately divided. The tall main flowering-stalks (peduncles) are leafless and bear the flower-heads, 1.5 to 4 cm wide. The flower-head, of numerous small flowers, is surrounded by an involucre of bracts each 8 to 15 mm long. The individual flowers have 8 sepals forming the calyx; they have awn-like bristles and form a cup. The tubular corolla has 4 to 5 lobes and is lilac-purple; the flowers on the margin of the flower-head are larger than the others. There are 4 stamens arising from inside the corolla-tube and appearing above it (extruded).

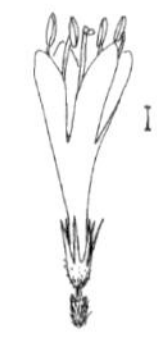

The gynoecium is inferior and turns into a hairy achene enveloped in an involucel and crowned by the persistent spiky calyx. Blue buttons is an introduced plant, found on waste ground, in fields and along roadsides.

NOTES: *The generic name* Knautia *honours Christian Knaut (1654-1716), a German botanist and physician from Saxony.*

Campanulaceae HAREBELL FAMILY

Harebells are perennial herbaceous plants with alternate leaves. The attractive, showy flowers are usually blue; they may be solitary or growing in a raceme or panicle. Each flower has 5 united sepals with well-marked "teeth," and 5 petals forming a bell-shaped corolla (funnelform in *Campanula uniflora*). There are 5 stamens attached to the base of the corolla but free from the bell-shaped corolla-tube. The gynoecium in the centre consists of 3 to 5 carpels; it is inferior (below the rest of the floral structures), so the flower is epigynous. There is a long style with a trifid stigma, often a conspicuous feature of the flower. The fruit is a many-seeded, papery capsule, which opens by pores near the base. Members of this family are often called "bluebells."

Flowers 1 per stem; alpine slopes . . . . *C. uniflora*
Flowers several per stem; dry open areas of low elevations . . . . *C. rotundifolia*

Campanula rotundifolia L.

HAREBELL, BLUEBELL

Harebell is a dainty plant with large, bell-shaped flowers that hang downwards or at right angles to the stem. The flowers are a delicate shade of blue. The basal leaves are round or heart-shaped (hence the specific name *rotundifolia*), but as these leaves fall off very soon, people sometimes wonder why the plant was named "round-leaved" because the stem leaves (cauline) are long, narrow and alternately borne on the delicate stem. Each flower has a calyx with 5 pointed lobes, and the corolla has 5 united petals. There are 5 stamens, distended at the base, and the ovary has a long style ending in a trifid stigma—this can be clearly seen inside the "bell." The fruit is a papery, many-seeded capsule.

NOTES: *Harebell is found in a wide variety of habitats, from low elevations to the alpine region.*

The Latin word campanula *means "little bell," referring not only to the generic name, but to the family name.*

Campanula uniflora L.

ALPINE HAREBELL

Alpine harebell is a tiny plant: it grows to 1 dm high or less, and usually has 1 flower, hence the specific name *uniflora.* The corrolla of the flower is not bell-shaped but funnel-shaped. The 2 harebell species are easily distinguished. Alpine harebell is exclusively an alpine species.

Lobeliaceae LOBELIA FAMILY

The members of this family are usually herbs, occasionally shrubs or small trees, with simple, alternate leaves. The flowers are irregular (zygomorphic) with parts in 5s; the fruit is a capsule. The genus name *Lobelia* honours the botanist L'Obel, who was researching with roses in 1581.

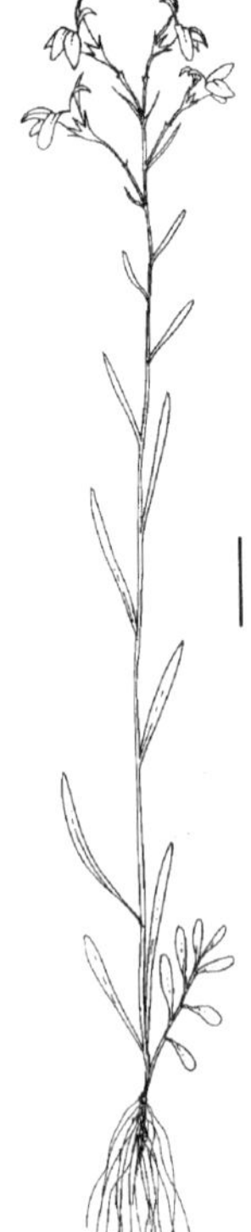

Lobelia kalmii L.

BROOK LOBELIA, KALM'S LOBELIA

Kalm's lobelia is a biennial with leafy stems arising from a basal rosette of spoon-shaped (spatulate) leaves; it is 1 to 5 dm high. The stem-leaves are lance-shaped, and the stem bears a few flowers in a loose raceme with slender stalks. Each flower has 5 sepals, partly united, and 5 blue petals, united, forming a corolla-tube with 2 lips (bi-labiate). The upper lip has 2 erect lobes, and the lower lobe is 3-cleft. The flower has a white throat and occasionally the corolla is all white. There are 5 stamens. The fruit is a 2-lobed capsule which splits to release many seeds. Kalm's lobelia is found in wet, calcareous meadows and bogs and on pond margins.

NOTES: *The specific name* kalmii *honours Peter Kalm, a student of Linnaeus. Kalm travelled widely in the English colonies of North America and in New France during the eighteenth century, in search of new flowers, many of which he sent back to Europe.*

Asteraceae (Compositae) DANDELION FAMILY

Asteraceae is the most highly developed family in the class Dicotyledoneae and has the most complex flower-structure. It is the most successful family, distributed worldwide, and members of the family are found in a wide variety of habitats. 2 factors affect its success: an extremely effective method of pollination and an equally effective method of fruit dispersal.

They are herbs, or, rarely, shrubs, with variable growth-habits. The individual flowers are always arranged in a cluster called a *capitulum*. The flowers arise from a common receptacle that is either flat or curved; the receptacle is surrounded by an involucre of bracts. Scales or tiny bracts are often present on the receptacle between the individual flowers. These scales are often referred to as "chaff." There are 2 distinct types of flowers (usually called "florets") in this family: the tubular or disc florets, and the ligulate or ray florets.

DISC OR TUBULAR FLORETS

The tubular or disc florets each have 5 petals, united to form a tubular corolla . The petals are free at the tips. The corolla is often surrounded by a ring of bristles, hairs or scales, which represent a modified calyx; these later develop into a pappus. There are 5 stamens with delicate filaments, which rise from inside the corolla-tube and bear elongated anthers which are usually united into a ring. These stamens are epipetalous. The gynoecium consists of 2 united carpels, with a well-marked style bearing 2 stigmatic lobes; the carpels are inferior (below the tubular corolla) so the flower is epigynous. The carpels form the fruit, a dry, 1-seeded inferior achene (called a *cypsela*). The fruit is often crowned by a feathery pappus, as in the thistles.

LIGULATE OR RAY FLORETS

Ligulate florets look quite different from tubular florets because the 5 petals are united to form a flat, strap-shaped portion which is only fused into a tube at the base. The florets of the dandelion show this development well because this flower-head has no tubular florets, only ligulate florets. The tip of the strap may be lobed or toothed (e.g., *Gaillardia*, where the flower-heads have 3-cleft ligulate florets). Ligulate florets are bilaterally symmetrical, and at the base the narrow tube is surrounded by a ring of tiny bracts or scales—a modified calyx. There are 5 stamens attached to the inside of the narrow tube-like portion of the corolla. These have delicate filaments and elongated anthers, joined in a ring, similar to the stamens in the tubular florets. The gynoecium is also similar; it consists of 2 united carpels, which are inferior and develop into a dry, 1-seeded achene. The tiny bracts or scales of the modified calyx often develop into a feathery pappus; the familiar dandelion fruit is a good example.

TYPES OF FLOWER-HEADS

There are 3 types of flower-heads in Asteraceae. The discoid ones have only disc or tubular florets (e.g., thistles). Ligulate flower-heads have only ligulate florets (e.g., dandelion). Radiate heads have disc florets in the centre surrounded by ligulate or ray florets (e.g., gaillardia).

INVOLUCRAL BRACTS

Involucral bracts are characteristic of Asteraceae. The involucre is a ring of bracts encircling the flower-head; these bracts are modified leaves and sometimes look like the sepals of a flower. They perform the same function as sepals because they protect the flower-head when in bud. The bracts are often useful in helping to determine a genus, and there are several different types: they may be overlapping (imbricate), as in the asters, or they may form two distinct series, an outer and an inner, as in hawksbeard *(Crepis)*. The bracts may be hooked, spiny or hairy: there is a range of variations. The colour may also vary: they may be black-tipped, as in groundsel (*Senecio lugens*), white or sometimes rose-coloured, as in pussytoes (*Antennaria*).

POLLINATION

Members of Asteraceae are so successful because they have excellent pollination mechanisms. In thistles, for example, the disc florets develop in 2 stages. In the first stage, the 2-lobed stigma, still tightly closed, grows up through the ring of stamens. These shed their pollen inwards so that the stigmas, still tightly closed and not receptive, become covered with pollen and emerge at the top of the corolla tube. In the second stage, the stigma-lobes open and become receptive. Any insect hovering or walking over the flower-head picks up pollen from the florets in the first stage and deposits the pollen grains on the receptive stigma-lobes in the second stage, a practically fool-proof method to ensure cross-pollination. If cross-pollination fails to take place, the stigmatic lobes grow downwards and collect pollen from the stamens of the same floret, ensuring self-pollination. Cross-pollination takes place not only in the discoid flower-heads but also in the ligulate florets.

FRUIT DISPERSAL

Members of this family have also developed an efficient method of fruit dispersal. Many genera have fruits with a feathery pappus, which are dispersed by the wind. There are numerous examples of this trait: goatsbeard (*Tragopogon*), false dandelion (*Agoseris*), sow thistle (*Sonchus*), groundsel (*Senecio*), daisy (*Erigeron*) and asters.

1. Leaves basal 2
   Leaves alternate or opposite 8
2. Plants with milky juice 3
   Plants without milky juice 6
3. Leaves lobed 4
   Leaves not lobed . . . . *Agoseris*
4. Flowering-stalk hollow 5
   Flowering-stalk not hollow . . . . *Crepis*
5. Involucral bracts reflexed; upper surface of leaf glabrous . . . . *Taraxacum*
   Involucral bracts not reflexed; upper surface of leaf hairy . . . . *Agoseris*
6. Flowering-stalk appearing before the leaves . . . . *Petasites*
   Flowering-stalk appearing at the same time or after the leaves 7
7. Edges of involucral bracts irregularly toothed . . . . *Townsendia*
   Edges of involucral bracts entire . . . . *Erigeron*

8. Leaves opposite 9
Leaves alternate 10
9. Leaves and stems with stiff, rough hairs; fruits without silky hairs . . . . . *Helianthus*
Leaves and stems with soft, silky hairs; fruits with silky hairs . . . . . *Arnica*
10. Plants with milky juice 11
Plants without milky juice 14
11. Stems with solitary flower-heads 12
Stems with many flower-heads 13
12. Involucral bracts 5 to 13 (usually 13); fruits with feathery hairs . . . *Tragopogon*
Involucral bracts more than 13 . . . . . *Hieracium*
13. Involucral bracts in 1 to 2 series . . . . . *Crepis*
Involucral bracts in several series . . . . . *Sonchus*
14. Leaves deeply lobed 15
Leaves entire 20
15. Flowers yellow 16
Flowers not yellow 17
16. Leaves bristly-hairy; disc florets reddish purple . . . . . *Gaillardia*
Leaves not bristly-hairy; disc florets yellow . . . . . *Senecio*
17. Flowers white 18
Flowers green 19
18. Leaves and stems woolly, fragrant . . . . . *Achillea*
Leaves and stems glabrous . . . . . *Matricaria*
19. Crushed leaves have a sage odour . . . . . *Artemisia*
Crushed leaves have a pineapple odour . . . . . *Matricaria*
20. Flowers yellow 21
Flowers not yellow 24
21. Plants glabrous 22
Plants hairy 23
22. Flower-heads sticky . . . . . *Grindelia*
Flower-heads not sticky . . . . . *Solidago*
23. Plants with soft hairs . . . . . *Haplopappus*
Plants with stiff, rough hairs . . . . . *Heterotheca*
24. Leaves prickly . . . . . *Cirsium*
Leaves not prickly 25
25. Flowers white or blue 26
Flowers pink or purple 29
26. Stems with large, well-developed leaves 27
Stems with small leaves 28
27. Leaves densely woolly . . . . . *Anaphalis*
Leaves hairy but not woolly . . . . . *Aster*
28. Involucral bracts white . . . . . *Antennaria*
Involucral bracts green . . . . . *Erigeron*
29. Stems leafy . . . . . *Aster*
Stems not leafy; leaves mostly basal 30
30. Flowers pink or cream-coloured 31
Flowers purple 32

31. Involucral bracts pink, white or black-tipped .... *Antennaria*
Involucral bracts green .... *Erigeron*
32. Involucral bracts in a series of 1 or 2; leaves without cobweb-like hairs .... *Townsendia*
Involucral bracts in a series of 4 or 5; leaves with cobweb-like hairs......... .... *Saussurea*

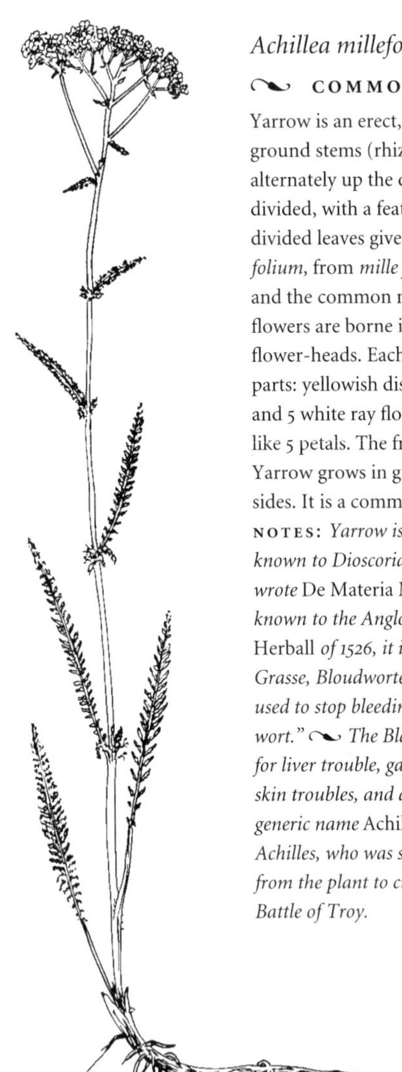

Achillea millefolium L.

COMMON YARROW, MILFOIL

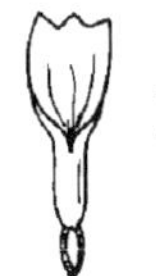

Yarrow is an erect, aromatic perennial with underground stems (rhizomes). The leaves grow alternately up the densely hairy stem and are finely divided, with a feathery appearance. These finely divided leaves give rise to the specific name *millefolium*, from *mille feuilles* ("a thousand leaves") and the common name "milfoil." The white flowers are borne in flat clusters of numerous flower-heads. Each flower-head is made up of 2 parts: yellowish disc florets (3 to 10) in the centre and 5 white ray florets on the outside, which look like 5 petals. The fruit is a dry, 1-seeded achene. Yarrow grows in grassy places and along roadsides. It is a common prairie plant.

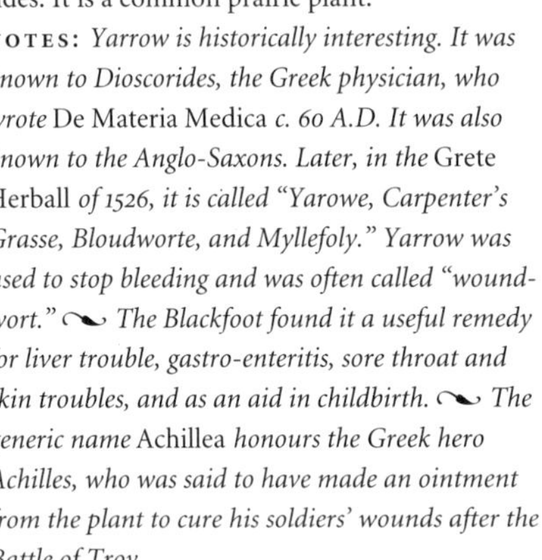

NOTES: *Yarrow is historically interesting. It was known to Dioscorides, the Greek physician, who wrote* De Materia Medica *c. 60 A.D. It was also known to the Anglo-Saxons. Later, in the* Grete Herball *of 1526, it is called "Yarowe, Carpenter's Grasse, Bloudworte, and Myllefoly." Yarrow was used to stop bleeding and was often called "woundwort."* ~ *The Blackfoot found it a useful remedy for liver trouble, gastro-enteritis, sore throat and skin troubles, and as an aid in childbirth.* ~ *The generic name* Achillea *honours the Greek hero Achilles, who was said to have made an ointment from the plant to cure his soldiers' wounds after the Battle of Troy.*

Agoseris Raf. FALSE DANDELION

These are perennial, herbaceous plants rising from a strong taproot with tufted basal leaves, entire or pinnatifid. The solitary flower-heads grow on leafless stalks, sometimes more than 1 per plant, made up of yellow or orange ligulate florets. The involucral bracts are imbricated. The fruit is a cypsela (inferior achene) with a well-developed white pappus. These species contain a milky latex.

Leaves green; flowers burnt-orange . . . . *A. aurantiaca*
Leaves blue-grey (glaucous); flowers yellow . . . . *A. glauca*

Agoseris aurantiaca (Hook.) Greene

ORANGE FALSE DANDELION

Orange false dandelion is a perennial plant with a tap-root. At first sight it resembles a dandelion: both have basal leaves growing in a tuft, both have solitary flower-heads borne on long leafless stalks (peduncles) and both contain a sticky, milky juice. The orange florets of *A. aurantiaca* are ray-florets—there are no disc florets. The bracts beneath the flower-head are in several series and overlapping (imbricate). In the true dandelion, the bracts are in 2-series; the second series is markedly reflexed. The lanceolate leaves are usually entire. Orange false dandelion is found in meadows, prairies, and open habitats from the montane to the alpine zone.

NOTES: *This species has been found in Kananaskis Country near Marmot Creek.*

Agoseris glauca (Pursh) Raf.

YELLOW FALSE DANDELION

The false dandelion species can be difficult to distinguish, apart from the colour of the florets. The leaves in the yellow species are blue-grey (glaucous, hence the specific name), while the leaves of the orange species are not. Both have achenes with beaks to which the pure white pappus bristles are attached, forming a parachute. In the orange species, the beak is at least as long as the fruit itself; in the yellow species, the beak is gradually tapered and is not as long as the fruit. Yellow false dandelion grows on prairies, in meadows and on subalpine and alpine slopes. It is found in Kananaskis Country near Barrier Mountain along an old forestry road, altitude 1400 m, and near the main road at the base of Plateau Mountain. It has also been found near Marmot Creek.

Anaphalis margaritacea (L.) Benth. and Hook.

PEARLY EVERLASTING

Pearly everlasting is a perennial, 25 to 30 cm tall, growing from rhizomes. The lance-shaped leaves grow alternately up the stem; they are green above and white-woolly underneath (tomentose). The flower-heads are pearly white, hence the common name, but the colour is actually due to the "everlasting" involucral bracts which encircle the composite flowers. There are no ray florets, only disc florets, which may be either male or female. The pappus bristles are dirty-white. The fruit is *papillose*: covered with papillae (minute projections). Pearly everlasting is found in open woods at low to middle altitudes, but is not found in the alpine zone.

NOTES: *The specific name* margaritacea *means "pearly."* *Pearly everlasting is sometimes used for dried-flower arrangements.*

Antennaria Gaertn. PUSSYTOES, EVERLASTING

All the pussytoes have a similar growth-habit: they are perennials, often mat-forming, with short, creeping branches. The flowering stalks (peduncles) usually arise from a basal leaf-rosette. The leaves are usually entire, simple and alternate, chiefly basal. The flower-heads have disc florets only, usually in a dense inflorescence; the involucral bracts are thin, papery and "everlasting," hence the common name. The small flowers are clustered together at the top of the flower-stalk and may be either male or female. Staminate flowers are either white or pink, with a tubular corolla; carpellary flowers have a filiform corolla and 2 inferior carpels, united, with a 2-cleft style. The fruits are cypselas, each with a dense pappus. The name "pussytoes" comes from a fancied resemblance between the flower-heads of some species of *Antennaria* and a kitten's paws.

1. Involucral bracts pink . . . . *A. rosea*
   Involucral bracts white or black 2
2. Plants less than 20 cm tall; alpine and subalpine meadows . . . . *A. lanata*
   Plants more than 20 cm tall; dry areas of lower elevations . . . . *A. pulcherrima*

Antennaria lanata (Hook.) Greene

WOOLLY PUSSYTOES, WOOLLY EVERLASTING

Woolly pussytoes grows only 1 to 2 dm tall and gets its common name because all its green parts are densely felty-hairy—even the involucral bracts are hairy at the base, a characteristic that helps to distinguish it from the other species of *Antennaria.* The bracts are brown or greenish black below, paler at the tips. Woolly pussytoes is found in alpine and subalpine meadows.

Antennaria pulcherrima (Hook.) Greene

SHOWY EVERLASTING, SHOWY PUSSYTOES

Showy everlasting is similar to woolly everlasting, but several differences help to distinguish them. Showy everlasting grows 2 to 5 dm tall. Its stem-leaves are less than half as long as the basal leaves. Woolly pussytoes is much "woollier" than showy everlasting, and the latter has a dark spot at the base of each involucral bract, missing in woolly pussytoes. Showy everlasting is found in moist grassy areas and in open woods.

NOTES: *This species has been found in Kananaskis Country at Barrier Creek, on Marmot Creek at 1600 m and elsewhere.*

Antennaria rosea Greene

ROSY PUSSYTOES, PINK PUSSYTOES

Pink pussytoes can be distinguished from the other everlasting species because the involucral bracts are tinged reddish pink or occasionally deep red. The basal leaves are much smaller than those of the other species, and the stem-leaves are also smaller. The whole plant is also less woolly.

NOTES: *Pink pussytoes is a prairie species, but is also found in montane areas. These plants often have redder bracts: a specimen collected near Picklejar Creek in 1962 still shows deep red bracts. A specimen was found on Kent Ridge, above the Smith-Dorrien Highway, altitude 1900 m; it has been seen in other places in Kananaskis Country.* ~ *The Blackfoot used the leaves of this plant in a tobacco mixture.*

Arnica L. ARNICA

These species are perennials, arising from rhizomes, with erect stems bearing opposite, lance-shaped to heart-shaped leaves; the leaves are usually entire. The flower-heads may be solitary or several, with disc florets and conspicuous ray florets, both types yellow. The involucral bracts grow in 1 or 2 series. The fruits are slender achenes (cypselas), bearing a white or brownish pappus.

1. Lower leaves heart-shaped . . . . *A. cordifolia*
   Lower leaves not heart-shaped 2
2. Flower-heads 1; stem-leaves 2 to 6; alpine and subalpine habitats . . . . *A. angustifolia*
   Flower-heads 1 to 9; stem-leaves 4 to 6; lower elevations . . . . . *A. lonchophylla*

Arnica angustifolia M. Vahl

ALPINE ARNICA

(*A. alpina* (L.) Olin)

Alpine arnica is a small plant, less than 2 dm tall, found on rocky ridges and exposed sites at high elevations. The plant is white-hairy, growing from a stout rootstock, and often minutely glandular, with 1 to 3 pairs of leaves. The solitary flower-head bears ray florets and disc florets. The ray florets are brilliant yellow, surrounded by an involucre of white-hairy bracts. There is often a ring of white hairs just below the involucre.

NOTES: *Alpine arnica has been found on the summit of Plateau Mountain, at 2500 m, and also on Plateau Mountain slope, at 2100 m. It has been found near the summit of Hailstone Butte at 2300 m and near Pasque Mountain at 2100 m. It has also been recorded from a "fjaeld field" (an area covered with large stones), at 2400 m, below the summit of Plateau Mountain and elsewhere.*

Arnica cordifolia Hook.

HEART-LEAVED ARNICA

Heart-leaved arnica grows from a rhizome, 2 to 6 dm tall, and its basal leaves are heart-shaped, hence the common name. The basal leaves have long petioles and are distinctive. The stem-leaves grow in pairs (2 to 4 pairs); the upper leaves are small and have no petioles. There are 1 to 3 large, showy flower-heads. The ray florets are 2 to 3 cm long and brilliant yellow; the disc florets are also yellow. The pappus is white or nearly so. The involucre of bracts surrounding the flower is usually glandular-hairy. Heart-leaved arnica can be distinguished from alpine arnica by its size and its heart-shaped leaves, but also by its habitat: it is a woodland plant and is not found in high alpine situations.

NOTES: *Heart-leaved arnica has been found in spruce woods, in a moist, shady location near Marmot Basin and also in spruce woods near Elbow Falls. It was also common on Mount Baldy, in pine-spruce forest, elevation 2000 m, near Barrier Reservoir at 1400 m and many other places.*

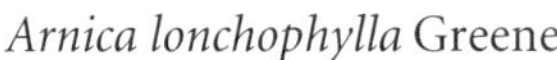

Arnica lonchophylla Greene

SPEAR-LEAVED ARNICA

Spear-leaved arnica is not as tall as heart-leaved arnica—it is only 2 to 4 dm tall—but it too grows from a branching rhizome and spreads easily. It has a single stem that often branches; there may be 1 to 9 flower-heads. It has pairs of leaves along the stem. Spear-leaved arnica can be distinguished from alpine arnica because it is much taller and is not found on high alpine slopes, and from heart-leaved arnica because the basal leaves are not heart-shaped but lanceolate and toothed. Spear-leaved arnica is found on montane slopes.

Artemisia L. SAGEWORT, WORMWOOD

Sageworts are aromatic, herbaceous plants that may grow as annuals, biennials or perennials. They are sometimes mat-forming. The leaves are finely divided, sometimes tomentose, as in *A. fridgida.* The flower-heads are small and clustered, bearing disc florets only; flowers are tubular. The involucral bracts are dry and imbricate; the fruits are achenes with no pappus.

1. Leaf divisions about 1 mm wide 2
   Leaf divisions more than 2 mm wide .... *A. biennis*
2. Plants silver-grey; less than 40 cm tall .... *A. frigida*
   Plants greyish green; 30 to 60 cm tall .... *A. campestris*

Artemisia biennis Willd.

WORMWOOD, BIENNIAL SAGEWORT

Biennial sagewort grows 3 to 10 dm high from a taproot. The alternate leaves are pinnately divided, and the lobes are sharply toothed. The discoid heads are arranged in dense spikes in the form of a leafy panicle, which trait helps distinguish the species from other sageworts. It is found in disturbed areas, on shores and in moist habitats.

NOTES: *Biennial sagewort is usually a biennial plant, hence its name, but sometimes grows as an annual.* Artemisia *may come from Queen Artemisia, the wife of Mausolus, whose tomb at Halicarnassus was one of the seven wonders of the ancient world. The name likely comes from Artemis (Diana), goddess of hunting.*

Artemisia campestris L.

PLAINS WORMWOOD, CUT-LEAF WORMWOOD

The stems of cut-leaf wormwood are 3 to 6 dm tall and arise from a taproot. The basal leaves are pinnately lobed, each lobe further divided into 3. The basal leaves grow in a tuft and have distinct petioles, while the stem leaves lack them—they are often sessile. The discoid flower-clusters are arranged in a panicle which is not leafy. Cut-leaf wormwood is a non-aromatic prairie plant that is also found on sandy shores and in woods.

NOTES: *This plant has been recorded from an open site, elevation 1996 m, near Hailstone Butte and at other places in Kananaskis Country. The common name cut-leaf wormwood is misleading because several* Artemisia *species have dissected leaves.*

Artemisia frigida Willd.

PASTURE SAGEWORT

Pasture sagewort can be distinguished from the other *Artemisia* species because of its small size (1 to 4 dm tall) and its silver-grey appearance. The whole plant is silver-grey, and the basal leaves are tufted and deeply dissected, giving the base a "fluffy" appearance. The tuft of fluffy basal leaves is quite different from the tuft of leaves in cut-leaf wormwood, and biennial sagewort has no tuft of leaves. The flowers are in discoid clusters, arranged in leafy racemes.

NOTES: *This plant is strongly fragrant.* ~ *The Blackfoot used the leaves for making a "tea," used by the women for menstrual problems; they also used it for colds and coughs.* ~ *Pasture sagewort was important for ceremonial use. A related species, pasture sage (*A. ludoviciana*), was also important for ceremonial use, in sweat lodges and for the Sun Dance. Both species are found in Kananaskis Country.* ~ *The species name* frigida *means "of cold regions," but the plant is not confined to these regions: it is found on the prairie and on open dry slopes, as well as in alpine areas.*

Aster L. — ASTER

Asters are perennials. They bear entire or toothed simple, alternate leaves. There are usually several flower-heads (solitary in *A. sibiricus*), with ligulate and disc florets. Ligulate florets are blue-purple. The involucral bracts are in 2 or more series and imbricate. The fruits form as achenes with a pappus of numerous bristles.

1.	Involucral bracts densely hairy	*A. sibiricus*
	Involucral bracts not hairy	2
2.	Leaves 3 to 5 cm across; upper surface rough	*A. conspicuus*
	Leaves glabrous, sometimes glaucous	*A. laevis*

Aster conspicuus Lindl.

SHOWY ASTER

Showy aster is well named, because it is a tall plant (2 to 10 dm) that has large, toothed leaves up to 18 cm long and numerous, large bluish violet flowers. The flowers are arranged in a broad cluster like a corymb—very "showy." Each flower-head has blue-violet ray florets and numerous disc florets in the centre. The flower-heads each have an involucre of overlapping glandular bracts; each has a papery base and spreading, pointed, green tips. The fruit is an achene crowned with a pappus of capillary bristles. The plant is found in open woodlands.

NOTES: *Showy aster has been recorded from a spruce-pine forest north of the Kananaskis Field Stations on Highway 40. It has also been found in a pine wood on the west side of Spray Lake, 11.3 km from the dam, and many other places.*

Aster laevis L.

SMOOTH ASTER

Smooth aster is a perennial, 4 to 10 dm tall. The leaves and stem are glabrous, hence the common name. The lower leaves have winged petioles, and the upper leaves are more or less clasping; they are sessile and are often auriculate (with ear-shaped lobes). The flower-heads grow in panicles; the ray florets are blue or purple, and there are numerous disc florets in the centre. The fruit is a glabrous achene with a tawny pappus. The plant is found on prairies and in open meadows at low elevations. It can be distinguished from showy aster, because the involucre is not densely glandular (this is well marked in showy aster). The flowers of smooth aster are smaller than those of showy aster. The leaves are also larger and broader in showy aster.

NOTES: *Smooth aster has been found in the Lower Marmot Creek, near the Kananaskis Field Stations in a dry meadow, and many other places.*

Aster sibiricus L.

SIBERIAN ASTER

Siberian aster is different from the other *Aster* species, because it is only 1 to 2 dm high and often forms loose mats. The flower-head is usually solitary. Siberian aster has purplish involucral bracts, not found in the other 2 species. The ray florets are purple or blue, and the fruit is a dry achene, crowned by a pappus. Siberian aster is found in diverse habitats, ranging from river flats to alpine elevations.

NOTES: *This species has been found in Kananaskis Country in Lineham Creek on an alluvial fan, in a rocky streamside by the bridge on Sheep River, along the Sheep Forestry Road and elsewhere.*

Cirsium Mill. — THISTLE

Thistles are coarse biennial or perennial plants rising from taproots or rhizomes, with very spiny leaves. The flower-heads vary in number, bearing only purple or whitish disc florets. The involucral bracts grow in several series. The fruits form as spiny achenes, each with a pappus of numerous plumose bristles.

1. Flowers white . . . . *C. hookerianum*
   Flowers purple 2
2. Upper leaf surface covered in woolly hairs . . . . *C. flodmanii*
   Upper leaf surface not hairy . . . . *C. arvense*

Cirsium arvense (L.) Scop.

CANADA THISTLE, CREEPING THISTLE

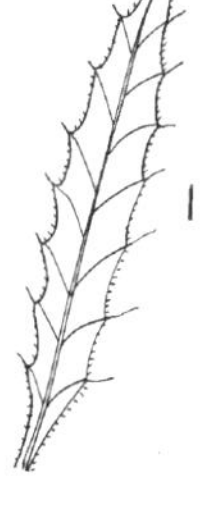

Canada thistle grows 3 to 10 dm tall and is a prickly, leafy plant. The plants are dioecious. The numerous flower-heads consist of tubular disc florets only and are subtended by spiny, overlapping involucral bracts. The disc florets are pinkish purple, sometimes white. The fruits are achenes with a pappus of plumose bristles. Canada thistles are found on waste ground, roadsides and cultivated fields.

NOTES: *"Canada thistle" is a misnomer because this species is not native to Canada—it comes from Europe, Asia and North Africa.* *The species name* arvense *means "of cultivated fields" and is unfortunately accurate: thistle is a noxious weed. Canada thistle grows from deep-seated, spreading rhizomes. When these are cut up by a plough or farm machine, the weed spreads.*

Cirsium flodmanii (Rydb.) Arthur

FLODMAN'S THISTLE

Flodman's thistle can be distinguished from Canada thistle because the deep purple flower-heads are much larger (up to 2 cm wide) and they are often solitary or few. The lower leaves are also different in the two species. In Flodman's thistle, the first-year rosettes are seldom spiny; they are often only spinose-ciliate, and the undersurfaces are white-tomentose. When the plants are in flower, the basal leaves have often disappeared, making Flodman's thistle difficult to distinguish from wavy-leaved thistle (*C. undulatum*). Flodman's thistle is found along roadsides, on dry prairies and in moist grassland.

Cirsium hookerianum Nutt.

HOOKER'S THISTLE, WHITE THISTLE

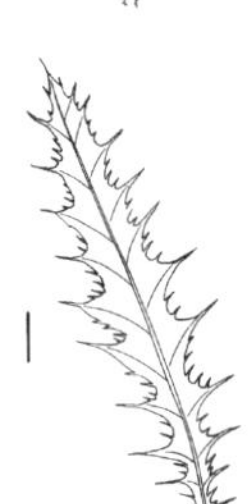

Hooker's thistle is generally hoary with dense, white hairs. Its flowers are large and creamy-white; the flower-heads measure 3 to 5 cm across. As in all *Cirsium* species, there are only disc florets present, no ray florets. The involucre is cup-shaped, tipped with spines and covered with cobwebby hairs. Hooker's thistle is found in open woods and meadows, up to subalpine elevations. It is also found on moist soils in valley bottoms.

NOTES: *This species has been found at an open, windy site near Hailstone Butte and near the junction of Sheep River and Gorge Creek, altitude 1500 m. It has also been found along roadsides and streams in the Marmot Basin, on the west side of Spray Lake, at 1700 m, and in other places.* ~ *The taproots and peeled stems are edible and can be eaten raw or cooked; they are said to be tasty and were eaten by the Blackfoot.* ~ *Hooker's thistle is named for the botanist Sir William Hooker, whose name will always be associated with the Royal Botanic Gardens at Kew, London.*

Crepis L. — HAWKSBEARD

These are perennial species containing milky juice. The leaves are often clustered at the base and may be spatulate, entire or dentate. The flower-heads are yellow and may be borne among the basal leaves or on an erect, branched stem. The involucral bracts grow in 1 to 2 series, and the fruits are achenes, each with a pappus of whitish bristles.

Plants 2 to 8 cm tall; alpine slopes . . . . *C. nana*
Plants 15 to 25 cm tall; streambanks and lakeshores . . . . *C. elegans*

Crepis elegans Hook.

HAWKSBEARD, YOUNGIA

Hawksbeard is a perennial with milky juice, 1.5 to 2.5 dm high, with several erect, branching stems and tufted leaves at the base. The stems bear numerous clusters of flower-heads. The flower-heads are bright yellow and are made up of numerous ray florets; there are no disc florets. The involucres beneath the flower-heads have 2 kinds of bracts: the main ones are long and narrow, usually in 2 series, with a few short ones below. The fruit is a slender achene with a delicate beak and a white pappus. Hawksbeard is found on mountain slopes, riverbanks and sandbars.

Crepis nana Rich.

DWARF HAWKSBEARD

C. nana can be easily distinguished from *C. elegans* because it rises only 2 to 8 cm high and grows from a tap root. The leaves grow in a dense tuft from which the flower-stalks rise, bearing yellow flower-heads. The flower-stalks are short and slender—they are often shorter than the tuft of leaves—so the flower-heads look like bright yellow tufts inside the leaves. Dwarf hawksbeard is an alpine plant. The fruits are slender achenes, each with a white pappus.

NOTES: *Dwarf hawksbeard has been recorded from the summit of Moose Mountain, 2400 m, from a ridge near Marmot Creek, at 2575 m, and elsewhere.*

Erigeron L. **FLEABANE, DAISY**

Fleabanes are perennials growing from a taproot, stout caudex or rhizome. The basal leaves are ovate, elliptical or oblanceolate; the cauline leaves are reduced upward in size and number. Flower-heads are few or solitary, and may be yellow, bluish or rose-purple. The involucral bracts are narrow and often greenish. The fruits are achenes, with a pappus bearing white bristles.

1. Ray florets yellow . . . . *E. aureus*
   Ray florets not yellow. 2
2. Ray florets purple; moist, wooded areas. . . . . *E. peregrinus*
   Ray florets white, blue or pink; dry, open areas . . . . *E. caespitosus*

Erigeron aureus Greene

GOLDEN FLEABANE

Golden fleabane is a dwarf perennial plant (only 2 to 15 cm tall). The golden flower-heads seem disproportionately large because they are 2 to 3 cm across. Such uneven balance between the flower-head and plant seems to be characteristic of some alpine plants. The oval basal leaves are deep green with short leaf-stalks and grow in a rosette; stem-leaves are usually absent. The single flower-head has 25 to 70 bright yellow ray florets on the outside, with yellow disc florets in the centre. The involucral bracts below the florets are purplish or have purplish tips, and are often covered with woolly hairs. The fruits are achenes with a double pappus, the outer one of narrow scales and the inner one of coarse bristles. Golden fleabane grows on rocky mountain slopes, scree slopes and alpine meadows.

NOTES: *This species has been found near the summit of Plateau Mountain at 2400 m, in an alpine meadow; also in alpine meadows on a ridge in the upper Evan-Thomas Creek, at 2300 m. It has been found, at 2100 m, in an alpine meadow on a north-east-facing slope near the upper terminus of Moose Mountain Lookout Road, and in other alpine and subalpine places.*

Erigeron caespitosus Nutt.

TUFTED FLEABANE, TUFTED DAISY

Tufted fleabane can be fairly easily separated from golden fleabane because it is taller (1 to 2.5 dm). Tufted fleabane has several decumbent stems and pale purple flower-heads. Its leaves are narrower and are found on the flowering stems; the flower-stalks of golden fleabane are not leafy. The involucre of bracts below the flower-head in tufted fleabane is glandular. Tufted fleabane is found on prairies and dry, open areas.

Erigeron peregrinus (Pursh) Greene

TALL PURPLE FLEABANE

Tall purple fleabane is a striking plant with its large purple flowers: there are 20 to 80 ray florets in each flower-head, and the disc florets are yellow. It is much taller than the other fleabanes (1 to 7 dm high) and its flowers are deep blue or purple. Its leaves resemble those of tufted fleabane but are larger (17 by 3 cm). It grows on dry, open, montane slopes, in damp meadows and along open trails up to 2700 m.

NOTES: *This species has been found in an alpine meadow at 2200 m, at Storm Creek, Highwood Pass, near the summit of Moose Mountain at 2000 m, on the summit of Plateau Mountain, at the head of Cabin Fork Creek at 2200 m, in the Marmot Basin and elsewhere.* ~ *The species name* peregrinus *means "exotic or foreign," and the plant is sometimes called "aster fleabane" because it looks like an aster.*

Gaillardia aristata Pursh

GAILLARDIA, BROWN-EYED SUSAN

Brown-eyed susan is a striking prairie plant bearing golden blooms with orange-brown centres, hence the common name. These blooms may be up to 6.5 cm across; they are borne on leafy stalks, 3 to 6 dm tall. The leaves are alternate. The basal leaves have leaf-stalks and are usually entire, but the stem-leaves are toothed and sessile. The flower-heads are solitary, at the tops of the stalks, and have numerous yellow ray florets, which are 3-cleft. The bases of these ray florets are often reddish. The disc florets in the centre are densely hairy. The involucral bracts are in 2 or 3-series and are often reflexed. After the ray florets drop, the rounded receptacle continues to develop until the disc florets form a nearly globular head. The fruit is an achene with a pappus of 5 to 10 scales. Brown-eyed susan is found on roadsides and dry, open areas, as well as forming sheets of colour on prairie grassland.

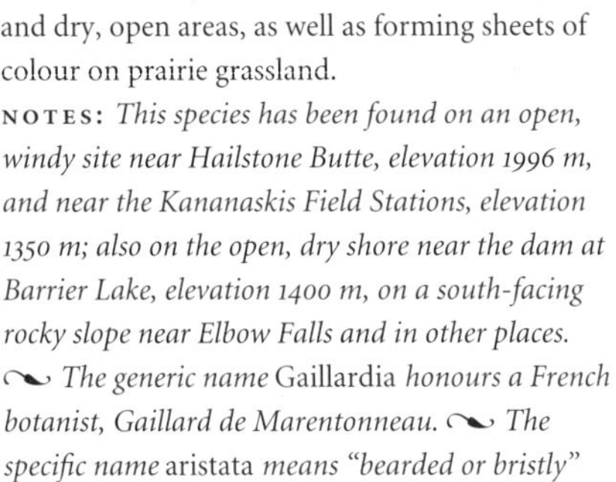

NOTES: *This species has been found on an open, windy site near Hailstone Butte, elevation 1996 m, and near the Kananaskis Field Stations, elevation 1350 m; also on the open, dry shore near the dam at Barrier Lake, elevation 1400 m, on a south-facing rocky slope near Elbow Falls and in other places.* ~ *The generic name* Gaillardia *honours a French botanist, Gaillard de Marentonneau.* ~ *The specific name* aristata *means "bearded or bristly" and refers to the flower-head after the fruits have been dispersed, when the numerous bristles that cover the receptacle can be clearly seen.* ~ *An infusion of this plant was used by some Stoney women for menstrual problems.*

Grindelia squarrosa (Pursh) Dunal

GUMWEED

Gumweed is a biennial or perennial with purple stems arising from a woody taproot; these stems are 3 to 6 dm high. There are glandular hairs on the leaves, which grow alternately up the stem. The lower leaves have leaf-stalks, but the stem-leaves clasp the stem and are often toothed. Numerous flower-heads grow on leafy axillary branches. Each flower-head is 2 to 3 cm across and surrounded by an involucre of bracts. The bracts are strongly curved, resinous and sticky, and covered with glandular hairs. The ray florets (ligulate) are bright yellow, 10 mm long and 1 to 2 mm wide; they are carpellary (pistillate) and fertile. The disc florets are yellow and also fertile. The fruits are brown, inferior, 4-angled achenes (cypselas), crowned by a pappus of a few bristly awns which soon fall off. Gumweed is found in dry, disturbed places, prairie grassland and saline flats.

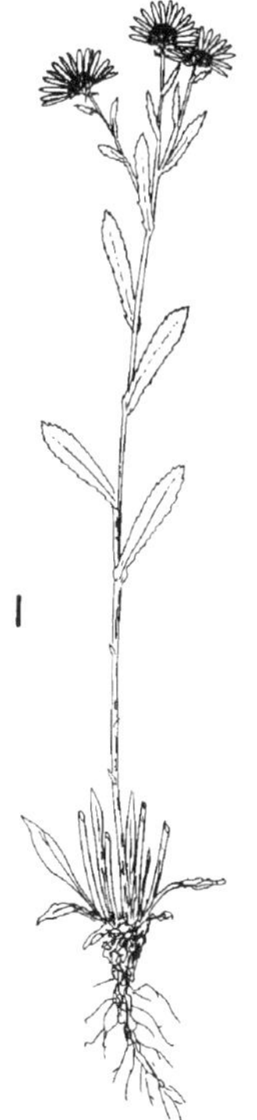

NOTES: *The plant was named in honour of David Grindel (1766-1836), a Russian botanist.* ~ *The species name* squarrosa *comes from "squarrose" (curled back) and refers to the sticky involucral bracts. These bracts also suggest the common name "gumweed."*

Haplopappus lyallii A. Gray

LYALL'S IRON PLANT

Lyall's iron plant is a dwarf perennial usually less than 10 cm tall, with creeping rhizomes. The basal leaves are bunched together; they are hairy and often glandular. There are several stems, each supporting a single flower-head which seems too large for a small plant. Each flower-head is surrounded by involucral bracts covered with glandular hairs; the bracts are narrowly lanceolate. Both the ray florets and disc florets are yellow, and the fruit is an achene with a whitish pappus. Lyall's iron plant is a mountain dweller, forming small clumps on alpine slopes above treeline.

NOTES: *Lyall's iron plant might be confused with golden fleabane* (Erigeron aureus), *but fleabane's leaves have leaf stalks, while the iron plant's leaves are sessile.* ~ *The species name* lyallii *honours David Lyall (1817-1895), a Scottish botanist and physician who made extensive botanical collections in North America while serving with the North American Boundary Commission.*

Helianthus nuttallii Torr. and Gray

COMMON TALL SUNFLOWER

Common tall sunflower is a tall, slender plant, up to 2 m in height, with tuberous creeping rootstocks. The largest leaves are in pairs and are up to 15 cm long; they are lance-shaped and sharply pointed. The flower-heads are 1 to several, bright yellow and quite striking, especially when a colony has arisen from the tuberous rootstocks—they make a bright splash of colour in prairie grasslands and on roadsides. The flower-heads have both ray florets and disc florets, and the involucral bracts below the flower-heads are in 4 to 6 series, overlapping. The fruit is an achene with a dirty-white pappus.

NOTES: *The generic name* Helianthus *comes from 2 Greek words:* helios, *meaning "sun," and* anthos, *meaning "flower."* *The specific name* nuttallii *honours Thomas Nuttall, who emigrated to the United States from England in 1808. Nuttall made many collections and published his* Genera of North American Plants *in 1818.*

Heterotheca villosa (Pursh) Shinners

GOLDEN ASTER, HAIRY GOLDEN ASTER

(*Chrysopsis villosa* (Pursh) Nutt.)

This plant is a much-branched perennial, growing from a tap-root; it often has a sprawling habit. The plant is covered with soft, white hairs. There are several flower-heads, each on a separate stalk, and these have 10 to 25 yellow ray florets and many orange-brown disc florets. The flower-heads are about 2.5 cm across, and the involucral bracts are narrow and overlapping. The fruit is a flat, hairy achene with a double pappus, with short bristles on the outside and long capillary bristles on the inside. Hairy golden aster is a typical prairie plant; it grows in dry, open, sandy areas and may be a weed in disturbed, sandy ground.

NOTES: *This species has been found on the Sibbald Creek Road in a sunny gravelly site, on the roadside near the base of Plateau Mountain and elsewhere.* *The common name is well chosen in one respect, but is also a misnomer. The plant is certainly hairy, the flower-heads are golden, but it is not an aster: Alberta asters are not yellow.*

Hieracium umbellatum L.

NARROW-LEAVED HAWKWEED

(H. canadense Michx.)

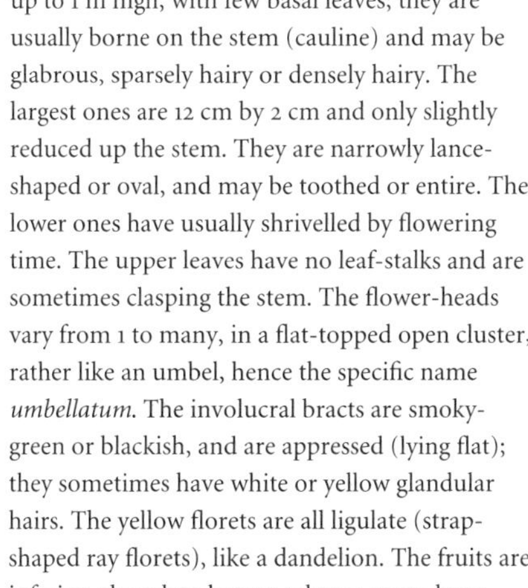

Narrow-leaved hawkweed is a coarse, leafy plant up to 1 m high, with few basal leaves; they are usually borne on the stem (cauline) and may be glabrous, sparsely hairy or densely hairy. The largest ones are 12 cm by 2 cm and only slightly reduced up the stem. They are narrowly lance-shaped or oval, and may be toothed or entire. The lower ones have usually shrivelled by flowering time. The upper leaves have no leaf-stalks and are sometimes clasping the stem. The flower-heads vary from 1 to many, in a flat-topped open cluster, rather like an umbel, hence the specific name *umbellatum*. The involucral bracts are smoky-green or blackish, and are appressed (lying flat); they sometimes have white or yellow glandular hairs. The yellow florets are all ligulate (strap-shaped ray florets), like a dandelion. The fruits are inferior, chocolate-brown achenes 3 mm long. Narrow-leaved hawkweed is found in woodland sites, open meadows and disturbed ground.

Matricaria matricarioides (Less.) Porter

PINEAPPLEWEED

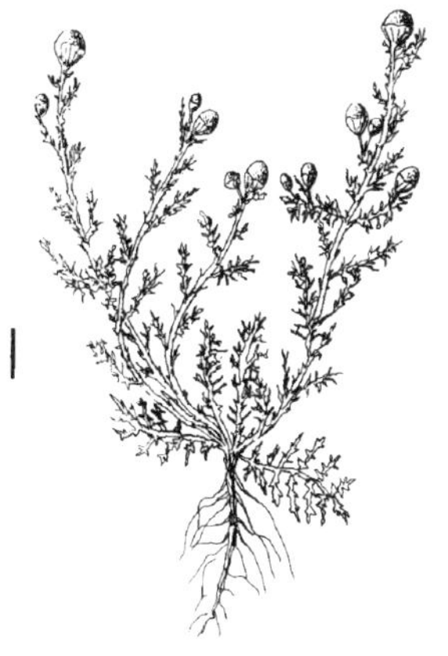

Pineappleweed is a small, leafy annual plant with dense, fibrous roots and a pineapple scent; it grows 1 to 4 dm high. The numerous leaves are 1 to 5 cm long and feathery, pinnately divided into filamentous portions. The flower-heads are small, green and conical; the involucral bracts are brownish green, 4 mm long, with a membraneous margin. There are no ray florets (ligulate). The disc florets are greenish yellow, with a 4-lobed corolla, and grow on a conical hypanthium. The fruits are small achenes, inferior, brownish and less than 2 mm long. The plant is an introduced weed, found on waste ground, roadsides and disturbed places.

NOTES: *Pineappleweed gets its common name from its pineapple scent and the resemblance of its conical flower-head to a pineapple.*

Petasites Mill. COLTSFOOT

These are perennials, rising from stout rhizomes. The basal leaves may be tomentose, cordate, sagittate, or nearly orbicular, and deeply divided. The flowering stem has large sheathing bracts. The flower-heads are usually numerous, growing in racemes or corymbs. The ray florets creamy white, but may be absent in some species. The disc florets are whitish. The involucral bracts are in 1 series. Fruits are achenes with a pappus of soft white bristles.

1. Leaves arrowhead-shaped . . . . *P. sagittatus*
   Leaves divided into 5 to 7 lobes 2
2. Leaves almost divided to the midrib . . . . *P. palmatus*
   Leaves divided half-way to the midrib . . . . *P. vitifolius*

Petasites palmatus (Ait.) A. Gray

PALMATE-LEAVED COLTSFOOT

(*P. frigidus* (L.) Fries var. *palmatus* (Ait.) Cronq.)

Palmate-leaved coltsfoot is a moisture-loving perennial plant, with stout creeping rhizomes and large basal leaves, 5 to 20 cm wide, with slender leaf-stalks. The leaves are deeply cleft, palmately, into 5 or 7 pointed lobes, with palmate veins. The upper surface of each leaf is green, but the under-surface is woolly-tomentose, that is, covered with white, felty hairs. The main flowering-stalks, 1 to 8 dm tall, appear early in the spring, before the leaves, and have large sheathing bracts; the flowers soon reach the fruiting stage and die down, then the leaves appear. This creates a problem of identification because the flowers of the 4 Alberta *Petasites* species are similar although the leaves are distinctive: in the flowering stage, it can be difficult to determine species. Usually there are some dead leaves of the year before, which help to determine the species. The whitish flower-heads are numerous, in dense terminal clusters, loosely racemose or corymbose. The involucre of each flower is cup-shaped, and the bracts are in 1 series; they are up to 8 mm long and are often purple-margined. Fertile plants have carpellary (pistillate) florets, which may be ligulate or tubular. Substerile plants have numerous staminate flowers in the centre, surrounded by a row of carpellary ray florets. The fruits are linear, inferior achenes, 5 to 10-ribbed and 2 mm long, with a soft white pappus. Palmate-leaved coltsfoot is found in moist woods and swamps.

Petasites sagittatus (Pursh) A. Gray

ARROW-LEAVED COLTSFOOT

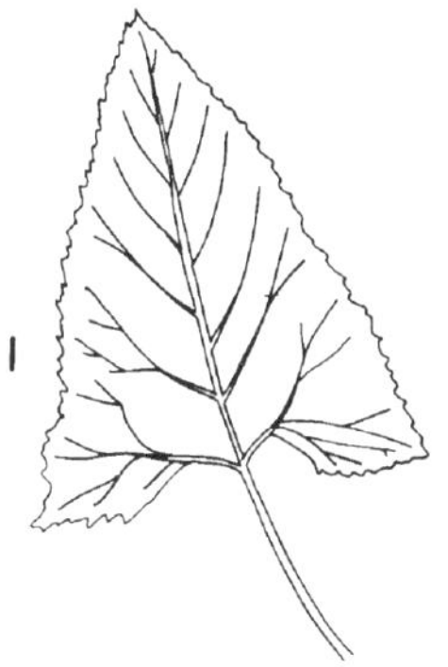

Although this coltsfoot is called "arrow-leaved" and the specific name *sagittatus* means "arrow-shaped," this is not a good description of the leaf: it is too rounded to be truly sagittate. But its leaves are easy to distinguish from palmate-leaved coltsfoot. It is also a stouter plant with a larger inflorescence. The large leaf-blades are 1 to 3 dm long and 1 to 2 dm broad, with green upper surfaces and velvety-tomentose below. The fruits are achenes, 2 to 3.5 cm long. Arrow-leaved coltsfoot is found in wet meadows and boggy areas.

Petasites vitifolius Greene

VINE-LEAVED COLTSFOOT

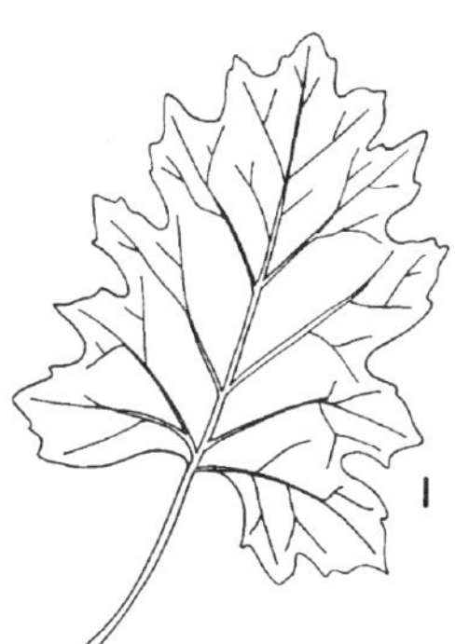

Vine-leaved coltsfoot is probably a hybrid between the other species: it has characteristics of both. The leaf-blades are deeply cordate, with 5 to 7 lobes. The flowering-heads and the bracts growing along the axis of the flowering-stalk are more numerous than in the other species. Like the other coltsfoot plants, vine-leaved coltsfoot is a moisture-loving plant and is found in wet woods.

Saussurea nuda Ledeb.

DWARF SAW-WORT

(*S. densa* (Hook.) Hult.)

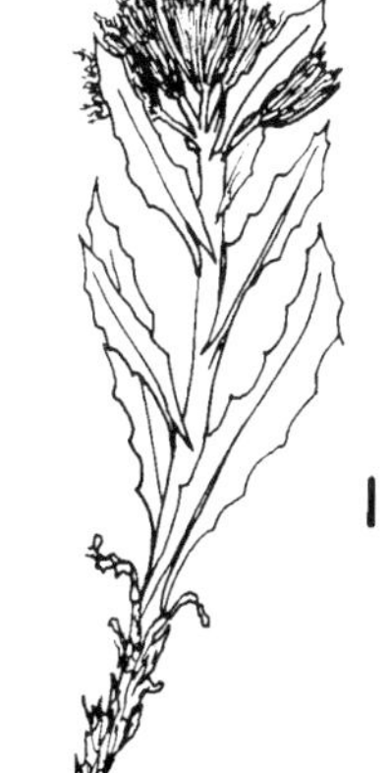

Dwarf saw-wort is an alpine plant that grows on rocky slopes. It reaches 1 to 2 dm and has hairy, rather cobwebby, stems with characteristic lanceolate leaves which are markedly saw-toothed, hence the common name. The leaves help to identify the plant; the lower ones have leaf-stalks, and the upper ones are sessile. The flower-heads are crowded at the top of the stem and look like thistle-heads because they have only disc florets and are blue-purple. The involucres are cobwebby (arachnoid), and the involucral bracts are in 4 to 5 series. Each disc floret has a blue-purple corolla, and the fruits are inferior achenes (cypselas), each crowned by a pappus of bristles, 1 to 2 mm long.

Senecio L. GROUNDSEL, RAGWORT

Groundsels are perennials. The leaves may be entire, toothed, lobed or arrow-shaped; the flower-heads grow in a loose terminal cluster; both the ligulate florets and disc florets are yellow. The involucral bracts usually appear in 1 series. The fruits are slender achenes, with 5 to 10 ribs and a pappus of numerous soft bristles.

1. Leaves silvery-grey . . . . *S. canus*
   Leaves green 2
2. Leaves triangular . . . . *S. triangularis*
   Leaves not triangular 3
3. Involucral bracts black-tipped . . . . *S. lugens*
   Involucral bracts not black-tipped . . . . *S. pauperculus*

Senecio canus Hook.

PRAIRIE GROUNDSEL

Prairie groundsel is a perennial covered with white, woolly hairs, especially on the undersides of the leaves. It has 2 types of leaves. The basal leaves are lanceolate, tapering into long petioles, and entire. The stem-leaves are smaller, have no petioles and are often toothed. Prairie groundsel grows up to 4 dm tall, and is an attractive prairie plant with many bright yellow, star-like flowers. The flower-heads grow at the tops of the stems (i.e., they are terminal), and have 5 to 12 yellow ray florets. These are narrow and separated, giving the flower-head a star-like appearance, typical of the *Senecio* genus. The disc florets in the centre are also yellow. The involucral bracts are bright green. The fruit is a slender achene, crowned with a pappus of soft white bristles. Prairie groundsel is found in dry, open, rocky areas, as well as on prairie grassland, and is occasionally found in subalpine elevations.

NOTES: *In Kananaskis Country, this species is found in heavily grazed grassland in the Sheep River Wildlife Sanctuary, on dry slopes below the Barrier Dam and in many other places.*

Senecio lugens Richards.

BLACK-TIPPED GROUNDSEL

Black-tipped groundsel is floccose-woolly when young, but as the plant matures it loses most of the woolly hairs. An erect, solitary flower-stalk, 2 to 6 dm high with very few leaves, rises from the basal leaves. The flowering-stem bears 5 to 15 flower-heads, all bright yellow and similar to those of prairie groundsel. The involucral bracts below the flower-head are black-tipped—a characteristic feature and hence the common name. Black-tipped groundsel is a plant of the alpine meadows and moist subalpine and alpine slopes.

NOTES: *The species name* lugens *is a Latin word meaning "to mourn," and the reason explorer John Richardson gave it this name is historically interesting. He collected the plant at Bloody Falls on the Coppermine River, the site of a 1771 massacre of unsuspecting Inuit by Indian warriors who accompanied Samuel Hearne on his expedition to the Arctic coast.*

Senecio pauperculus Michx.

BALSAM GROUNDSEL, RAGWORT

This groundsel is a perennial plant, with basal leaves more or less tufted and slender stems 1 to 6 dm tall. There are often tiny woolly tufts in the leaf-axils. The basal leaves have slender leaf-stalks, and the blades vary in shape from lanceolate to spatulate or elliptic. The leaf margin may be scalloped, toothed or practically entire; the base is usually wedge-shaped. The stem-leaves have no leaf-stalks, while the leaf-stalks of the basal leaves are long, sometimes longer than the blade. The stem-leaves are greatly reduced upwards, sometimes clasping the stem, and usually pinnatifid. The flower-heads number up to 15, and all except the central ones are usually long-stalked, resulting in a corymbose inflorescence. The ray florets are yellow, as are the disc florets. The fruit is a slender, inferior achene, crowned with a pappus of soft, white bristles. It is found in various habitats, in open woods, moist meadows, and roadsides.

NOTES: *Ragwort can be distinguished from triangle-leaved ragwort* (S. triangularis) *because its leaves are not triangular.*

Senecio triangularis Hook.

GIANT RAGWORT, TRIANGLE-LEAVED RAGWORT, BROOK RAGWORT

Triangle-leaved ragwort is a tall plant (3 to 15 dm), bearing characteristic leaves shaped like a triangle and drawn out into a long point. Its flowers are similar to those of black-tipped groundsel, with the narrow ray florets producing a star-like flower-head. Triangle-leaved ragwort is associated with streams and is found in subalpine meadows.

NOTES: *This species has been recorded from Storm Creek, at the Highwood Pass, on the banks of streamlets crossing the road 1 km east of Elbow Lake and elsewhere.*

I

Solidago L. — GOLDENROD

These species are perennials with erect stems bearing alternate, often lanceolate, leaves which may be entire or toothed. The numerous small, yellow flower-heads grow in a large inflorescence. The involucral bracts are imbricate, in several series; the fruits are achenes, more or less cylindrical, with a pappus of white bristles.

1. Plants 5 to 20 dm tall; stems leafiest at the middle .... *S. gigantea*
   Plants less than 50 cm tall; stems leafiest at the base 2
2. Stalk of basal leaves hairy .... *S. multiradiata*
   Stalk of basal leaves not hairy 3
3. Basal leaves 3-nerved .... *S. missouriensis*
   Basal leaves 1-nerved; stems often reddish .... *S. spathulata*

Solidago gigantea Ait.

LATE GOLDENROD, TALL GOLDENROD

Tall goldenrod is a leafy perennial growing from a creeping rootstock, with erect stems bearing alternate leaves. The leaves are narrowly lanceolate, sharply pointed and toothed in the upper half of the leaf-blade. The leaves associated with the inflorescence are not toothed. None of the leaves have leaf-stalks. The terminal flowering-heads grow in a pyramidal panicle, and the heads are often secund, that is, they grow along only 1 side of the branching flowering-stalks. Each flower-head has narrow involucral bracts, 3 to 6 mm high and overlapping (imbricate). They protect the bright yellow ligulate florets (about 15) in bud, and the yellow disc florets in the centre. The fruit is an inferior achene (cypsela), usually ribbed, more or less cylindrical and crowned with a pappus of whitish bristles. Tall goldenrod is found in damp habitats, moist open woods, thickets and alluvial flats.

NOTES: *This species can be distinguished from the other goldenrods because of its height and its large pyramidal inflorescence.*

Solidago missouriensis Nutt.

LOW GOLDENROD, MISSOURI GOLDENROD

Missouri goldenrod grows from 1 to 6 dm tall from creeping rhizomes and often forms colonies. The basal leaves are often tufted and are lanceolate, tapering into long petioles. From the basal leaves arises a solitary, leafy flower-stalk bearing a large inflorescence of tiny flower-heads, golden in colour. The flower-heads are usually secund (arranged along 1 side of the flowering-stalk). Each flower has short yellow ray florets in a ring on the outside, and disc florets, also yellow, in the centre. The shape of the inflorescence is variable: it is usually pyramidal but is sometimes a flat-topped panicle. The fruit is an achene with a pappus of capillary bristles. Missouri goldenrod is a typical plant of prairie grasslands.

NOTES: *Missouri goldenrod has been found in Kananaskis Country at Elbow Falls, elevation 1500 m, at Sandy McNabb Recreation Area, 1400 m, and in other places.*

Solidago multiradiata Ait.

ALPINE GOLDENROD

Alpine goldenrod has erect, leafy stems, grows only 0.5 to 5 dm tall, and its basal leaves are usually tufted, with white, hairy-margined petioles. The plant is often red-tinged. The flower-heads are in a dense cluster at the top of the stem and are larger than those of the Missouri goldenrod; there are 8 ray florets on the outside ring, about 6 mm long, and 13 or more disc florets. Alpine goldenrod, as its common name implies, is usually found in subalpine or alpine situations, although it can be found in moist open areas at lower elevations. It has been found near the summit of Plateau Mountain, at 2400 m, and in many other places.

Solidago spathulata D.C.

MOUNTAIN GOLDENROD

(*S. decumbens* Greene)

Goldenrod is an apt name for this species of *Solidago* because the flower-heads usually form a dense, cylindrical raceme and look like a "rod." The flower-heads are not secund. Mountain goldenrod lacks the hairy-margined petioles of alpine goldenrod. It is found in prairie grassland and in open woods, sometimes reaching higher elevations; the "mountain" in the common name is not very appropriate.

Sonchus arvensis L.

PERENNIAL SOW THISTLE

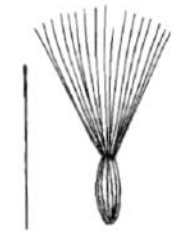

Perennial sow thistle is a tall plant (4 to 20 dm), contains milky juice and has a leafy stem. The stems rise from a semi-taproot, but the plant has creeping rhizomes, which make it a troublesome weed. The leaves are pinnately lobed and prickly margined; the lobes are often pointed, with the points bending backwards (runcinate). The lower ones tend to have clasping bases, but the leaves higher up the stem are definitely clasping and are often auriculate (i.e., with little "ears"). The flower-heads are bright yellow, and there are several of them in an open corymbose cluster. The involucral bracts below the flower-head have characteristic coarse glandular hairs. There are only ray florets, no disc florets, unlike the true thistles (*Cirsium*). These flower-heads produce large quantities of fruits, achenes crowned with a white pappus of capillary bristles. Perennial sow thistle is found in waste ground, fields and gardens.

NOTES: *Perennial sow thistle, like Canada thistle, is an unwelcome invader from Eurasia: it competes in fields with crops.* ~ *The specific name* arvensis *means "of the fields."*

Taraxacum Weber — DANDELION

Dandelions are perennials which grow from fleshy taproots; they contain milky juice. The stems are scapose, and the leaves may be entire, deeply toothed or lobed. The flower-heads are solitary, bearing only yellow ligulate florets. The involucral bracts appear in 2 series, the outer series usually shorter than the inner and often reflexed. The achenes are spindle-shaped, with a long beak, crowned with a pappus of numerous bristles, and are white or whitish in colour.

1. Involucral bracts closely appressed . . . . *T. ceratophorum*
   Involucral bracts spreading 2
2. Leaf blade divided to the midrib (deeply lobed) . . . . *T. laevigatum*
   Leaf blade divided half-way to the midrib . . . . *T. officinale*

Taraxacum ceratophorum (Ledeb.) D.C.

~ **NORTHERN DANDELION, MOUNTAIN DANDELION**

This dandelion is similar to the common dandelion but is only 2 to 25 cm tall. Unlike the common dandelion, mountain dandelion is not a weed; it is found at alpine elevations, sometimes on exposed summits. Its leaves are 2 to 20 cm long and pinnately lobed; the lobes are pointed and bend backwards, and the bases of the petioles are often winged. The flower-head consists of numerous yellow ray florets (there are no disc florets); the inner bracts beneath the flower-head often have hooded tips. The fruit is a yellowish brown achene, with a "parachute" of pappus hairs.

Taraxacum laevigatum (Willd.) D.C.

RED-SEEDED DANDELION

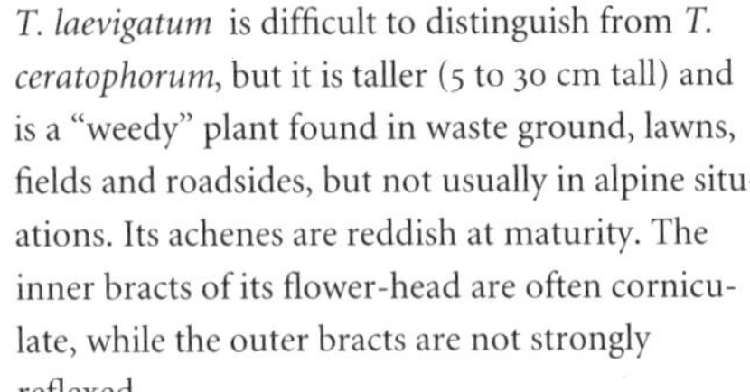

T. laevigatum is difficult to distinguish from *T. ceratophorum*, but it is taller (5 to 30 cm tall) and is a "weedy" plant found in waste ground, lawns, fields and roadsides, but not usually in alpine situations. Its achenes are reddish at maturity. The inner bracts of its flower-head are often corniculate, while the outer bracts are not strongly reflexed.

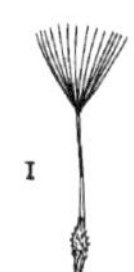

Taraxacum officinale Weber

COMMON DANDELION

Common dandelion, with its familiar flower-head of yellow ray florets and the characteristic shape of the basal leaves, would appear to be so characteristic that it should be easy to distinguish it from other flowers of the same genus. But it is similar to red-seeded dandelion and care is needed to distinguish the species. Common dandelion and red-seeded dandelion both have deeply lobed leaves, but in common dandelion the terminal lobe is larger than the others. The flower-head of common dandelion has 2 rows of involucral bracts; those in the inner row are only slightly corniculate (horned), while those in the outer row are strongly reflexed. Common dandelion has a grey-brown achene; the achene of red-seeded dandelion is reddish and has a shorter beak.

Townsendia Hook. **TOWNSENDIA**

Members of this genus are dwarf or short perennials rising from taproots. Their leaves are linear or narrowly lanceolate, often basal. The flower-heads are either sessile amongst the basal leaves or solitary at the tips of the flowering stems. The ligulate florets may be blue or purple, while the disc florets are yellow. The involucral bracts are lanceolate, sharply pointed and imbricate. The achenes of the disc florets have a pappus of rigid, scabrous bristles; the pappus of the ray florets is similar or composed of small scales.

Low caespitose perennial; large flower-heads among leaves, ray florets white to pinkish . . . . *T. exscapa*

Stout perennial, 0.5 to 2.5 dm tall; several flowering stems from a cluster of basal leaves, ray florets violet or bluish purple . . . . *T. parryi*

Townsendia exscapa (Richards.) Porter

LOW TOWNSENDIA

Low townsendia flowers in early spring; it is an almost stemless plant and grows from a taproot. The lance-shaped leaves grow in a rosette and are covered with silvery hairs—a useful protection against the frosts of early spring. The large flower-heads have no flowering stalks but arise from amongst the leaves. Each flower-head has white or pink ray florets and yellow disc florets in the centre; the heads are 2 cm across. The fruit is an achene with a white pappus "parachute." Low townsendia is found on dry hillsides and prairie grassland.

NOTES: *The root of low townsendia was boiled by the Blackfoot and the resulting infusion was used as horse medicine.*

Townsendia parryi D.C. Eat.

PARRY'S TOWNSENDIA

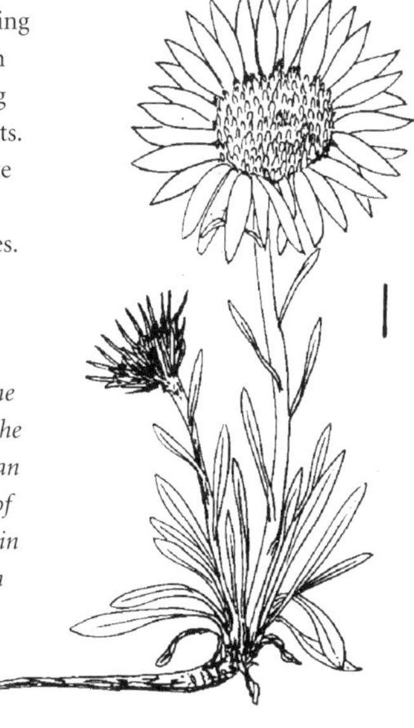

Parry's townsendia is quite distinct from low townsendia. It is a perennial, 0.5 to 2.5 dm in height, and is quite striking, with several flowering stems arising from a cluster of basal leaves. Each stem has usually a single flower-head, consisting of purplish blue ray florets and yellow disc florets. The involucral bracts are covered with stiff white hairs. The fruits are flattened achenes, usually hairy and crowned with a pappus of rigid bristles. The achenes of the ray florets sometimes have small scales. This plant's usual habitats are dry hills, banks and roadsides.

NOTES: *When Parry's townsendia grows in alpine situations it often becomes a small plant, though the flowers are large—for example, on the Mount Allan Centennial Trail at 2575 m and near the summit of Hailstone Butte at 3300 m. It has also been found in Kananaskis Country in a small grassy ravine 8 km south of Cat Creek and elsewhere.*

Tragopogon dubius Scop.

YELLOW GOAT'S-BEARD

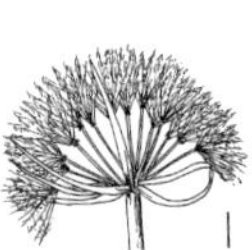

Yellow goat's-beard is a stiffly erect plant, 4 to 10 dm tall, bluish green, with a milky juice. The leaves are long (up to 20 cm) and narrow, rather grass-like; they are sessile. The plant has a taproot. The flower-heads are terminal and possess only bright yellow ray florets. The long, lanceolate bracts of the involucre enclose the flower-head; they are 2.5 to 4 cm long in the fruit. The flowering stem is swollen just below the flower-head. The fruits of goat's-beard are unique: each consists of a long, thin achene, crowned with a delicate pappus of interwoven white, plumose bristles. Goat's-beard is a weed and is common in disturbed ground, roadsides and gardens.

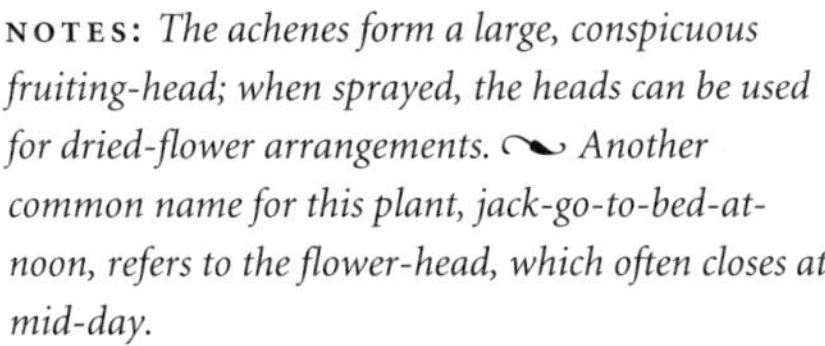

NOTES: *The achenes form a large, conspicuous fruiting-head; when sprayed, the heads can be used for dried-flower arrangements.* ~ *Another common name for this plant, jack-go-to-bed-at-noon, refers to the flower-head, which often closes at mid-day.*

Appendix 1

Other Botanical Investigations in Kananaskis Country

JOHN MACOUN collected many lichen and vascular species in Kananaskis Country in 1902. C.D. Bird of the University of Calgary examined Macoun's collections in the Ottawa Herbaria while researching the distribution of lichens in the family Parmeliaceae in southwestern Alberta; Bird found many that had been collected in Kananaskis Country, in Jumping Pound Creek, on Forget-me-not Mountain and near Elbow Lake. He documented Macoun's species in an article in the *Canadian Journal of Botany*, 1973.

NORMAN BETHUNE SANSON, a pioneer botanist, was born in Toronto in 1862 and came west with the King's Rifles to fight in the Riel Rebellion. He later made his home in Banff and was Curator of the Banff Museum from 1896 to 1931. He collected vascular plant species from many parts of the Rockies, not only in Alberta but in British Columbia as well. He made several forays into the area now called Kananaskis Country and collected from various locations, including the North Kananaskis Pass, Mud Lake and the Burstall Lakes. After he retired in 1931, he continued collecting in his previous locations and made two extended visits to the Arctic, collecting specimens there. By the time he died in 1948, his collection contained 6000 sheets. In 1971 the whole collection was presented to the University of Calgary Herbarium; it was catalogued and annotated between 1978 and 1982 by Beryl Hallworth. There were 136 species from Kananaskis Country, from 8 locations.

ALFRED HENRY BRINKMAN (1873-1945), who lived in Craigmyle, Alberta, collected liverworts (*Hepaticae*), bryophytes (mosses) and lichens in the Rocky Mountains, including the area now called Kananaskis Country, from the Highwood River and Jumping Pound Creek. The lichen species were annotated by C.D. Bird and deposited in the Herbarium of the University of Calgary.

In 1957, A.J. BREITUNG made an extensive collection of lichens in the Kananaskis area. These were also deposited in the Herbarium.

C.D. BIRD collected many species of vascular plants and lichens in the Kananaskis area from 1960 to 1975. All the specimens he collected are deposited in the Herbarium of the University of Calgary.

J.R. SEABORN, a graduate student at the University of Calgary, completed his thesis, *The Bryophyte Ecology of Marl Bogs in Alberta*, in 1966. Two of the marl bogs he studied were in Kananaskis Country. All his specimens are deposited in the University of Calgary Herbarium.

In 1968, J.P. BRYANT completed his M.Sc. thesis, *Alpine vegetation and frost activity in an alpine fellfield on the summit of Plateau Mountain, Alberta*, which was largely concerned with the vegetation of patterned ground. Plateau Mountain lies in southern Kananaskis Country.

In 1969, C.L. KIRBY and R.T. OGILVIE conducted an extensive study on the flora of the Kananaskis area. Ogilvie collected several thousand vascular specimens and deposited them in the Canadian Forestry Service Herbarium, Edmonton.

In 1970, J.P. BRYANT did more field work in the fellfield on the summit of Plateau Mountain with E. Scheinberg. Their joint research was published in the *Canadian Journal of Botany* under the title "Vegetation and frost activity in an alpine fellfield on the summit of Plateau Mountain, Alberta."

C. WALLIS and C. WERSHLER carried out an ecological survey of the Bow Valley Provincial Park, in the north of Kananaskis Country, in 1972. In the same year, they carried out a similar survey of Bragg Creek Provincial Park, also in Kananaskis Country.

Also in 1972, G. TROTTIER, a graduate student at the University of Calgary, completed his M.Sc. thesis, *The ecology of the alpine vegetation of Highwood Pass, Alberta*. Highwood Pass, in the southern part of Kananaskis Country, has the highest paved road in western Canada. Trottier found two species that were new botanical records for Alberta, the dwarf willow (*Salix stolonifera*) and a whitlow grass (*Draba porsildii*), which was described by Mulligan (Ottawa).

Trottier described ten alpine plant associations in the area. These were as follows:

a. ***Dryas octopetala*** **association**
Found on snow-free, wind-exposed ridge-slopes and moraines. The following plants form colonies on stone-stripes: *Arenaria obtusiloba*, *Oxytropis deflexa* and *Potentilla diversifolia*. Elsewhere, *Dryas octopetala* forms an almost continuous cover. There is a bryophyte-lichen layer.

b. ***Kobresia myosuroides*** **association**
Found on wind-exposed, snow-free slopes and ridge-crests. *Kobresia* forms robust tussocks on the ridge-crests; these grow close together and provide a continuous cover. Steep slopes have a low cover of *Kobresia*. This association has two distinct layers: a sedge-graminoid-herb stratum and a bryophyte-lichen stratum. Three dwarf shrubs, *Salix arctica*, *Salix nivalis* and *Arctostaphylos uva-ursi* occur occasionally. *Kobresia* is the dominant species, associated with *Dryas octopetala*, *Potentilla nivea*, *Silene acaulis* and *Antennaria alpina*. Graminoids include *Poa alpina*, *Bromus pumpellianus*, *Festuca saximontanus*, *Luzula spicata* and *Carex rupestris*.

c. ***Salix nivalis*** **association**
Found on high-elevation "saddle" areas exposed to strong winds with very little snow. *Salix nivalis*, *Salix arctica*, *Dryas octopetala*, *Potentilla diversifolia*, *Silene acaulis*, *Polygonum viviparum* and *Astragalus alpinus* occur. The bryophyte-lichen layer is well developed in some areas.

d. ***Phyllodoce*** **association**
Found at timberline and low alpine zones, in openings among tree islands and as an understory in the subalpine *Picea-Abies-Larix* forest on gentle alpine slopes. *Phyllodoce empetriformis* and *Phyllodoce glanduliflora* are dominant. *Vaccinium scoparium* is a co-dominant, and *Salix arctica* is a sub-dominant. The bryophyte-lichen layer is poorly developed.

e. ***Thalictrum occidentale*** **association**
This association is an avalanche-meadow type, found on steep, well-drained slopes around *2290 m (7500 ft)* elevation, subject to downslope snow movement. Although *Thalictrum occidentale* is the dominant species, there are several other important herbaceous species: *Valeriana sitchensis*, *Erigeron peregrinus*, *Achillea millefolium*, *Fragaria virginiana*, *Hedysarum sulphurescens* and *Aquilegia flavescens*. There are also many grasses in this association, but the bryophyte-lichen layer is poorly developed.

f. ***Salix arctica*** **association**

Found on wet or mesic alluvial fans and along high mountain streams. All stands receive melt water. The shrub layer is dominated by *Salix arctica*; *Salix glauca* and *Salix nivalis* are also found. Very rich herb layer dominated by *Erigeron peregrinus*, *Fragaria virginiana*, *Achillea millefolium*, *Potentilla diversifolia*, *Anemone parviflora* and *Pedicularis groenlandica*. Some graminoids are present: *Poa alpina*, *Poa ampla*, *Deschampsia caespitosa* and *Phleum alpinum*. There are few bryophytes; lichens are not common.

g. ***Salix barrattiana*** **association**

Occurs above treeline, along creeks and in gullies, also in the valley bottom. *Salix farrae* is co-dominant with *Salix barrattiana* and is associated with *Valeriana septentrionalis*, *Erigeron peregrinus*, *Deschampsia caespitosa* and *Potentilla diversifolia*. In stands dominated by *Salix barrattiana*, *Achillea millefolium*, *Erigeron peregrinus* and *Trisetum spicatum* occur. All understorey herbs suffer from reduced light conditions. The bryophyte-lichen layer is poorly developed.

h. ***Cassiope tetragona*** **association**

Well-developed stands of this association are rare in the Highwood Pass area. The stands occur on cool, north-facing slopes with deep snow cover. Snow accumulates in "bowls," typical habitats for *Cassiope tetragona*. There are three vegetative areas: shrub, herb and bryophyte-lichen. Sub-dominant shrubs are *Salix arctica*, *Salix nivalis* and *Dryas octopetala*. The herb layer is dominated by *Astragalus alpinus*, *Carex atrosquama* and *Poa longipila*. There is a well-developed bryophyte-lichen layer.

i. ***Deschampsia caespitosa*** **association**

Only one stand was found at the Highwood Pass, situated in the valley bottom at the summit of the Pass—a well-protected meadow type. Associated plants are *Poa pratensis*, *Agropyron latiglume*, *Carex haydeniana*, *Achillea millefolium*, *Fragaria virginiana*, *Veronica alpina*, *Veronica serpillifolia*, *Sibbaldia procumbens* and *Ranunculus eschscholtzia*. The tussocks of *Deschampsia caespitosa* grow very close together and form a thick cover. There are very few bryophyte species and no lichens.

j. *Carex nigricans* association
Occurs in bowl-shaped snow beds on the lee side of exposed ridges. The snow melts by mid-July. There are two vegetation layers: a graminoid-herb layer and a bryophyte layer. *Carex nigricans* is the dominant species, forming an almost pure cover in the very late-melting snowbed stands. Species associated with *Carex nigricans* in the stands free from snow earlier include *Veronica alpina, Antennaria lanata, Ranunculus eschscholtzia, Sibbaldia procumbens, Juncus drummondii, Poa alpina* and *Deschampsia atropurpurea.* The bryophyte layer is poorly developed.

In the early 1970s, personnel at the Kananaskis Centre (now the Kananaskis Field Stations), situated near Barrier Lake, worked on a detailed and extensive study of the Kananaskis Valley, *Mountain Environment and Urban Society, Kananaskis Pilot Study.* Gordon Hodgson, the Director of the Centre, supervised the work and wrote the chapter "Organization and Structure." This was the first comprehensive study of the Kananaskis area and dealt in detail with environmental history, geology, hydrology, climatology, botany and wildlife, and the influence of urban society on a mountain system. The section on vegetation was covered by D.R. Jacques and A.H. Legge, assisted by J. Corbin; they collected 4000 sheets of 1350 specimens. This study was a review of all the different aspects of the valley and was illustrated by numerous maps and appendices of relevant facts. The survey was completed in 1974. This pilot study was an excellent review, but it is now more than twenty years old and would need a vast amount of revision to update it.

In 1977, the Government of Alberta established Kananaskis Country. Kananaskis Provincial Park was also established at that time; in 1986, its name was changed to Peter Lougheed Provincial Park. In 1977, D.R. Jacques produced *A preliminary check-list of plants found in Kananaskis Provincial Park, with special reference to rare and disjunct species.* In 1978, N. Kondla carried out a detailed vegetation survey of the same area.

In 1981, C. WALLIS and C. WERSHLER prepared a natural history and assessment in Bow Valley Provincial Park, this time in the Many Springs area. This area is well known for the orchids found there.

In 1983 and 1988, JOHN CORBIN did field work in Kananaskis Country and produced two check-lists of plants deposited in the Kananaskis Herbarium at the Kananaskis Field Stations. These specimens were later given to the University of Calgary Herbarium.

E.H. MOSS had written his pioneer *Flora of Alberta* in 1958, but by the 1980s it was out of date. It was completely revised and updated by John G. Packer, with distribution maps and many new keys, in 1983. All the plants known at that time to be found in Kananaskis Country were included in the revised text.

In 1984, C. MACGREGOR carried out an ecological land classification and evaluation of Kananaskis Country for Alberta Forestry, Lands and Wildlife, Edmonton.

In 1988, JOAN WILLIAMS produced a vegetation inventory of Bow Valley Provincial Park, and in 1989 she carried out a monitoring program of the Many Springs area in that Park. In 1990, she worked in Peter Lougheed Provincial Park and carried out a vegetative inventory of the Opal, Kananaskis and Misty Ranges in that Park. In the same year, she produced an annotated vegetative species check-list.

In 1990, C.D. BIRD published a check-list of "Alpine plants on Plateau Mountain" in *Pica* 10 (3). In 1991, John Corbin published a check-list of "Vascular Plants of the Kananaskis Valley" in *Pica*; Stuart Harris published a "Vascular Plant-List of the Plateau Mountain area" in the same publication later that year.

In 1992, JOAN WILLIAMS worked in Peter Lougheed Provincial Park on a wetlands classification system, habitat inventory and description of wetlands. These included the two marl bogs described by J.R. Seaborn. Williams described five types of bogs: shore bogs, floating bogs, blanket bogs, flat bogs and basin bogs.

The diagram on page 281 illustrates two of these, a shore bog and a floating bog. The plant succession from south to north is as follows:

1. **Upland forest of pine and spruce**
2. **Shore bog zone**

   This zone is characterized by *Sphagnum* moss and Labrador tea; there are many hummocks, usually 1 m high. Small

PROFILE DIAGRAM

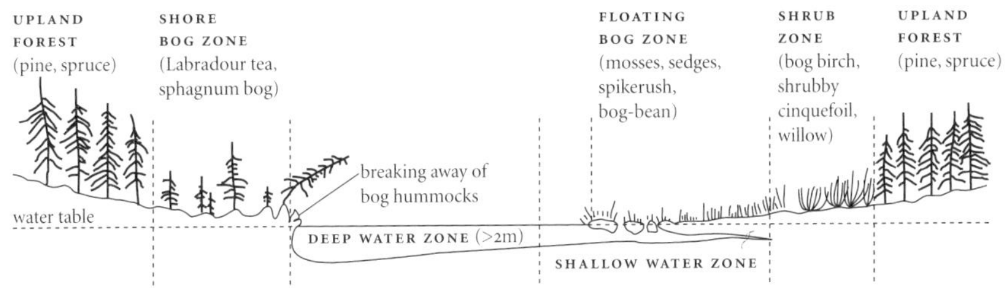

VERTICAL DIAGRAM

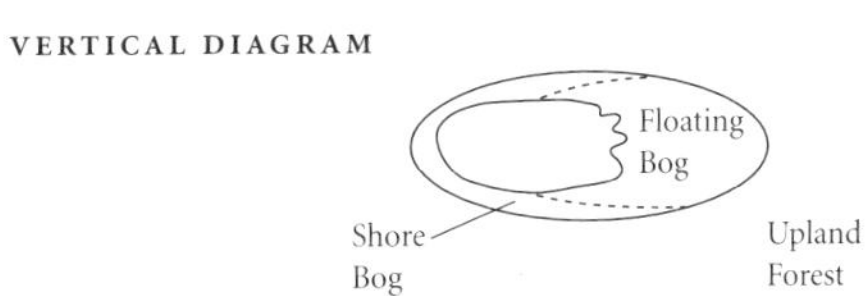

saplings of spruce or pine grow on the tops of these hummocks, and mosses, lichens, *Kalmia* spp., crowberry and bog cranberry grow on the lower parts.

3. **Deep-water zone, >2 m deep**
   Vegetation is often restricted to stonewort.
4. **Floating bog zone**
   This is found in the shallow-water area and includes mosses, sedges, spike-rush, cottongrass, rushes and buckbean.
5. **Shrub zone**
   Bog birch is found here, as well as willow and shrubby cinquefoil.
6. **Upland forest of pine and spruce**
   There are several examples of lakes or ponds with floating bogs and/or shore bogs in the Park, e.g., Sounding Lake.

Seaborn described eight study areas, two of which are in Peter Lougheed Provincial Park; these were Kananaskis Fen Number 1 and Kananaskis Fen Number 2.

Marl bogs are not true bogs. The term "marl" refers to a granular adhesive coating, composed chiefly of calcium carbonate. This coating inhibits the growth of *Sphagnum* moss, which is characteristic of bogs. Seaborn refers to all eight study areas as "fens." Marl bogs exhibit a hummock-hollow topography; the hollows form pools, and the hummocks form a network between them.

Kananaskis Fen Number 1 is 1.6 km (1 mi.) northeast of Lower Kananaskis Lake, elevation 1650 m (5400 ft.). It occupies 3000 m^2 and is surrounded by white spruce forest. Large pools of water occur throughout the fen, and the bottoms of these pools are heavily coated with marl; mosses, bladderwort and stonewort are found here. The central portion of the fen is dominated by sedges, spike-rush and mosses; willow and birch are also present. *Sphagnum* moss is found on the fen margins, as are other mosses and plants, such as bog cranberry. Crowberry is also found there, with Labrador tea and shrubby cinquefoil. Many other vascular plants were found in other parts of the fen.

Kananaskis Fen Number 2 lies 60 metres east of Fen Number 1 and is similar. It has extensive marl accumulation. Seaborn found several additional species here, such as tufted hair grass, hairy wild rye, a sedge (*Carex capillaris*), slender bog orchid, ladies'-tresses orchid, balsam poplar, willow (*Salix barrattiana*), aspen, dwarf raspberry and red-osier dogwood. Seaborn included detailed maps of both these marl bogs.

Also in 1992, as a graduate student of the University of Calgary, JOAN WILLIAMS presented her M.Sc. thesis, *Vegetative Classification using Landset T.M. and SPOT HRV Imagery in Mountainous Terrain, Kananaskis Country, Southwestern Alberta.* In 1993, she returned to the wetlands of Peter Lougheed Provincial Park to work on "Phytoplankton/Algal Sampling of Selected Wetlands," and in 1994, she made "An assessment of recreational impact on vegetation of the Visitor Services Meadow in Peter Lougheed Provincial Park."

NEIL EMERY, also a graduate student of the University of Calgary, submitted his Ph.D. thesis in 1994 on *Plasticity in Ecotypes of* Stellaria longipes. Part of his field work was carried out on Hailstone Butte and Plateau Mountain, and he wrote a check-list of plants he had found on Plateau Mountain. Also in 1994, Ian Macdonald produced an annotated list of vascular plants found in the immediate vicinity of Camp Adventure at Sibbald Lake.

Other botanists have worked in different parts of Kananaskis Country, but it is not possible to detail them all, and of course, this work is ongoing.

Appendix 2

Plants of Kananaskis Country Species List

Arranged in Alphabetical Order of Scientific Family Name

Aceraceae MAPLE FAMILY

Acer glabrum Torr. . . . . mountain maple
Acer negundo L. . . . . Manitoba maple, box elder

Alismataceae WATER-PLANTAIN FAMILY

Alisma gramineum Lej. . . . . narrow-leaved water-plantain
Alisma plantago-aquatica L. . . . . broad-leaved water-plantain
Sagittaria cuneata Sheld. . . . . arrowhead, wapato

Amaranthaceae AMARANTH FAMILY

Amaranthus albus L. . . . . tumbleweed
Amaranthus graecizans L. . . . . prostrate amaranth
Amaranthus retroflexus L. . . . . red-root pigweed

Apiaceae [Umbelliferae] CARROT FAMILY

Angelica arguta Nutt. . . . . white angelica
Angelica dawsonii S. Wats. . . . . yellow angelica
Carum carvi L. . . . . caraway
Cicuta maculata L. . . . . water-hemlock
Heracleum lanatum Michx. . . . . cow parsnip
Lomatium dissectum (Nutt.) Mathias & Constance . . . . mountain wild parsnip, prairie parsley
Lomatium macrocarpum (Hook. & Arn.) Coult. & Rose . . . . . . long-fruited parsley
Lomatium sandbergii Coult. & Rose . . . . Sandberg's wild parsley
Lomatium triternatum (Pursh) Coult. & Rose . . . . western wild parsley
Musineon divaricatum (Pursh) Nutt. . . . . leafy musineon
Osmorhiza chilensis Hook. & Arn. . . . . blunt-fruited sweet cicely
Osmorhiza depauperata Philippi . . . . spreading sweet cicely
Perideridia gairdneri (Hook. & Arn.) Mathias . . . . squawroot
Sanicula marilandica L. . . . . snakeroot
Sium sauve Walt. . . . . water parsnip
Zizia aptera (A. Gray) Fern. . . . . heart-leaved alexanders

Apocynaceae — DOGBANE FAMILY

Apocynum androsaemifolium L. . . . . spreading dogbane, Indian hemp

Araliaceae — GINSENG FAMILY

Aralia nudicaulis L.. . . . . wild sarsaparilla

Asclepiadaceae — MILKWEED FAMILY

Asclepias speciosa Torr. . . . . showy milkweed

Asteraceae [Compositae] — SUNFLOWER FAMILY

Achillea millefolium L. . . . . common yarrow
Achillea millefolium L var. *lanulosa* (Nutt.) Piper . . . . milfoil
Agoseris aurantiaca (Hook.) Greene . . . . orange false dandelion
Agoseris glauca (Pursh) Raf. . . . . yellow false dandelion
Agoseris glauca (Pursh) Raf. var. *dasycephala* (T. & G.) Jepson. . . . . yellow false dandelion
Anaphalis margaritacea (L.) Benth. & Hook.. . . . . pearly everlasting
Antennaria alpina (L.) Gaertn. . . . . alpine everlasting
Antennaria anaphaloides Rydb. . . . . tall everlasting
Antennaria aprica Greene . . . . low everlasting
Antennaria corymbosa E. Nels. . . . . corymbose everlasting
Antennaria lanata (Hook.) Greene . . . . woolly everlasting
Antennaria luzuloides T. & G.. . . . . silvery everlasting
Antennaria microphylla Rydb. . . . . small-leaved everlasting
Antennaria monocephala DC. . . . . one-headed everlasting
Antennaria neglecta Greene . . . . broad-leaved everlasting
Antennaria parvifolia Nutt.. . . . . small-leaved everlasting
Antennaria pulcherrima (Hook.) Greene. . . . . showy everlasting
Antennaria pulvinata Greene . . . . cushion everlasting
Antennaria racemosa Hook.. . . . . racemose everlasting
Antennaria rosea Greene. . . . . rosy everlasting
Antennaria umbrinella Rydb. . . . . brown-bracted mountain everlasting
Arnica amplexicaulis Nutt. . . . . stem-clasping arnica
Arnica angustifolia M. Vahl. ssp. *tomentosa* (Macoun) Douglas . . . . . . alpine arnica
Arnica chamissonis Less. . . . . leafy arnica
Arnica chamissonis Less. ssp. *foliosa* (Nutt.) Maguire . . . . leafy arnica
Arnica cordifolia Hook. . . . . heart-leaved arnica
Arnica diversifolia Greene . . . . vari-leaved arnica, lawless arnica
Arnica fulgens Pursh . . . . shining arnica
Arnica gracilis Rydb. . . . . graceful arnica
Arnica latifolia Bong.. broad-leaved arnica, mountain arnica
Arnica lonchophylla Greene . . . . spear-leaved arnica
Arnica louiseana Farr . . . . Lake Louise arnica, rock arnica
Arnica mollis Hook. . . . . cordilleran arnica
Arnica rydbergii Greene. . . . . Rydberg's arnica, narrow-leaved arnica

Arnica sororia Greene . . . . . twin arnica
Artemisia absinthium L.. . . . . absinthe, wormwood
Artemisia biennis Willd.. . . . . biennial sagewort
Artemisia borealis Pallas. . . . . northern wormwood
Artemisia campestris L. . . . . plains wormwood
Artemisia frigida Willd.. . . . . pasture sagewort
Artemisia ludoviciana Nutt. . . . . prairie sagewort
Artemisia michauxiana Bess.. . . . . Michaux's sage
Artemisia norvegica Fries . . . . . mountain sage
Aster alpinus L. . . . . alpine aster
Aster borealis (T. & G.) Prov. . . . . rush aster, marsh aster
Aster ciliolatus Lindl. . . . . Lindley's aster
Aster conspicuus Lindl.. . . . . showy aster
Aster ericoides L. ssp. *pansus* (Blake) A. G. Jones. . . . . tufted white prairie aster
Aster falcatus Lindl. . . . . creeping white prairie aster
Aster hesperius A. Gray . . . . . western willow aster
Aster laevis L. . . . . smooth aster
Aster modestus Lindl. . . . . large northern aster
Aster puniceus L. . . . . purple-stemmed aster
Aster sibiricus L. . . . . Siberian aster, arctic aster
Aster subspicatus Nees. . . . . leafy-bracted aster
Balsamorhiza sagittata (Pursh) Nutt. . . . . balsamroot
Centaurea maculosa Lam.. . . . . spotted knapweed
Chrysanthemum leucanthemum L.. . . . . ox-eye daisy
Cirsium arvense (L.) Scop. . . . . Canada thistle, creeping thistle
Cirsium flodmanii (Rydb.) Arthur . . . . . Flodman's thistle
Cirsium hookerianum Nutt. . . . . white thistle
Cirsium undulatum (Nutt.) Spreng.. . . . . wavy-leaved thistle
Cirsium vulgare (Savi) Ten.. . . . . bull thistle
Crepis elegans Hook. . . . . hawk's-beard, youngia
Crepis nana Richards. . . . . dwarf hawk's-beard
Crepis runcinata (James) T. & G.. . . . . scapose hawk's-beard
Erigeron acris L. spp. *debilis* (A. Gray) Piper . . . . . northern daisy, fleabane
Erigeron acris L. spp. *politus* (E. Fries) Schinz & Keller . . . . northern daisy, fleabane
Erigeron annus (L.) Pers. . . . . fleabane, whitetop
Erigeron aureus Greene. . . . . golden fleabane, yellow daisy
Erigeron caespitosus Nutt.. . . . . tufted fleabane
Erigeron canadensis L. . . . . horseweed
Erigeron compositus Pursh var. *discoideus* A. Gray . . . . . cut-leaf fleabane
Erigeron compositus Pursh var. *glabratus* Macoun . . . . . cut-leaf fleabane
Erigeron elatus Hook. . . . . tall fleabane
Erigeron flagellaris A. Gray. . . . . creeping fleabane
Erigeron glabellus Nutt. ssp. *pubescens* (Hook.) Cronq. . . . . smooth fleabane
Erigeron grandiflorus Hook. . . . . large-flowered daisy, large-flowered fleabane
Erigeron humilis Grah. . . . . purple fleabane, purple daisy
Erigeron lanatus Hook.. . . . . woolly fleabane, woolly daisy
Erigeron lonchophyllus Hook.. . . . . hirsute fleabane
Erigeron ochroleucus Nutt.. . . . . yellow-bracted alpine daisy
Erigeron peregrinus (Pursh) Greene . . . . . tall purple fleabane, wandering daisy

Erigeron philadelphicus L. . . . . Philadelphia fleabane
Erigeron pumilus Nutt. . . . . hairy daisy
Erigeron purpuratus Greene . . . . pale alpine fleabane, pale daisy
Erigeron radicatus Hook. . . . . dwarf fleabane
Erigeron speciosus (Lindl.) DC. . . . . showy fleabane
Gaillardia aristata Pursh. . . . . brown-eyed susan, gaillardia
Grindelia squarrosa (Pursh) Dunal . . . . gumweed
Haplopappus lyallii A. Gray . . . . Lyall's ironplant, sticky stenotus
Helianthus nuttallii T. & G. . . . . common tall sunflower
Helianthus subrhomboideus Rydb. rhombic-leaved sunflower
Heterotheca villosa (Pursh) Shinners . . . . golden aster
Hieracium albiflorum Hook. . . . . white hawkweed
Hieracium cynoglossoides Arr.-Touv. . . . . woolly hawkweed
Hieracium triste Willd. ssp. *gracile* (Hook.) Calder & Taylor . . . .
. . . . slender hawkweed, alpine hawkweed
Hieracium umbellatum L. . . . . narrow-leaved hawkweed
Matricaria matricarioides (Less.) Porter . . . . pineapple-weed
Matricaria perforata Mérat . . . . scentless chamomile
Petasites palmatus (Ait.) A. Gray . . . . palmate-leaved coltsfoot, sweet coltsfoot
Petasites sagittatus (Pursh) A. Gray . . . . arrow-leaved coltsfoot
Petasites vitifolius Greene . . . . vine-leaved coltsfoot
Rudbeckia hirta L. . . . . black-eyed susan
Saussurea nuda Ledeb. var. *densa* (Hook.) Hult. . . . . dwarf saw-wort
Senecio canus Hook. . . . . prairie groundsel
Senecio congestus (R. Br.) DC. . . . . marsh ragwort
Senecio conterminus Greenm. . . . . arctic butterweed
Senecio cymbalaroides Buek. . . . . ragwort
Senecio eremophilus Richards. . . . . cut-leaved ragwort
Senecio fremontii T. & G. . . . . . . dwarf mountain groundsel, mountain butterweed
Senecio indecorus Greene . . . . rayless ragwort
Senecio integerrimus Nutt. . . . . entire-leaved groundsel
Senecio lugens Richards. . . . . black-tipped groundsel
Senecio pauciflorus Pursh . . . . few-flowered ragwort
Senecio pauperculus Michx. . . . . balsam groundsel
Senecio pseudaureus Rydb. . . . . thin-leaved groundsel
Senecio streptanthifolius Greene . . . . northern ragwort
Senecio triangularis Hook. . . . . triangular-leaved ragwort, brook ragwort
Senecio vulgaris L. . . . . common groundsel
Solidago canadensis L. . . . . Canada goldenrod
Solidago gigantea Ait. var. *serotina* (Ait.) Cronq. . . . . late goldenrod, tall goldenrod
Solidago missouriensis Nutt. . . . . Missouri goldenrod, low goldenrod
Solidago multiradiata Ait. . . . . alpine goldenrod
Solidago nemoralis Ait. . . . . showy goldenrod
Solidago spathulata DC. . . . . mountain goldenrod
Sonchus arvensis L. . . . . perennial sow thistle
Sonchus asper (L.) Hill . . . . prickly sow thistle
Sonchus oleraceus L. . . . . annual sow thistle
Sonchus uliginosus Bieb. . . . . smooth perennial sow thistle
Taraxacum ceratophorum (Ledeb.) DC. . . . . northern dandelion

Taraxacum laevigatum (Willd.) DC. . . . . red-seeded dandelion
Taraxacum officinale Weber . . . . common dandelion
Townsendia exscapa (Richards.) Porter. . . . . low townsendia
Townsendia parryi D. C. Eat. . . . . Parry's townsendia
Tragopogon dubius Scop. . . . . yellow goat's-beard

Betulaceae BIRCH FAMILY

Alnus crispa (Ait.) Pursh ssp. *sinuata* (Regel) Hult.. . . . . green alder
Alnus tenuifolia Nutt. . . . . river alder
Betula glandulosa Michx. . . . . bog birch
Betula occidentalis Hook. . . . . water birch, black birch
Betula pumila L. . . . . dwarf birch

Boraginaceae BORAGE FAMILY

Echium vulgare L. . . . . viper's-bugloss, blue devil
Hackelia floribunda (Lehm.) I.M. Johnston . . . . large-flowered stickseed
Hackelia jessicae (McGreg.) Brand . . . . Jessica's stickseed
Lappula occidentalis (S. Wats.) Greene. . . . . western bluebur
Lappula squarrosa (Retz.) Dumort. . . . . bluebur
Lithospermum incisum Lehm.. . . . . narrow-leaved puccoon, incised puccoon
Lithospermum ruderale Lehm. . . . . woolly gromwell
Mertensia paniculata (Ait.) G. Don . . . . tall lungwort
Myosotis alpestris Schmidt ssp. *asiatica* Vestergr. . . . . alpine forget-me-not
Myosotis micrantha Pallas ex Lehm.. . . . . forget-me-not
Onosmodium molle Michx.. . . . . western false gromwell

Brassicaceae [Cruciferae] MUSTARD FAMILY

Arabis divaricarpa A. Nels.. . . . . purple rock cress
Arabis drummondii A. Gray . . . . Drummond's rock cress
Arabis glabra (L.) Bernh. . . . . tower mustard
Arabis hirsuta (L.) Scop. . . . . hairy rock cress
Arabis holboellii Hornem. . . . . Holboell's rock cress, reflexed rock cress
Arabis lemmonii S. Wats.. . . . . Lemmon's rock cress
Arabis lyallii S. Wats.. . . . . Lyall's rock cress
Arabis nuttallii Robinson . . . . Nuttall's rock cress
Barbarea vulgaris R. Br.. . . . . yellow rocket
Braya humilis (C.A. Mey.) Robins. . . . . leafy braya
Capsella bursa-pastoris (L.) Medic. . . . . shepherd's-purse
Cardamine pensylvanica Muhl. . . . . bitter cress
Cardamine pratensis L.. . . . . meadow bitter cress
Cardamine umbellata Greene . . . . mountain cress
Descurainia pinnata (Walt.) Britt. . . . . green tansy mustard
Descurainia richardsonii (Sweet) O.E. Schulz . . . . grey tansy mustard
Descurainia sophia (L.) Webb. . . . . flixweed
Diplotaxis muralis (L.) DC. . . . . sand rocket
Draba aurea Vahl. . . . . golden whitlow grass

Draba borealis DC. . . . . northern whitlow grass
Draba cana Rydb. . . . . whitlow grass
Draba crassifolia R. Grah. . . . . thick-leaved whitlow grass
Draba incerta Payson . . . . whitlow grass
Draba kananaskis Mulligan . . . . Kananaskis whitlow grass
Draba lonchocarpa Rydb. . . . . white whitlow grass
Draba macounii O.E. Schulz . . . . Macoun's whitlow grass
Draba nemorosa L. . . . . annual whitlow grass
Draba nivalis Liljebl. . . . . whitlow grass
Draba oligosperma Hook. . . . . few-seeded whitlow grass
Draba paysonii Macbr. . . . . Payson's whitlow grass
Draba porsildii Mulligan. . . . . Porsild's whitlow grass
Draba praealta Greene. . . . . low alpine whitlow grass
Draba stenoloba Ledeb. . . . . slender whitlow grass
Draba ventosa A. Gray . . . . whitlow grass
Erucastrum gallicum (Willd.) Schulz . . . . dog mustard
Erysimum cheiranthoides L. . . . . wormseed mustard
Erysimum inconspicuum (S. Wats.) MacM. . . . . small-flowered rocket
Lepidium campestre (L.) R. Br. . . . . cow cress, pepperwort or peppergrass
Lepidium densiflorum Schrad. . . . . common peppergrass
Lepidium ramosissimum A. Nels.. . . . . branched peppergrass
Lesquerella arctica (Wormskj. ex Hornem.) S. Wats. . . . . northern bladderpod
Lesquerella arenosa (Richards.) Rydb. . . . . sand bladderpod
Physaria didymocarpa (Hook.) A. Gray. . . . . double bladderpod
Raphanus raphanistrum L. . . . . jointed charlock, wild radish
Rorippa palustris (L.) Besser . . . . marsh yellow cress
Rorippa sylvestris (L.) Besser . . . . creeping yellow cress
Rorippa tenerrima Greene . . . . slender cress
Sisymbrium altissimum L.. . . . . tumbling mustard
Sisymbrium loeselii L. . . . . tall hedge mustard
Smelowskia calycina (Stephen) C.A. Mey . . . . silver rock cress
Thlaspi arvense L. . . . . stinkweed

Callitrichaceae WATER-STARWORT FAMILY

Callitriche verna L.. . . . . vernal water-starwort

Campanulaceae BLUEBELL FAMILY

Campanula rapunculoides L.. . . . . garden bluebell
Campanula rotundifolia L. . . . . harebell, bluebell
Campanula uniflora L. . . . . alpine harebell

Caprifoliaceae HONEYSUCKLE FAMILY

Linnaea borealis L. . . . . twinflower
Lonicera dioica L.. . . . . twining honeysuckle
Lonicera involucrata (Richards.) Banks. bracted honeysuckle

Sambucus racemosa L. var. *melanocarpa* (Gray) McMinn .... elderberry
Sambucus racemosa L. var. *pubens* (Michx.) Koehne.............. red elderberry
Symphoricarpos albus (L.) Blake .... snowberry
Symphoricarpos occidentalis Hook.................................. buckbrush
Viburnum edule (Michx.) Raf. .... low-bush cranberry, mooseberry

Caryophyllaceae — PINK FAMILY

Arenaria capillaris Poir. .... linear-leaved sandwort
Cerastium arvense L.............................. field mouse-ear chickweed
Cerastium beeringianum Cham. & Schlecht.......... alpine mouse-ear chickweed
Cerastium nutans Raf...................... long-stalked mouse-ear chickweed
Cerastium vulgatum L......................... common mouse-ear chickweed
Dianthus plumarius L... garden pink
Minuartia austromontana Wolf & Packer .... green alpine sandwort
Minuartia biflora (L.) Schinz & Thell.................... dwarf alpine sandwort
Minuartia dawsonensis (Britt.) House .... Dawson's sandwort
Minuartia elegans (Cham. & Schlecht.) Schischk.......... purple alpine sandwort
Minuartia nuttallii (Pax) Briq. .... Nuttall's sandwort
Minuartia obtusiloba (Rydb.) House .... arctic sandwort
Minuartia rubella (Wahl.) Graebn........................ red-seeded sandwort
Moehringia lateriflora (L.) Fenzl........................ blunt-leaved sandwort
Sagina saginoides (L.) Karst............................ mountain pearlwort
Silene acaulis L. .... moss campion
Silene cserei Baumg. .... smooth catchfly
Silene drummondii Hook........... Drummond's cockle, Drummond's campion
Silene menziesii Hook. .... Menzies' catchfly
Silene noctiflora L............................... night-flowering catchfly
Silene parryi (S. Wats.) C.L. Hitchc. & Maguire .... Parry's campion
Silene uralensis (Rupr.) Bocq................................ nodding cockle
Stellaria borealis Bigel.............................. northern stitchwort
Stellaria crassifolia Ehrh.................................. fleshy stitchwort
Stellaria crispa Cham. & Schlecht. .... wavy-leaved chickweed
Stellaria longifolia Muhl............................. long-leaved chickweed
Stellaria longipes Goldie .... long-stalked chickweed
Stellaria umbellata Turcz. .... chickweed

Chenopodiaceae — GOOSEFOOT FAMILY

Axyris amaranthoides L. .... Russian pigweed
Chenopodium album L...................................... lamb's-quarters
Chenopodium capitatum (L.) Aschers. .... strawberry blite
Chenopodium gigantospermum Aellen.................. maple-leaved goosefoot
Kochia scoparia (L.) Schrad..................... summer cypress, burning bush
Monolepis nuttalliana (Schultes) Greene .... spear-leaved goosefoot
Salsola kali L. .... Russian thistle

Cornaceae DOGWOOD FAMILY

Cornus canadensis L. . . . . bunchberry
Cornus stolonifera Michx. . . . . red-osier dogwood

Crassulaceae STONECROP FAMILY

Sedum lanceolatum Torr. . . . . lance-leaved stonecrop
Sedum stenopetalum Pursh . . . . narrow-petalled stonecrop
Tolmachevia integrifolia (Raf.) Löve & Löve . . . . rose-root

Cupressaceae CYPRESS FAMILY

Juniperus communis L.. . . . . ground juniper
Juniperus horizontalis Moench . . . . creeping juniper
Juniperus scopulorum Sarg. . . . . Rocky Mountain juniper
Thuja plicata D. Don . . . . western red cedar

Cyperaceae SEDGE FAMILY

Carex adusta Boott . . . . browned sedge
Carex aenea Fern. . . . . silvery-flowered sedge
Carex albo-nigra Mack. . . . . sedge
Carex aperta Boott . . . . open sedge
Carex aquatilis Wahlenb.. . . . . water sedge
Carex atherodes Spreng. . . . . awned sedge
Carex atrosquama Mack. . . . . dark-scaled sedge
Carex aurea Nutt.. . . . . golden sedge
Carex bebbii Olney ex Fern. . . . . Bebb's sedge
Carex brunnescens (Pers.) Poir. . . . . brownish sedge
Carex buxbaumii Wahlenb. . . . . brown sedge
Carex capillaris L. . . . . hair-like sedge
Carex capitata L.. . . . . capitate sedge
Carex chordorrhiza L.f.. . . . . prostrate sedge
Carex concinna R. Br. . . . . beautiful sedge
Carex concinnoides Mack.. . . . . low northern sedge
Carex crawei Dewey . . . . Crawe's sedge
Carex curta Good.. . . . . short sedge
Carex deflexa Hornem. . . . . bent sedge
Carex deweyana Schwein. . . . . Dewey's sedge
Carex diandra Schrank.. . . . . two-stamened sedge
Carex disperma Dewey . . . . two-seeded sedge
Carex eburnea Boott . . . . bristle-leaved sedge
Carex enanderi Hult. . . . . sedge
Carex epapillosa Mack. . . . . sedge
Carex filifolia Nutt.. . . . . thread-leaved sedge
Carex flava L.. . . . . yellow sedge
Carex gynocrates Wormsk. . . . . northern bog sedge
Carex haydeniana Olney . . . . Hayden's sedge

Carex heleonastes Ehrh. . . . . Hudson Bay sedge
Carex hoodii Boott . . . . Hood's sedge
Carex illota Bailey . . . . sedge
Carex interior Bailey . . . . sedge
Carex kelloggii Boott. . . . . Kellogg's sedge
Carex lacustris Willd.. . . . . lakeshore sedge
Carex lanuginosa Michx. . . . . woolly sedge
Carex lasiocarpa Ehrh.. . . . . hairy-fruited sedge
Carex leptalea Wahlenb. . . . . bristle-stalked sedge
Carex limosa L.. . . . . mud sedge
Carex livida (Wahlenb.) Willd.. . . . . livid sedge
Carex macloviana D'Urv. . . . . thick-spike sedge
Carex microglochin Wahlenb. . . . . short-awned sedge
Carex microptera Mack. . . . . sedge
Carex nardina Fries . . . . fragrant sedge
Carex nebraskensis Dewey . . . . Nebraska sedge
Carex nigricans C.A. Meyer . . . . blackening sedge
Carex norvegica Retz. . . . . Norway sedge
Carex obtusata Lilj. . . . . blunt sedge
Carex oligosperma Michx. . . . . few-fruited sedge
Carex pachystachya Cham.. . . . . sedge
Carex parryana Dewey . . . . Parry's sedge
Carex paysonis Clokey . . . . Payson's sedge
Carex pensylvanica Lam. var. *digyna* Boeckl. . . . . sun-loving sedge
Carex petricosa Dewey . . . . stone sedge
Carex phaeocephala Piper . . . . head-like sedge
Carex platylepis Mack. . . . . sedge
Carex podocarpa R. Br. . . . . alpine sedge
Carex praegracilis W. Boott . . . . graceful sedge
Carex praticola Rydb. . . . . meadow sedge
Carex preslii Steud. . . . . Presl sedge
Carex pyrenaica Wahlenb. . . . . spiked sedge
Carex raymondii Calder . . . . Raymond's sedge
Carex richardsonii R. Br. . . . . Richardson's sedge
Carex rossii Boott . . . . Ross' sedge
Carex rostrata Stokes . . . . beaked sedge
Carex rupestris All. . . . . rock sedge
Carex sartwellii Dewey . . . . Sartwell's sedge
Carex saxatilis L. . . . . rocky ground sedge
Carex scirpoidea Michx. . . . . rush-like sedge
Carex scopulorum Holm. . . . . sedge
Carex siccata Dewey. . . . . hay sedge
Carex simulata Mack.. . . . . sedge
Carex spectabilis Dewey . . . . showy sedge
Carex sprengelii Dewey. . . . . Sprengel's sedge
Carex stenophylla Wahl. . . . . low sedge
Carex sychnocephala Carey. . . . . many-headed sedge, long-beaked sedge
Carex tincta Fern. . . . . tinged sedge
Carex umbellata Schk. . . . . umbellate sedge

Carex vaginata Tausch . . . . sheathed sedge
Carex viridula Michx. . . . . green sedge
Carex xerantica Bailey . . . . white-scaled sedge
Eleocharis palustris (L.) R. & S. . . . . creeping spike rush
Eleocharis quinqueflora (F.X. Hartm.) O. Schwarz. . . . . few-flowered spike rush
Eriophorum brachyantherum Trautv.. . . . . close-sheathed cotton grass
Eriophorum polystachion L. . . . . tall cotton grass
Eriophorum scheuchzeri Hoppe. . . . . one-spike cotton grass
Eriophorum viridi-carinatum (Engelm.) Fern. . . . . thin-leaved cotton grass
Kobresia myosuroides (Vill.) Fiori & Paol.. . . . . bog sedge
Kobresia simpliciuscula (Wahlenb.) Mack. . . . . simple bog sedge
Scirpus acutus Muhl. ex Bigel. . . . . great bulrush
Scirpus caespitosus L. . . . . tufted bulrush
Scirpus clintonii A. Gray. . . . . Clinton's bulrush
Scirpus microcarpus Presl . . . . small-fruited bulrush
Scirpus pungens Vahl . . . . three-square rush
Scirpus validus Vahl . . . . common great bulrush

Dipsacaceae — TEASEL FAMILY

Knautia arvensis (L.) Duby . . . . blue buttons

Droseraceae — SUNDEW FAMILY

Drosera anglica Huds. . . . . oblong-leaved sundew
Drosera linearis Goldie . . . . slender-leaved sundew

Elaeagnaceae — OLEASTER FAMILY

Elaeagnus commutata Bernh. ex Rydb. . . . . wolf willow, silver-berry
Shepherdia canadensis (L.) Nutt. . . . . Canada buffaloberry

Empetraceae — CROWBERRY FAMILY

Empetrum nigrum L. . . . . crowberry

Equisetaceae — HORSETAIL FAMILY

Equisetum arvense L.. . . . . common horsetail
Equisetum fluviatile L. . . . . swamp horsetail
Equisetum hyemale L.. . . . . common scouring rush
Equisetum laevigatum A. Br. . . . . smooth scouring rush
Equisetum palustre L.. . . . . marsh horsetail
Equisetum pratense Ehrh. . . . . meadow horsetail
Equisetum scirpoides Michx. . . . . dwarf scouring rush
Equisetum sylvaticum L. . . . . woodland horsetail
Equisetum variegatum Schleich.. . . . . variegated horsetail

Ericaceae HEATH FAMILY

Arctostaphylos rubra (Rehder & Wils.) Fern. .... alpine bearberry
Arctostaphylos uva-ursi (L.) Spreng. .... common bearberry, kinnikinnick
Cassiope mertensiana (Bong.) D. Don................ western mountain heather
Cassiope tetragona (L.) D. Don .... white mountain heather
Kalmia microphylla (Hook.) Keller......................... mountain laurel
Kalmia polifolia Wang. .... northern laurel
Ledum groenlandicum Oeder .... common Labrador tea
Menziesia ferruginea J. E. Smith .... false huckleberry, false azalea
Oxycoccus microcarpus Turcz............................ small bog cranberry
Phyllodoce empetriformis (Smith) D. Don........... red heather, purple heather
Phyllodoce glanduliflora (Hook.) Coville........................ yellow heather
Phyllodoce glanduliflora x intermedia (Hook.) Camp .... heather
Rhododendron albiflorum Hook.. .... white-flowered rhododendron
Vaccinium caespitosum Michx............................... dwarf bilberry
Vaccinium membranaceum Dougl. ex Hook. .... tall bilberry
Vaccinium myrtillus L. .... low bilberry
Vaccinium scoparium Leiberg.................................. grouseberry
Vaccinium vitis-idaea L.......................... bog cranberry, cow-berry

Euphorbiaceae SPURGE FAMILY

Euphorbia esula L. .... leafy spurge
Euphorbia glyptosperma Engelm.. .... thyme-leaved spurge
Euphorbia serpyllifolia Pers. .... thyme-leaved spurge

Fabaceae [Leguminosae] PEA FAMILY

Astragalus aboriginum Richards. .... Indian milk vetch
Astragalus alpinus L...................................... alpine milk vetch
Astragalus americanus (Hook.) M. E. Jones............... American milk vetch
Astragalus bisulcatus (Hook.) A. Gray................. two-grooved milk vetch
Astragalus bourgovii A. Gray .... Bourgov's milk vetch
Astragalus canadensis L. .... Canadian milk vetch
Astragalus crassicarpus Nutt.................... buffalo bean, ground plum
Astragalus dasyglottis Fisch. ex DC. .... purple milk vetch
Astragalus drummondii Dougl. ex Hook. .... Drummond's milk vetch
Astragalus eucosmus Robins. .... milk vetch
Astragalus flexuosus Dougl. ex G. Don .... slender milk vetch
Astragalus miser Dougl. ex Hook. .... timber milk vetch
Astragalus missouriensis Nutt. .... Missouri milk vetch
Astragalus purshii Dougl. ex Hook.. .... Pursh's milk vetch
Astragalus robbinsii (Oakes) A. Gray var. *occidentalis* S. Wats. ....
.... Robbins' milk vetch
Astragalus striatus Nutt.. .... ascending purple milk vetch
Astragalus tenellus Pursh .... loose-flowered milk vetch
Astragalus vexilliflexus Sheldon .... few-flowered milk vetch
Caragana arborescens Lam. .... common caragana

Hedysarum alpinum L. .... alpine hedysarum
Hedysarum boreale Nutt. .... northern hedysarum
Hedysarum boreale Nutt. var. *mackenzii* (Rich.) C. L. Hitchc. ....
.... Mackenzie's hedysarum
Hedysarum sulphurescens Rydb.. .... yellow hedysarum
Lathyrus ochroleucus Hook.. .... cream-coloured vetchling
Lotus corniculatus L. .... bird's-foot trefoil
Lupinus argenteus Pursh .... silvery perennial lupine
Lupinus sericeus Pursh .... silky perennial lupine
Medicago falcata L. .... yellow lucerne
Medicago lupulina L. .... black medick
Medicago sativa L. .... alfalfa, lucerne
Melilotus alba Desr.. .... white sweet clover
Melilotus officinalis (L.) Lam. .... yellow sweet clover
Onobrychis viciifolia Scop. .... sainfoin
Oxytropis cusickii Greenm.. .... alpine locoweed
Oxytropis deflexa (Pall.) DC. .... reflexed locoweed
Oxytropis monticola A. Gray. .... late yellow locoweed
Oxytropis podocarpa A. Gray .... inflated oxytrope
Oxytropis sericea Nutt. .... early yellow locoweed
Oxytropis splendens Dougl. ex Hook. .... showy locoweed
Oxytropis viscida Nutt. .... viscid locoweed
Thermopsis rhombifolia (Nutt.) Richards.. .... golden bean
Trifolium aureum Poll. .... yellow clover, hop clover
Trifolium hybridum L.. .... alsike clover
Trifolium pratense L. .... red clover
Trifolium repens L. .... white clover, Dutch clover
Vicia americana Muhl. .... wild vetch
Vicia cracca L.. .... tufted vetch

Fumariaceae — FUMITORY FAMILY

Corydalis aurea Willd. .... golden corydalis
Corydalis sempervirens (L.) Pers. .... pink corydalis

Gentianaceae — GENTIAN FAMILY

Gentiana affinis Griseb. .... prairie gentian
Gentiana prostrata Haenke .... moss gentian
Gentianella amarella (L.) Börner. .... northern gentian, felwort
Gentianella crinita (Froel.) G. Don ssp. *macounii* (Holm) Gillett. ....
.... fringed gentian
Gentianella propinqua (Richards.) J.M. Gillett .... four-parted gentian
Halenia deflexa (Sm.) Griseb. .... spurred gentian

Geraniaceae — GERANIUM FAMILY

Geranium bicknellii Britt. .... Bicknell's geranium, crane's-bill

Geranium carolinianum L. . . . . Carolina wild geranium
Geranium richardsonii Fisch. & Trautv. wild white geranium
Geranium viscosissimum Fisch. & Mey. . . . . sticky purple geranium

Grossulariaceae CURRANT/GOOSEBERRY FAMILY

Ribes americanum Mill. . . . . wild black currant
Ribes aureum Pursh . . . . golden currant
Ribes glandulosum Grauer . . . . skunk currant
Ribes hudsonianum Richards. . . . . northern black currant
Ribes lacustre (Pers.) Poir.. . . . . bristly black currant
Ribes oxyacanthoides L.. . . . . northern gooseberry

Haloragaceae WATER-MILFOIL FAMILY

Myriophyllum exalbescens Fern. . . . . spiked water-milfoil

Hippuridaceae MARE'S-TAIL FAMILY

Hippuris montana Ledeb. . . . . mountain mare's-tail
Hippuris vulgaris L.. . . . . common mare's-tail

Hydrophyllaceae WATERLEAF FAMILY

Phacelia franklinii (R. Br.) A. Gray . . . . Franklin's scorpion weed
Phacelia hastata Dougl. ex Lehm. . . . . silver-leaved scorpion weed
Phacelia sericea (Graham) A. Gray . . . . silky scorpion weed
Romanzoffia sitchensis Bong. mist maiden, Sitka romanzoffia

Iridaceae IRIS FAMILY

Sisyrinchium montanum Greene . . . . common blue-eyed grass
Sisyrinchium septentrionale Bicknell. . . . . pale blue-eyed grass

Juncaceae RUSH FAMILY

Juncus albescens (Lange) Fern. . . . . white rush
Juncus alpinoarticulatus Chaix. . . . . alpine rush
Juncus balticus Willd.. . . . . wire rush
Juncus biglumis Willd. . . . . two-glumed rush
Juncus bufonius L. . . . . toad rush
Juncus castaneus L. . . . . chestnut rush
Juncus drummondii E. Meyer . . . . Drummond's rush
Juncus ensifolius Wikstr. var. *montanus* (Engelm.) C.L. Hitchc. . . . .
. . . . equitant-leaved rush
Juncus longistylis Torr. . . . . long-styled rush
Juncus mertensianus Bong. . . . . slender-stemmed rush
Juncus nodosus L. . . . . knotted rush

Juncus parryi Engelm. . . . . Parry's rush
Juncus tracyi Rydb. . . . . mud rush
Luzula hitchcockii (L.) Hämet-Ahti . . . . smooth wood rush
Luzula multiflora (Retz.) Lej. . . . . field wood rush
Luzula parviflora (Ehrh.) Desv. . . . . small-flowered wood rush
Luzula piperi (Cov.) M.E. Jones . . . . mountain wood rush
Luzula spicata (L.) DC. . . . . spiked wood rush

Juncaginaceae — ARROW-GRASS FAMILY

Triglochin maritima L. . . . . seaside arrow-grass
Triglochin palustris L. . . . . slender arrow-grass

Lamiaceae [Labiatae] — MINT FAMILY

Dracocephalum parviflorum Nutt. . . . . American dragonhead
Mentha arvensis L. . . . . wild mint
Monarda fistulosa L. . . . . wild bergamot
Monarda fistulosa L. var. *menthaefolia* (Grah.) Fern.. . . . . horse mint
Prunella vulgaris L. . . . . heal-all
Stachys palustris L. . . . . marsh hedge nettle

Lemnaceae — DUCKWEED FAMILY

Lemna minor L.. . . . . common duckweed
Lemna trisulca L.. . . . . ivy-leaved duckweed

Lentibulariaceae — BLADDERWORT FAMILY

Pinguicula vulgaris L. . . . . common butterwort
Utricularia minor L. . . . . small bladderwort
Utricularia vulgaris L.. . . . . common bladderwort

Liliaceae — LILY FAMILY

Allium cernuum Roth. . . . . nodding onion
Allium schoenoprasum L. . . . . wild chives
Allium textile Nels. & Macbr. . . . . prairie onion
Disporum trachycarpum (S. Wats.) B. & H. . . . . fairy-bells
Erythronium grandiflorum Pursh . . . . glacier lily
Lilium philadelphicum L. . . . . western wood lily
Smilacina racemosa (L.) Desf.. . . . . false Solomon's-seal
Smilacina stellata (L.) Desf. . . . . star-flowered Solomon's-seal
Smilacina trifolia (L.) Desf. . . . . three-leaved Solomon's-seal
Stenanthium occidentale A. Gray. . . . . bronzebells
Streptopus amplexifolius (L.) DC. clasping-leaved twisted stalk
Tofieldia glutinosa (Michx.) Pers.. . . . . sticky false asphodel
Tofieldia pusilla (Michx.) Pers. . . . . bog false asphodel

Veratrum eschscholtzii (R. & S.) A. Gray .... green false hellebore
Zigadenus elegans Pursh .... white camas
Zigadenus venenosus S. Wats. var. *gramineus* (Rydb.) Walsh .... death camas

Linaceae FLAX FAMILY

Linum lewisii Pursh.. wild blue flax

Lobeliaceae LOBELIA FAMILY

Lobelia kalmii L.. Kalm's lobelia

Loranthaceae MISTLETOE FAMILY

Arceuthobium americanum Nutt............................. dwarf mistletoe

Lycopodiaceae CLUB-MOSS FAMILY

Lycopodium alpinum L. .... alpine club-moss
Lycopodium annotinum L. .... stiff club-moss
Lycopodium complanatum L.................................. ground-cedar

Malvaceae MALLOW FAMILY

Sphaeralcea coccinea (Pursh) Rydb............................ scarlet mallow

Menyanthaceae BUCK-BEAN FAMILY

Menyanthes trifoliata L.. buck-bean

Monotropaceae INDIAN PIPE FAMILY

Monotropa uniflora L. .... Indian pipe

Onagraceae EVENING PRIMROSE FAMILY

Epilobium anagallidifolium Lam............................ alpine willowherb
Epilobium angustifolium L. .... common fireweed, great willowherb
Epilobium ciliatum Raf. ssp. *glandulosum* (Lehm.) Hoch & Raven..............
.... northern willowherb
Epilobium clavatum Trel.. willowherb
Epilobium hornemannii Reichenb.................... Hornemann's willowherb
Epilobium lactiflorum Hausskn.................................. willowherb
Epilobium latifolium L. .... river beauty, broad-leaved fireweed
Epilobium palustre L.................................. marsh willowherb
Epilobium saximontana Hausskn.................. rocky mountain willowherb

Oenothera biennis L. . . . . yellow evening primrose

Ophioglossaceae ADDER'S-TONGUE FAMILY

Botrychium boreale Milde . . . . northern grape fern
Botrychium dusenii (Christ) Alston. . . . . grape fern
Botrychium lanceolatum (Gmel.) Angstr. . . . . lance-leaved grape fern
Botrychium lunaria (L.) Sw. . . . . moonwort
Botrychium virginianum (L.) Sw. . . . . Virginia grape fern

Orchidaceae ORCHID FAMILY

Calypso bulbosa (L.) Oakes. . . . . Venus'-slipper
Corallorhiza maculata Raf. . . . . spotted coral-root
Corallorhiza striata Lindl. . . . . striped coral-root
Corallorhiza trifida Châtelain . . . . pale coral-root
Cypripedium calceolus L.. . . . . yellow lady's-slipper
Cypripedium passerinum Richards. . . . .
. . . . sparrow's-egg lady's-slipper, northern lady's-slipper
Goodyera oblongifolia Raf. . . . . rattlesnake plantain
Goodyera repens (L.) R. Br. . . . . creeping plantain, lesser rattlesnake plantain
Habenaria dilatata (Pursh) Hook. . . . . tall white bog orchid
Habenaria hyperborea (L.) R. Br.. northern green bog orchid
Habenaria obtusata (Pursh) Richards. . . . . blunt-leaved bog orchid
Habenaria saccata Greene . . . . slender bog orchid
Habenaria viridis (L.) R. Br. var. *bracteata* (Muhl.) Gray . . . . . . bracted bog orchid
Listera borealis Morong . . . . northern twayblade
Listera cordata (L.) R. Br. . . . . heart-leaved twayblade
Orchis rotundifolia Banks ex Pursh . . . . round-leaved orchid
Spiranthes romanzoffiana Cham. & Schlecht.. . . . . hooded ladies'-tresses

Orobanchaceae BROOMRAPE FAMILY

Orobanche fasciculata Nutt. . . . . clustered broomrape
Orobanche uniflora L.. . . . . one-flowered cancer-root

Papaveraceae POPPY FAMILY

Papaver kluanensis D. Löve. . . . . alpine poppy
Papaver nudicaule L. . . . . Iceland poppy

Parnassiaceae GRASS-OF-PARNASSUS FAMILY

Parnassia fimbriata Konig . . . . fringed grass-of-Parnassus
Parnassia kotzebuei Cham. & Schlecht. . . . . small grass-of-Parnassus
Parnassia palustris L. . . . . northern grass-of-Parnassus

Parnassia parviflora DC. .... small northern grass-of-Parnassus

Pinaceae — PINE FAMILY

Abies balsamea (L.) Mill. .... balsam fir
Abies lasiocarpa (Hook.) Nutt. .... subalpine fir
Larix lyallii Parl. .... subalpine larch
Larix occidentalis Nutt. .... western larch
Picea engelmannii Parry ex Engelm. .... Engelmann spruce
Picea glauca (Moench) Voss .... white spruce
Picea mariana (Mill.) BSP. .... black spruce
Pinus albicaulis Engelm. .... whitebark pine
Pinus contorta Loudon .... lodgepole pine
Pinus flexilis James .... limber pine
Pseudotsuga menziesii (Mirb.) Franco .... Douglas fir

Plantaginaceae — PLANTAIN FAMILY

Plantago major L. .... common plantain, whiteman's-foot

Poaceae [Gramineae] — GRASS FAMILY

Agroelymus sp. G. Camus
Agropyron dasystachyum (Hook.) Scribn. .... northern wheat grass
Agropyron pectiniforme R. & S. .... crested wheat grass
Agropyron repens (L.) Beauv. .... couch grass, quack grass
Agropyron scribneri Vasey .... wheat grass
Agropyron smithii Rydb. var. *molle* (Scribn. & Smith) Jones ... western wheat grass
Agropyron spicatum (Pursh) Scribn. & Smith. .... bluebunch wheat grass
Agropyron trachycaulum (Link) Malte .... slender wheat grass
Agropyron trachycaulum (Link) Malte var. *glaucum* (Pease & Moore) Malte .... slender wheat grass
Agropyron trachycaulum (Link) Malte var. *unilaterale* (Cassidy) Malte .... slender wheat grass
Agropyron violaceum (Hornem.) Lange .... broad-glumed wheat grass
Agrostis exarata Trin. .... spike redtop
Agrostis scabra Willd. .... rough hair grass, tickle grass
Agrostis stolonifera L. .... redtop
Agrostis thurberiana A.S. Hitchc. .... bent grass
Agrostis variabilis Rydb. .... alpine redtop
Alopecurus aequalis Sobol. .... water foxtail
Beckmannia syzigachne (Steud.) Fern. .... slough grass
Bromus carinatus Hook. & Arn. .... brome
Bromus ciliatus L. .... fringed brome
Bromus inermis Leyss. .... awnless brome
Bromus tectorum L. .... downy chess

Calamagrostis canadensis (Michx.) Beauv.. . . . . bluejoint, marsh reed grass
Calamagrostis inexpansa A. Gray. . . . . northern reed grass
Calamagrostis montanensis Scribn.. . . . . plains reed grass
Calamagrostis purpurascens R. Br.. . . . . purple reed grass
Calamagrostis rubescens Buckl. . . . . pine reed grass
Calamagrostis stricta (Timm) Koeler . . . . narrow reed grass
Cinna latifolia (Trev.) Griseb. . . . . drooping wood reed
Dactylis glomerata L. . . . . orchard grass
Danthonia californica Boland. . . . . California oat grass
Danthonia parryi Scribn. . . . . Parry oat grass
Deschampsia caespitosa (L.) Beauv.. . . . . tufted hair grass
Deschampsia elongata (Hook.) Munro . . . . slender hair grass
Elymus glaucus Buckl. . . . . smooth wild rye
Elymus innovatus Beal . . . . hairy wild rye
Elymus piperi Bowden . . . . giant wild rye
Festuca baffinensis Polunin . . . . arctic fescue
Festuca brachyphylla Schultes . . . . alpine fescue
Festuca idahoensis Elmer . . . . bluebunch fescue
Festuca ovina L. . . . . sheep fescue
Festuca rubra L. . . . . red fescue
Festuca saximontana Rydb. . . . . sheep fescue
Festuca scabrella Torr. . . . . rough fescue
Glyceria grandis S. Wats. ex A. Gray. . . . . common tall manna grass
Glyceria pulchella (Nash) K. Schum. . . . . graceful manna grass
Glyceria striata (Lam.) A.S. Hitchc. . . . . fowl manna grass
Helictotrichon hookeri (Scribn.) Henr. . . . . Hooker's oat grass
Hierochloe odorata (L.) Beauv.. . . . . sweet grass
Hordeum jubatum L. . . . . foxtail barley
Koeleria macrantha (Ledeb.) J.A. Schultes . . . . June grass
Muhlenbergia cuspidata (Torr.) Rydb.. . . . . plains muhly
Muhlenbergia glomerata (Willd.) Trin. . . . . bog muhly
Muhlenbergia richardsonis (Trin.) Rydb. . . . . mat muhly
Oryzopsis asperifolia Michx. . . . . white-grained mountain rice grass
Oryzopsis exigua Thurb. . . . . little rice grass
Oryzopsis hymenoides (R.& S.) Ricker . . . . Indian rice grass
Oryzopsis pungens (Torr.) A.S. Hitchc. . . . . northern rice grass
Phalaris arundinacea L. . . . . reed canary grass
Phleum commutatum Gaudin . . . . mountain timothy
Phleum pratense L. . . . . timothy
Poa alpina L. . . . . alpine bluegrass
Poa annua L. . . . . annual bluegrass
Poa arctica R. Br. . . . . arctic bluegrass
Poa canbyi (Scribn.) Piper . . . . Canby bluegrass
Poa compressa L. . . . . Canada bluegrass
Poa cusickii Vasey . . . . early bluegrass
Poa epilis Scribn. . . . . skyline bluegrass
Poa glauca Vahl . . . . bluegrass
Poa gracillima Vasey . . . . pacific bluegrass
Poa interior Rydb. . . . . wood bluegrass

Poa juncifolia Scribn. . . . . alkali bluegrass
Poa leptocoma Trin. . . . . bog bluegrass
Poa nemoralis L. . . . . wood bluegrass
Poa nervosa (Hook.) Vasey . . . . Wheeler's bluegrass
Poa palustris L. . . . . fowl bluegrass
Poa pattersonii Vasey . . . . Patterson's bluegrass
Poa pratensis L. . . . . Kentucky bluegrass
Poa sandbergii Vasey. . . . . Sandberg's bluegrass
Puccinellia distans (L.) Parl. . . . . slender salt-meadow grass
Schizachne purpurascens (Torr.) Swallen . . . . purple oatgrass, false melic
Setaria glauca (L.) Beauv. . . . . yellow foxtail
Sphenopolis intermedia (Rydb.) Rydb. . . . . slender wedge grass
Stipa columbiana Macoun . . . . Columbia needle grass
Stipa comata Trin. & Rupr. . . . . needle-and-thread, spear grass
Stipa richardsonii Link . . . . Richardson's needle grass
Stipa spartea Trin. . . . . porcupine grass
Stipa viridula Trin. . . . . green needle grass
Trisetum spicatum (L.) Richt. . . . . spike trisetum
Vahlodea atropurpurea (Vahl.) Fries . . . . mountain hair grass

Polemoniaceae — PHLOX FAMILY

Collomia linearis Nutt. . . . . narrow-leaved collomia
Phlox alyssifolia Greene . . . . blue phlox
Phlox hoodii Richards. . . . . moss phlox
Polemonium pulcherrimum Hook. . . . . showy Jacob's-ladder
Polemonium viscosum Nutt. . . . . skunkweed

Polygalaceae — MILKWORT FAMILY

Polygala senega L. . . . . seneca snakeroot

Polygonaceae — BUCKWHEAT FAMILY

Eriogonum androsaceum Benth. . . . . cushion umbrella-plant
Eriogonum flavum Nutt. . . . . yellow umbrella-plant
Eriogonum ovalifolium Nutt. . . . . silver-plant
Eriogonum umbellatum Torr.. . . . . subalpine umbrella-plant
Fagopyrum tartaricum (L.) Gaertn. . . . . tartary buckwheat
Oxyria digyna (L.) Hill. . . . . mountain sorrel
Polygonum amphibium L. . . . . water smartweed
Polygonum arenastrum Jord. ex Bor. . . . . common knotweed, yard knotweed
Polygonum bistortoides Pursh . . . . western bistort
Polygonum coccineum Muhl. . . . . water smartweed
Polygonum convolvulus L.. . . . . wild buckwheat, black bindweed
Polygonum engelmannii Greene . . . . slender knotweed
Polygonum erectum L. . . . . striate knotweed
Polygonum persicaria L. . . . . lady's-thumb
Polygonum ramosissimum Michx.. . . . . bushy knotweed

Polygonum viviparum L. . . . . alpine bistort
Rumex acetosa L. ssp. *alpestris* (Scop.) Löve . . . . green sorrel
Rumex acetosella L. . . . . sheep sorrel
Rumex maritimus L. . . . . golden dock
Rumex occidentalis S. Wats. . . . . western dock
Rumex triangulivalvis (Dans.) Rech. f. . . . . narrow-leaved dock

---

Polypodiaceae — FERN FAMILY

Asplenium viride Huds. . . . . green spleenwort
Athyrium filix-femina (L.) Roth. . . . . lady fern
Cheilanthes feei Moore . . . . slender lip fern
Cryptogramma stelleri (S.G. Gmel.) Prantl . . . .
. . . . Steller's rock brake, fragile rock brake
Cystopteris fragilis (L.) Bernh. . . . . fragile bladder fern
Dryopteris assimilis S. Walker . . . . broad spinulose shield fern
Gymnocarpium dryopteris (L.) Newm.. . . . . oak fern
Pellaea atropurpurea (L.) Link. . . . . purple cliff brake
Pellaea glabella Mett. ex Kuhn . . . . smooth cliff brake, purple cliff brake
Polystichum lonchitis (L.) Roth . . . . northern holly fern
Woodsia glabella R. Br. . . . . smooth woodsia
Woodsia oregana D.C. Eat. . . . . Oregon woodsia
Woodsia scopulina D.C. Eat. . . . . mountain woodsia

Portulacaceae — PURSLANE FAMILY

Claytonia lanceolata Pursh. . . . . western spring beauty
Claytonia megarhiza (A. Gray) Parry. . . . . alpine spring beauty

---

Potamogetonaceae — PONDWEED FAMILY

Potamogeton alpinus Balbis . . . . alpine pondweed
Potamogeton filiformis Pers.. . . . thread-leaved pondweed, narrow-leaved pondweed
Potamogeton gramineus L. . . . . various-leaved pondweed
Potamogeton obtusifolius Mert. & Koch. . . . . blunt-leaved pondweed
Potamogeton pectinatus L. . . . . sago pondweed
Potamogeton praelongus Wulf. . . . . white-stem pondweed
Potamogeton richardsonii (Benn.) Rydb. . . . . clasping-leaf pondweed

---

Primulaceae — PRIMROSE FAMILY

Androsace chamaejasme Host . . . . sweet-flowered androsace
Androsace occidentalis Pursh . . . . western fairy candelabra
Androsace septentrionalis L. . . . . northern fairy candelabra
Dodecatheon conjugens Greene . . . . mountain shooting star
Dodecatheon pulchellum (Raf.) Merr. . . . . saline shooting star

Lysimachia ciliata L. .... fringed loosestrife
Primula mistassinica Michx. .... dwarf Canadian primrose

Pyrolaceae WINTERGREEN FAMILY

Chimaphila umbellata (L.) Bart. ssp. *occidentalis* (Rydb.) Hult. ....
.... prince's-pine, pipsissewa
Moneses uniflora (L.) A. Gray .... one-flowered wintergreen
Orthilia secunda (L.) House .... one-sided wintergreen
Pyrola asarifolia Michx. .... common pink wintergreen
Pyrola bracteata Hook.. .... large wintergreen
Pyrola chlorantha Sw.. .... greenish-flowered wintergreen
Pyrola elliptica Nutt. .... white wintergreen
Pyrola minor L. .... lesser wintergreen

Ranunculaceae CROWFOOT FAMILY

Actaea rubra (Ait.) Willd. .... red and white baneberry
Anemone canadensis L.. .... Canada anemone
Anemone cylindrica A. Gray. .... long-fruited anemone
Anemone lithophila Rydb. .... Drummond's anemone
Anemone multifida Poir. .... cut-leaved anemone
Anemone occidentalis S. Wats.. .... western anemone, chalice-flower
Anemone parviflora Michx.. .... small wood anemone
Anemone patens L.. .... prairie crocus, pasque flower
Anemone richardsonii Hook. .... yellow anemone
Aquilegia brevistyla Hook. .... blue columbine
Aquilegia flavescens S. Wats. .... yellow columbine
Clematis occidentalis (Hornem.) DC.. .... purple clematis
Clematis tangutica (Max.) Korsh. .... yellow clematis
Delphinium bicolor Nutt. .... low larkspur
Delphinium glaucum S. Wats.. .... tall larkspur
Delphinium nuttallianum Pritz. ex Walp. .... Nuttall's larkspur
Ranunculus abortivus L. ...... small-flowered crowfoot, smooth-leaved buttercup
Ranunculus acris L. .... tall buttercup
Ranunculus aquatilis L. .... large-leaved white water crowfoot
Ranunculus cardiophyllus Hook. .... heart-leaved buttercup
Ranunculus cymbalaria Pursh .... seaside buttercup, creeping buttercup
Ranunculus eschscholtzii Schlecht. .... mountain buttercup
Ranunculus glaberrimus Hook. .... early buttercup
Ranunculus gmelinii DC .... yellow water crowfoot
Ranunculus grayi Britt. .... Gray's buttercup, dwarf buttercup
Ranunculus inamoenus Greene ....
.... mountain meadow buttercup, graceful buttercup
Ranunculus macounii Britt. .... Macoun's buttercup
Ranunculus nivalis L.. .... snow buttercup
Ranunculus pedatifidus J.E. Smith .... northern buttercup

Ranunculus pensylvanicus L.f. . . . . bristly buttercup
Ranunculus pygmaeus Wahlenb. . . . . dwarf buttercup
Ranunculus repens L. . . . . creeping buttercup
Ranunculus reptans L. . . . . creeping spearwort
Ranunculus rhomboideus Goldie . . . . prairie buttercup
Ranunculus sceleratus L. . . . . cursed crowfoot, celery-leaved buttercup
Thalictrum occidentale A. Gray . . . . western meadow rue
Thalictrum venulosum Trel. . . . . veiny meadow rue
Trollius albiflorus (A. Gray) Rydb. . . . . globeflower

Rosaceae — ROSE FAMILY

Amelanchier alnifolia Nutt. . . . . saskatoon
Chamaerhodos erecta (L.) Bunge . . . . chamaerhodos
Dryas drummondii Richards. . . . . yellow mountain avens
Dryas integrifolia M. Vahl. . . . . northern white mountain avens
Dryas octopetala L. . . . . white mountain avens
Fragaria vesca L. . . . . woodland strawberry
Fragaria virginiana Duchesne ssp. *glauca* (S. Wats.) Staudt. . . . . wild strawberry
Geum aleppicum Jacq. . . . . yellow avens
Geum macrophyllum Willd. . . . . large-leaved yellow avens
Geum rivale L. . . . . purple avens
Geum triflorum Pursh. . . . old man's whiskers, three-flowered avens, prairie smoke
Potentilla anserina L. . . . . silverweed
Potentilla arguta Pursh. . . . . white cinquefoil
Potentilla concinna Richards. var. *macounii* (Rydb.) C. L. Hitchc. . . . . early cinquefoil
Potentilla diversifolia Lehm. . . . . . . mountain cinquefoil, smooth-leaved cinquefoil
Potentilla drummondii Lehm. . . . . Drummond's cinquefoil
Potentilla fruticosa L. . . . . shrubby cinquefoil
Potentilla glandulosa Lindl. . . . . sticky cinquefoil
Potentilla gracilis Dougl. ex Hook. . . . . graceful cinquefoil
Potentilla hippiana Lehm. . . . . woolly cinquefoil
Potentilla hookeriana Lehm. . . . . Hooker's cinquefoil
Potentilla hyparctica Malte. . . . . northern cinquefoil
Potentilla multisecta (S. Wats.) Rydb. . . . . smooth-leaved cinquefoil
Potentilla nivea L. . . . . snow cinquefoil
Potentilla norvegica L. . . . . rough cinquefoil
Potentilla ovina Macoun . . . . sheep cinquefoil
Potentilla palustris (L.) Scop. . . . . marsh cinquefoil
Potentilla pensylvanica L. . . . . prairie cinquefoil
Potentilla plattensis Nutt. . . . . low cinquefoil
Potentilla rivalis Nutt. . . . . brook cinquefoil
Potentilla uniflora Ledeb. . . . . one-flowered cinquefoil
Potentilla villosa Pallas ex Pursh . . . . hairy cinquefoil
Prunus pensylvanica L. . . . . pin cherry
Prunus virginiana L. . . . . choke cherry
Rosa acicularis Lindl. . . . . prickly rose

Rosa arkansana Porter . . . . prairie rose
Rosa woodsii Lindl. . . . . common wild rose
Rubus arcticus L. ssp. *acaulis* (Michx.) Focke . . . . dwarf raspberry
Rubus idaeus L. . . . . wild red raspberry
Rubus parviflorus Nutt. . . . . thimbleberry
Rubus pedatus J.E. Smith . . . . dwarf bramble
Rubus pubescens Raf.. . . . . dewberry, running raspberry
Sibbaldia procumbens L. . . . . sibbaldia
Sorbus scopulina Greene . . . . western mountain ash
Sorbus sitchensis Roemer . . . . Sitka mountain ash
Spiraea alba DuRoi. . . . . narrow-leaved meadowsweet
Spiraea betulifolia Pallas . . . . white meadowsweet
Spiraea densiflora Nutt.. . . . . pink meadowsweet

Rubiaceae MADDER FAMILY

Galium boreale L. . . . . northern bedstraw, cleavers
Galium trifidum L. . . . . small bedstraw
Galium triflorum Michx. . . . . sweet-scented bedstraw

Salicaceae WILLOW FAMILY

Populus balsamifera L. . . . . balsam poplar
Populus tremuloides Michx.. . . . . aspen
Salix arctica Pallas . . . . arctic willow
Salix athabascensis Raup . . . . Athabasca willow
Salix barclayi Anderss. . . . . Barclay's willow
Salix barrattiana Hook. . . . . Barratt's willow
Salix bebbiana Sarg. . . . . Bebb's willow, beaked willow
Salix brachycarpa Nutt. . . . . short-capsuled willow
Salix candida Fluegge ex Willd. . . . . hoary willow
Salix commutata Bebb. . . . . changeable willow
Salix discolor Muhl. . . . . pussy willow
Salix drummondiana Barr. ex Hook. . . . . Drummond's willow, satin willow
Salix exigua Nutt. . . . . sandbar willow
Salix farriae Ball . . . . Farr's willow
Salix glauca L. . . . . smooth willow
Salix lutea Nutt. . . . . yellow willow
Salix maccalliana Rowlee . . . . velvet-fruited willow
Salix melanopsis Nutt. . . . . willow
Salix myrtillifolia Anderss. . . . . myrtle-leaved willow
Salix pedicellaris Pursh . . . . bog willow
Salix petiolaris J.E. Smith . . . . basket willow
Salix planifolia Pursh . . . . flat-leaved willow
Salix prolixa Anderss. . . . . Mackenzie's willow
Salix pseudomonticola Ball. . . . . false mountain willow
Salix reticulata L. ssp. *nivalis* Löve, Löve & Kapoor . . . . dwarf willow, snow willow
Salix scouleriana Barr. ex Hook. . . . . Scouler's willow

Salix serissima (Bailey) Fern. . . . . autumn willow
Salix stolonifera Coville . . . . willow
Salix reticulata vestita Pursh . . . . rock willow

Santalaceae SANDALWOOD FAMILY

Comandra umbellata (L.) Nutt. var. *pallida* (DC.) M.E. Jones. . . . . bastard toadflax
Geocaulon lividum (Richards.) Fern. . . . . northern comandra

Saxifragaceae SAXIFRAGE FAMILY

Chrysosplenium iowense Rydb. . . . . golden saxifrage
Chrysosplenium tetrandrum (Lund) T. Fries . . . . golden saxifrage
Heuchera cylindrica Dougl. ex Hook. . . . . sticky alumroot
Heuchera parvifolia Nutt. ex T. & G. . . . . small-leaved alumroot
Heuchera richardsonii R. Br.. . . . . Richardson's alumroot
Leptarrhena pyrolifolia (D. Don) R. Br. . . . . leather-leaved saxifrage
Lithophragma glabrum Nutt. ex T. & G. . . . . rockstar, fringe-cup
Mitella breweri A. Gray . . . . Brewer's bishop's-cap, mitrewort
Mitella nuda L. . . . . bishop's-cap
Mitella pentandra Hook. . . . . bishop's-cap
Mitella trifida Graham . . . . bishop's-cap, three-toothed mitrewort
Saxifraga adscendens L.. . . . . wedge-leaved saxifrage
Saxifraga aizoides L. . . . . yellow mountain saxifrage
Saxifraga bronchialis L. . . . . common saxifrage, spotted saxifrage
Saxifraga caespitosa L. . . . . tufted saxifrage
Saxifraga cernua L.. . . . . nodding saxifrage
Saxifraga ferruginea Graham . . . . saxifrage
Saxifraga hyperborea R. Br.. . . . . brook saxifrage
Saxifraga lyallii Engler . . . . red-stemmed saxifrage
Saxifraga nivalis L.. . . . . alpine saxifrage
Saxifraga occidentalis S. Wats. . . . . . . western saxifrage, rhomboid-leaved saxifrage
Saxifraga oppositifolia L.. . . . . purple saxifrage
Tiarella trifoliata L.. . . . . laceflower
Tiarella unifoliata Hook.. . . . . sugar scoop, false mitrewort

Scrophulariaceae FIGWORT FAMILY

Besseya wyomingensis (A. Nels.) Rydb.. . . . . kittentails
Castilleja cusickii Greenm. . . . . yellow paintbrush
Castilleja hispida Benth. . . . . hispid paintbrush
Castilleja lutescens (Greenm.) Rydb. . . . . stiff yellow paintbrush
Castilleja miniata Dougl. ex Hook.. . . . . common red paintbrush
Castilleja occidentalis Torr.. . . . . . alpine yellow paintbrush, lance-leaved paintbrush
Castilleja rhexifolia Rydb.. . . . . alpine red paintbrush
Collinsia parviflora Lindl. . . . . blue-eyed mary
Euphrasia arctica Lange ex Rostrup . . . . eyebright
Linaria dalmatica (L.) Mill. . . . . broad-leaved toad-flax
Linaria vulgaris Hill . . . . butter-and-eggs

Mimulus lewisii Pursh .... red monkeyflower
Orthocarpus luteus Nutt. .... owl-clover
Pedicularis arctica R. Br. .... arctic lousewort
Pedicularis bracteosa Benth. .... bracted lousewort, western lousewort
Pedicularis contorta Benth. .... coiled-beak lousewort
Pedicularis flammea L. .... flame-coloured lousewort
Pedicularis groenlandica Retz. .... little red elephant, elephant's head
Pedicularis racemosa Dougl. ex Benth. .... leafy lousewort
Penstemon albertinus Greene. .... blue beardtongue
Penstemon confertus Dougl. .... yellow beardtongue
Penstemon ellipticus Coult. & Fisher .... creeping beardtongue
Penstemon eriantherus Pursh .... crested beardtongue
Penstemon fruticosus (Pursh) Greene. .... shrubby beardtongue
Penstemon gracilis Nutt.. .... lilac-flowered beardtongue
Penstemon lyallii A. Gray .... large-flowered beardtongue, Lyall's beardtongue
Penstemon nitidus Dougl. ex Benth. .... smooth blue beardtongue
Penstemon procerus Dougl. ex Grah. .... slender blue beardtongue
Rhinanthus minor L. .... yellow rattle
Veronica alpina L.. .... alpine speedwell
Veronica americana (Raf.) Schw.. .... American brooklime
Veronica peregrina L. .... hairy speedwell
Veronica scutellata L. .... marsh speedwell
Veronica serpyllifolia L. ssp. *humifusa* (Dickson) Syme ... thyme-leaved speedwell

---

Selaginellaceae LITTLE CLUB-MOSS FAMILY SPIKE-MOSS FAMILY

Selaginella densa Rydb. .... prairie selaginella
Selaginella rupestris (L.) Spring .... little club-moss
Selaginella selaginoides (L.) Link. .... little club-moss

---

Sparganiaceae BUR-REED FAMILY

Sparganium angustifolium Michx. .... narrow-leaved bur-reed
Sparganium minimum (Hartm.) Fries. .... slender bur-reed

---

Typhaceae CATTAIL FAMILY

Typha latifolia L. .... common cattail

Urticaceae NETTLE FAMILY

Urtica dioica L. .... common nettle

Valerianaceae VALERIAN FAMILY

Valeriana dioica L. ssp. *sylvatica* (Rich.) F. G. Mey. .... northern valerian
Valeriana sitchensis Bong. .... mountain valerian, mountain heliotrope

Violaceae VIOLET FAMILY

Viola adunca J.E. Smith . . . . early blue violet
Viola canadensis L. var. *rugulosa* (Greene) C. L. Hitchc. . . . . western Canada violet
Viola glabella Nutt.. . . . . yellow wood violet
Viola nephrophylla Greene . . . . bog violet
Viola nuttallii Pursh. . . . . yellow prairie violet
Viola orbiculata Geyer ex Hook. . . . . evergreen violet
Viola palustris L. . . . . marsh violet
Viola renifolia A. Gray. . . . . kidney-leaved violet

Glossary

ACAULESCENT
apparently stemless; the leaves and inflorescence arise from near ground-level

ACHENE
simple, dry, indehiscent fruit, sometimes winged or plumed; each contains a single seed

ACTINOMORPHIC *(flower)*
regular; radially symmetrical

ADVENTITIOUS *(roots)*
not arising from a radicle

AGGREGATE *(fruit)*
composed of a number of small fruits derived from numerous carpels of a single flower

ALTERNATE *(leaves and branches)*
not opposite each other

AMENT
catkin; a spike or raceme of tiny, usually unisexual, flowers

ANDROECIUM
collective noun for stamens, whose function is to produce pollen for fertilization

ANDROGYNOUS
bearing flowers in a spike, composed of carpellary (pistillate) and staminate flowers, with staminate flowers at the apex (see *Carex* spp.)

ANGIOSPERM
a plant that bears flowers possessing ovules in a ovary; the ovary becomes the fruit

ANNUAL
a plant that completes its life-cycle in one year

ANNULUS
ring of thin-walled cells partly surrounding the fern sporangium, which helps in spore-dispersal

ANTHER
pollen-bearing part of stamen

ANTHERIDIUM
multicellular structure from which sperms are produced

APETALOUS
without petals

APEX
tip

APOCARPOUS
free carpels

APPRESSED
lying close to an organ or parallel to it

ARACHNOID
cobwebby

ARCHEGONIUM
a multicellular structure, typically flask-shaped, from which ova (eggs) are produced

ARIL
a specialized outgrowth from the stalk of an ovule, which covers the mature seed; any fleshy thickening of the seed coat (e.g., mace is the aril of the nutmeg)

AURICLE
an ear-shaped lobe

AWN
a slender bristle, usually terminal

AXIL
angle between leaf and stem

AXILLARY
growing in an axil

BARBED
having short reflexed points or bristles

BEAKED *(fruit)*
bearing a short, stout terminal appendage

BERRY
a fleshy fruit with many seeds

BIENNIAL
a plant that lives for two years only, flowering in the second year

BI-LABIATE *(flower)*
with upper and lower lips

BIFID *(stigma)*
forked

BIPINNATE
twice pinnate

BLADE
expanded part of a leaf or a petal (limb)

BLOOM
a whitish, powdery coating

BRACT
leafy structure associated with a flower

BRACTLET
small bract

BRAIDED STREAM
a river that consists of a series of wide, shallow channels that join and divide in an interlacing pattern; the channels are separated by sand or gravel bars

BUD
undeveloped flower or leaf

BUD-SCALES
reduced leaves enclosing and protecting a bud

BULB
an underground bud with fleshy scales used for storing food; it bears buds for asexual reproduction

BULBLET
a small bulb, replacing flowers, e.g., alpine bistort (*Polygonum viviparum*)

CAESPITOSE
growing in tufts, sometimes forming mats

CALLUS
an abnormally thickened part at the base of a cutting

CALYX
collective noun for sepals, usually green and leaf-like; its function is to protect the petals, stamens and carpels in bud

CALYX-LOBE
part of a united calyx

CAMPANULATE
bell-shaped

CANESCENT
appearing pale or grey because of a fine, close pubescence; hairy

CAPILLARY
hair-like

CAPITULUM
inflorescence of dandelion family, Asteraceae (Compositae)

CAPSULE
a dry fruit that opens with splits or pores

CARBONATES
limestone or dolomite rocks composed of calcium carbonate or the double carbonates of calcium and magnesium; most are sedimentary

CARPEL
sometimes referred to as a *pistil*; each carpel usually has an ovary, style and stigma (the female reproductive parts of the flower)

CARPELLARY FLOWER *(pistillate)*
with carpels but no stamens

ASEXUAL REPRODUCTION

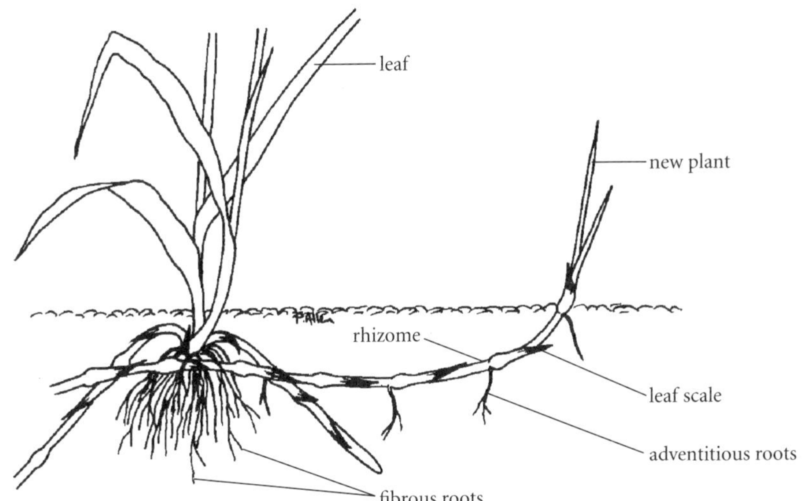

RHIZOME
QUACK GRASS (*AGROPYRON REPENS*)

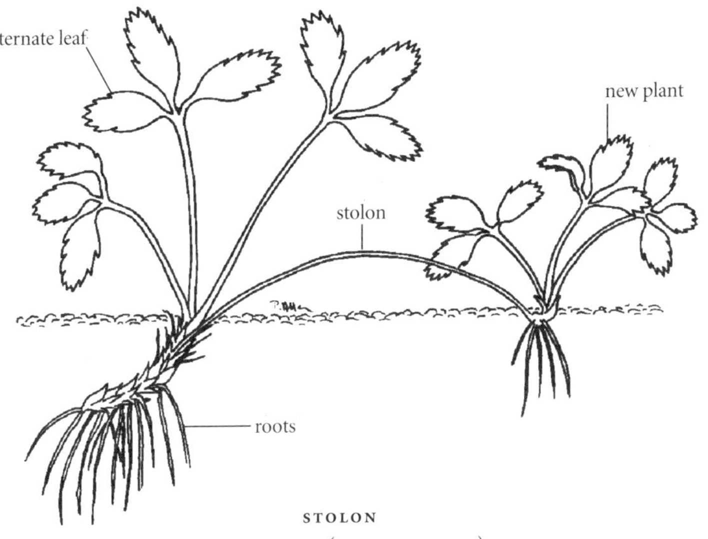

STOLON
STRAWBERRY (*FRAGARIA* SPP)

CARPOPHORE
part of a schizocarp fruit; each mericarp is suspended on a stalk, the carpophore

CARYOPSIS
the fruit of grasses; the seed-coat (testa) is joined to the pericarp (the fruitwall)

CATKIN
a spike or raceme of tiny flowers, usually unisexual

CAUDATE
having a tail-like terminal appendage

CAUDEX
the thickened base of a perennial plant

CAULINE *(leaves)*
borne on the stem

CILIATE
with a fringe of marginal hairs

CINEREOUS
ash-coloured

CIRCINATE
coiled in the bud

CIRCUMSCISSILE *(capsule)*
opening along a transverse circular line, the top of the capsule opening like a lid, e.g., plantain (*Plantago* sp.)

CLASPING *(leaf)*
with the base partly or completely surrounding the stem

CLASTIC ROCKS
sedimentary rocks composed of broken fragments of pre-existing rocks

CLAVATE
club-shaped

CLAW
the narrow base of some petals and sepals, e.g., *Silene* spp.

CLEISTOGAMOUS *(flower)*
fertilized without opening, e.g., violet (*Viola* sp.)

COLUMN *(in orchid)*
stamens united with the style

COMA
a tuft of hairs at the tip of a seed, as in fireweed (*Epilobium* sp.)

COMPOUND *(leaves)*
composed of two or more leaflets

CONNECTIVE
the part of a stamen which joins the two anther-lobes, clearly seen in shooting star (*Dodecatheon* sp.)

CORDATE
heart-shaped

CORM
the bulb-like, but solid, base of a stem, as in glacier lily (*Erythronium* sp.)

COROLLA
collective noun for petals; usually coloured, often perfumed. Its function is to attract insects to pollinate the flower.

COROLLA TUBE
petals united into a tube

CORONA
a flower-structure found in some plants between the corolla and the stamens, like a crown (as in daffodil)

CORYMB
a flat-topped flower cluster, where the central flowers bloom last

COTYLEDON
a leaf of the embryonic plant inside the seed

CREEPING
growing along the surface of the ground and often producing roots at the nodes

LEAF SHAPES

CORDATE

LANCEOLATE

LINEAR

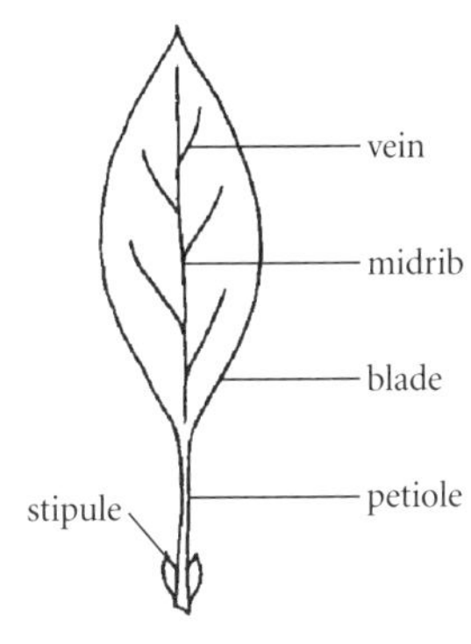

OVATE

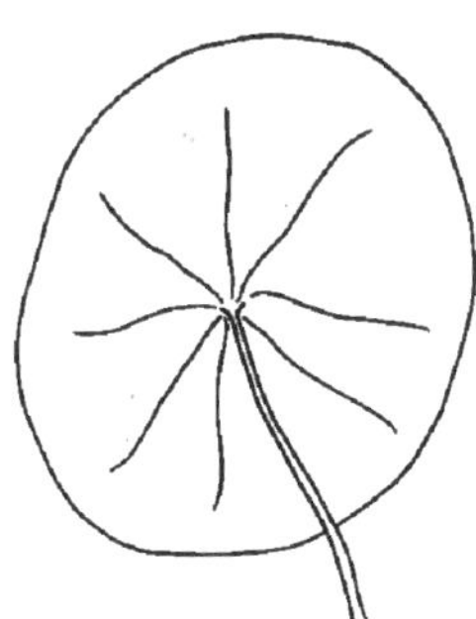

PELTATE

RENIFORM

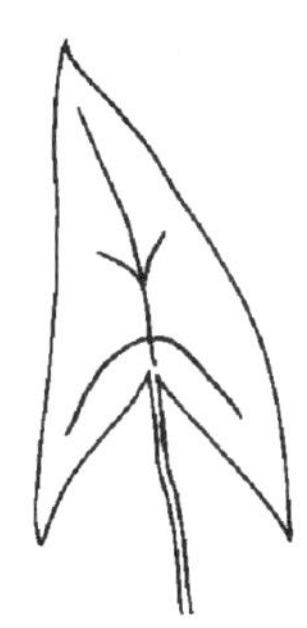

SAGITTATE

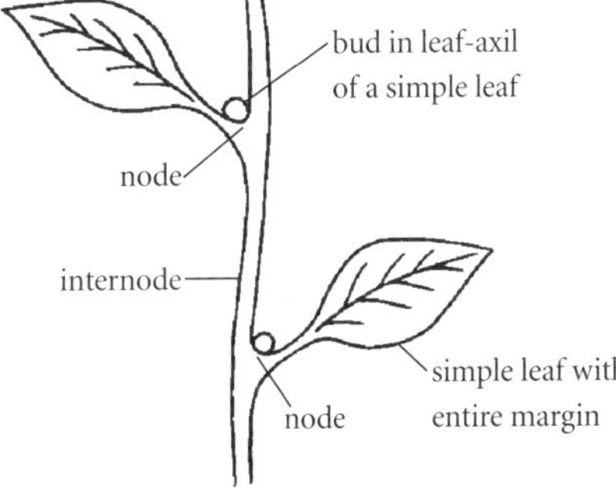

TWIG WITH ALTERNATE LEAVES

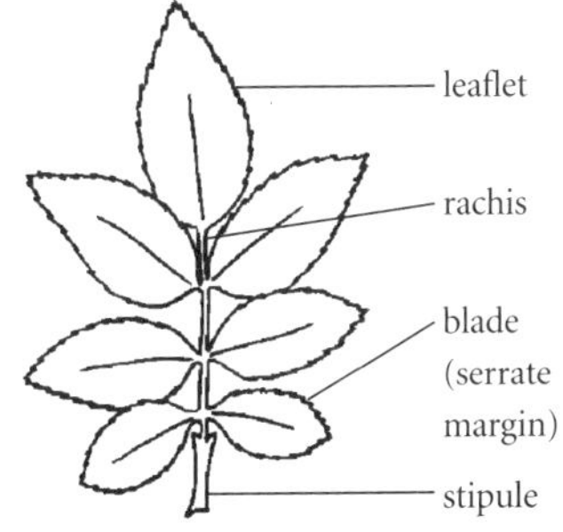

PINNATELY COMPOUND LEAF

CRENATE
scalloped

CRUCIFORM *(petals)*
arranged in the form of a cross, e.g., the mustard family, Brassicaceae (Cruciferae)

CULM
the stem of a grass or sedge

CUNEATE
wedge-shaped

CYME
a flat-topped or convex flower cluster, where the central flowers bloom first

CYPSELA
inferior, often plumed, achene, e.g., dandelion

DECIDUOUS *(leaves)*
not persistent; falling at the end of the growing season

DECUMBENT *(stems)*
prostrate at base, then erect or ascending

DEFLEXED
bent downwards

DEHISCENT
opening

DELTOID
shaped like an equilateral triangle

DENTATE
with teeth outwardly directed

DENUDATION
wearing away of the land by various agencies such as water erosion, wind erosion or glaciers

DIADELPHOUS *(stamens)*
arranged in two sets, often unequal, e.g., mustard family, Brassicaceae (Cruciferae)

DICHOTOMOUS
forking into two branches of about equal size

DIGITATE *(leaves)*
compound, with the leaflets all arising at the apex of the common stalk

DIOECIOUS
male and female flowers on separate plants

DISC FLORET
tubular flower in centre of capitulum, e.g., dandelion family, Asteraceae (Compositae)

DISSECTED
divided into narrow segments

DIVARICATE
widely divergent; spreading

DRUPE
fleshy, one-seeded fruit, with seed inside a stony endocarp, e.g., plum

DRUPELETS
small drupes

EMBRYO
the rudimentary, immature plant in the seed, with radicle, plumule and cotyledon

ENDOCARP
stony part enclosing seed of drupe, e.g., plum

EPICALYX
a ring of bracts alternating with sepals, (e.g., *Potentilla* sp.)

EPICARP
skin of a drupe

EPICONTINENTAL
shallow portions of the sea overlying the continental shelf

EPIGYNOUS
flower with inferior ovary

EPIPETALOUS
attached to petals

EXSERTED
projecting out, for example, stamens and style projecting beyond the corolla tube (e.g., *Rhododendron* sp.)

INFLORESCENCES (FLOWER ARRANGEMENTS)

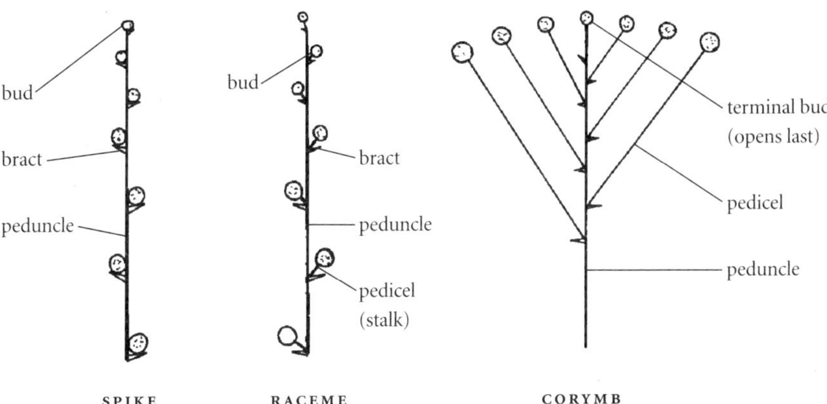

SPIKE RACEME CORYMB

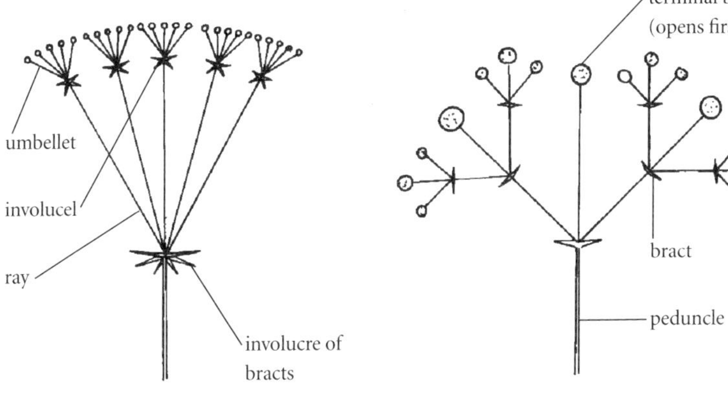

COMPOUND UMBEL COMPOUND CYME

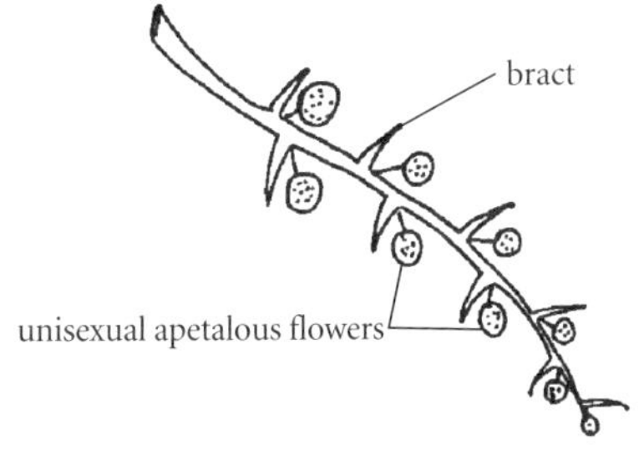

CATKIN (AMENT)

INFLORESCENCES (FLOWER ARRANGEMENTS)

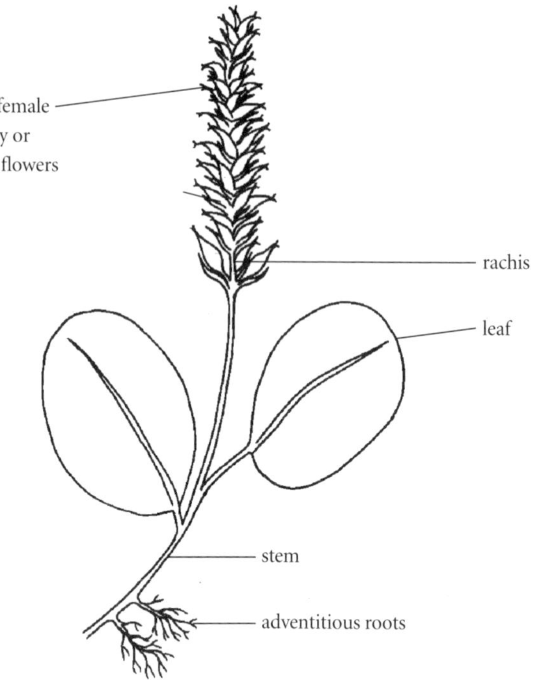

CATKIN INFLORESCENCE: SALICACEAE *(SALIX RETICULATA)*

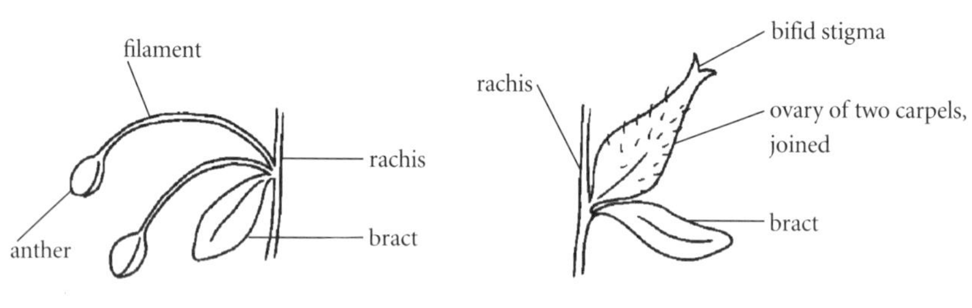

MALE FLOWER
(STAMINATE)

FEMALE FLOWER
(CARPELLARY OR PISTILLATE)

FASCICLE
a cluster

FELLFIELDS *(fjaeidfields)*
stony plateaus or ridges in high alpine zones, with sorted, patterned ground, often dominated by lichens

FEN
low, peaty land, covered wholly or partly with water

FIBRILLOSE
possessing small fibres

FILAMENT
stalk of stamen

FILIFORM
thread-like

FIMBRIATE
fringed

FLOCCOSE
covered with flocks or tufts of soft woolly hairs

FLORET
an individual flower of the flower-head of an Asteraceae plant

FOLLICLE *(fruit)*
like a legume, but opening only along one side, e.g., columbine fruit (*Aquilega* sp.)

FROND
fern leaf, curled in bud, hence called a "fiddlehead"

FRUIT
a ripened ovary

FUNNELIFORM
funnel-shaped

GALEA
the helmet-like upper lip of some bi-labiate corollas, e.g., *Castilleja* sp.

GAMETOPHYTE
the plant generation which produces gametes. In angiosperms, the female gametophyte is the embryo sac, and the male gametophyte is the pollen grain.

GAMOPETALOUS
united petals

GAMOSEPALOUS
united sepals

GENICULATE
abruptly bent or twisted

GLABROUS
smooth; without hairs

GLAND
a plant organ that secretes nectar or volatile oil

GLAUCOUS
covered with a whitish or grey-blue bloom

GLOBOSE
spherical

GLUME
one of a pair of bracts, growing at the base of a grass spikelet; the glumes do not subtend the flowers

GLUTINOUS
covered with a sticky substance

GYNOBASE
enlargement of hypanthium below the gynoecium

GYNOECIUM
collective noun for the carpels

HASTATE
like an arrowhead (sagittate), but with two divergent basal lobes

HERB
a plant with stems dying back to ground level at the end of the season; having no persistent woody tissue

HIRSUTE
hairy

HYPANTHIUM
saucer-shaped, cup-shaped or sometimes rod-shaped expansion of the flower stalk that bears sepals, petals, stamens and carpels; a modified *receptacle*

FLOWER TYPES, DEFINED BY THE OVARY POSITION

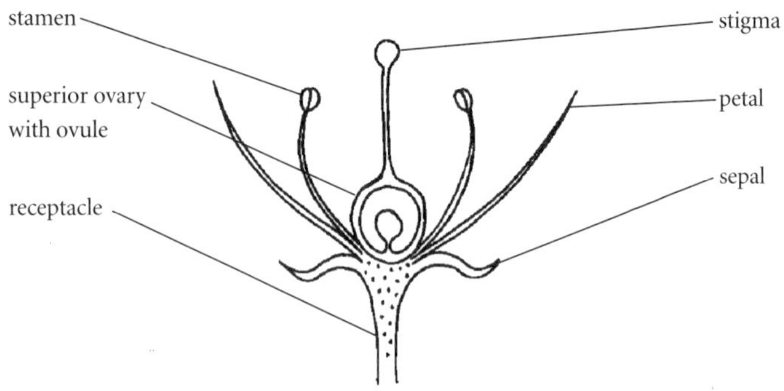

HYPOGYNOUS FLOWER WITH SUPERIOR OVARY

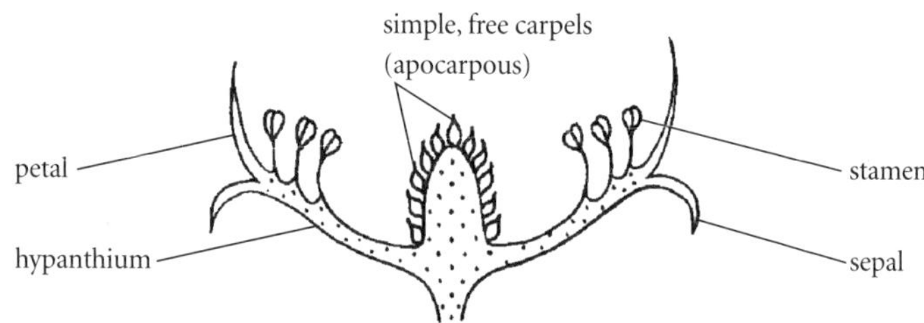

PERIGYNOUS FLOWER WITH EXPANDED RECEPTACLE (HYPANTHIUM)

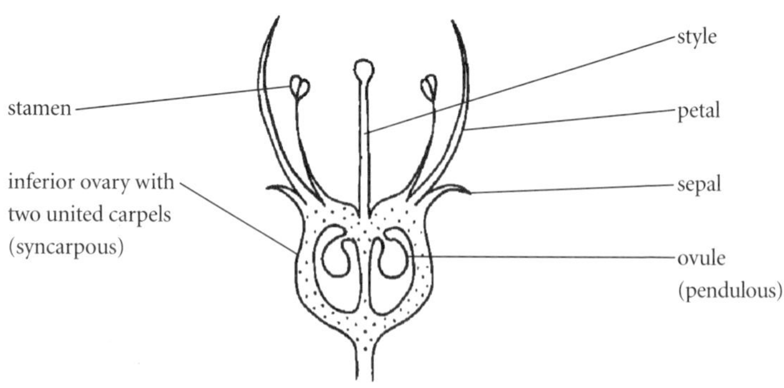

EPIGYNOUS FLOWER WITH INFERIOR OVARY

FLOWER TYPES, DEFINED BY THE OVARY POSITION

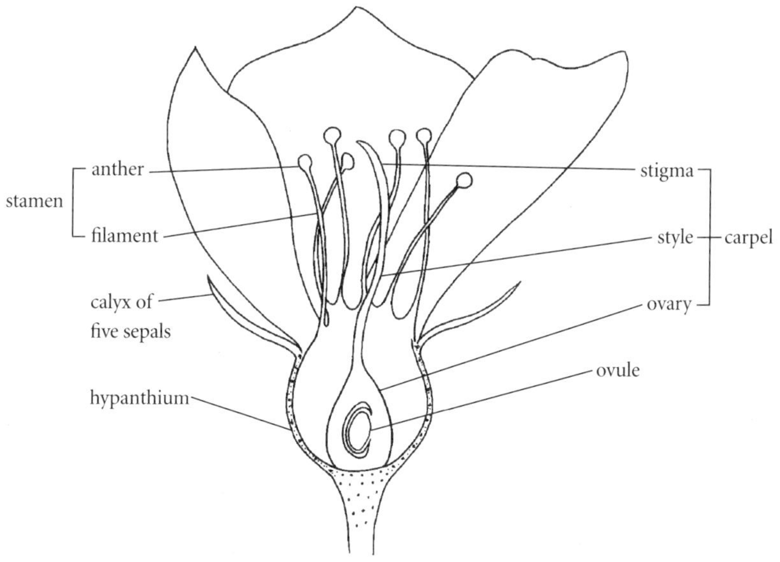

PERIGYNOUS FLOWER: *PRUNUS* (PLUM)

UNDERSURFACE OF POTENTILLA (CINQUEFOIL) FLOWER

FLOWER TYPES, DEFINED BY THE OVARY POSITION

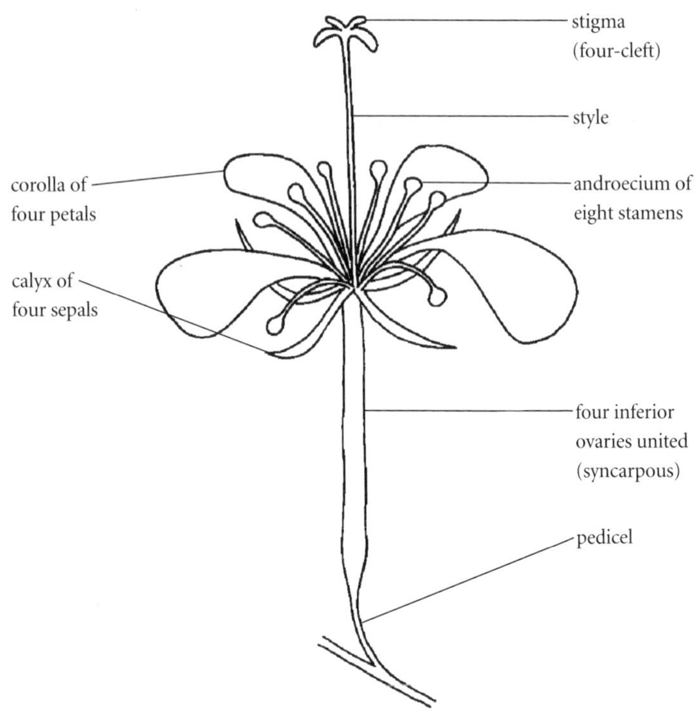

ACTINOMORPHIC POLYPETALOUS EPIGYNOUS FLOWER:
COMMON FIREWEED
(*EPILOBIUM ANGUSTIFOLIUM*)

HYPOGYNOUS
ovary superior; flower parts below the ovary

IMBRICATE
overlapping

INCLUDED *(stamens)*
not protruding from the corolla

INDEHISCENT
not opening

INDURATE
hardened

INDUSIUM
membrane covering a fern sorus

INFERIOR *(ovary)*
flower parts above the ovary

INFLORESCENCE
arrangement of flowers

INTERNODE
between the nodes

INTRORSE *(stamens)*
shedding pollen inwards

INVOLUCEL
small involucre

INVOLUCRE
circle of bracts beneath the flower or flower-head

INVOLUTE
rolled inward

IRREGULAR
bilaterally symmetrical

KEEL PETALS
two petals joined together in a Fabaceae (Leguminosae) flower

KRUMMHOLZ
dwarf trees at timberline (German for "crooked tree")

LACERATE
torn; with an irregularly jagged margin

LACINIATE
cut into narrow segments

LANCEOLATE
lance-shaped

LATEX
a milky juice found in some plants

LEAFLET
part of a compound leaf

LEAF-SCALE
protects bud in winter

LEAFBLADE
broad part of a leaf

LEMMA
lower (outer) of two bracts enclosing a flower, in Poaceae (Gramineae)

LIGULATE
bearing ligules

LIGULE
ear-like lobe, in Poaceae (Gramineae)

LINEAR
long and narrow

LOCULE
a cavity in an ovary or an anther

LOCULICIDAL
a fruit that opens (dehisces) along the middle of each locule

LODICULE
minute outgrowth in a Poaceae (Gramineae) flower; represents the missing perianth

LOMENT
a jointed legume, with constrictions between the seeds, e.g., milk vetch (*Hedysarum* spp.)

LYRATE
shaped like a lyre

MACROFOSSILS
remains of past life that can be seen and studied without the aid of a microscope

MACULATE
spotted

MERICARP
single-seeded fruit; part of a schizocarp

REGULAR AND IRREGULAR FLOWERS: SAMPLE SECTIONS

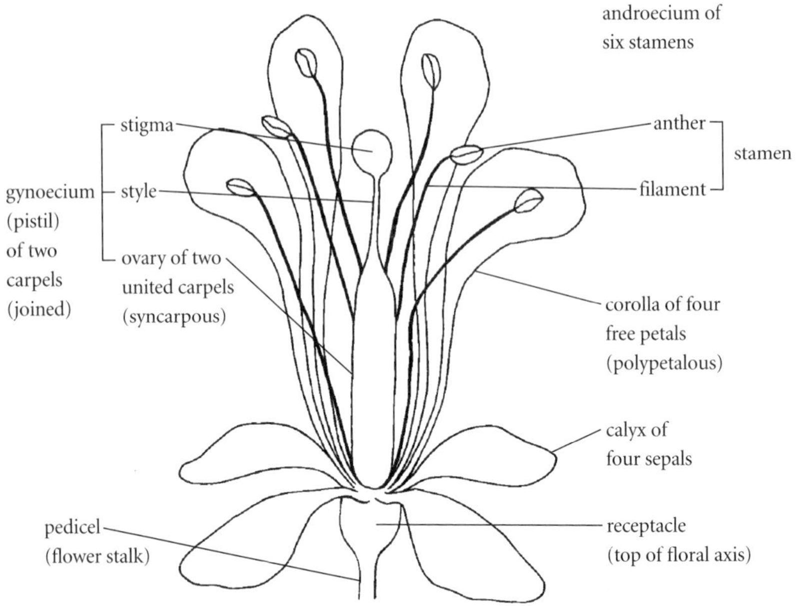

ACTINOMORPHIC (REGULAR) DICOTYLEDONOUS FLOWER: BRASSICACEAE (CRUCIFERAE)

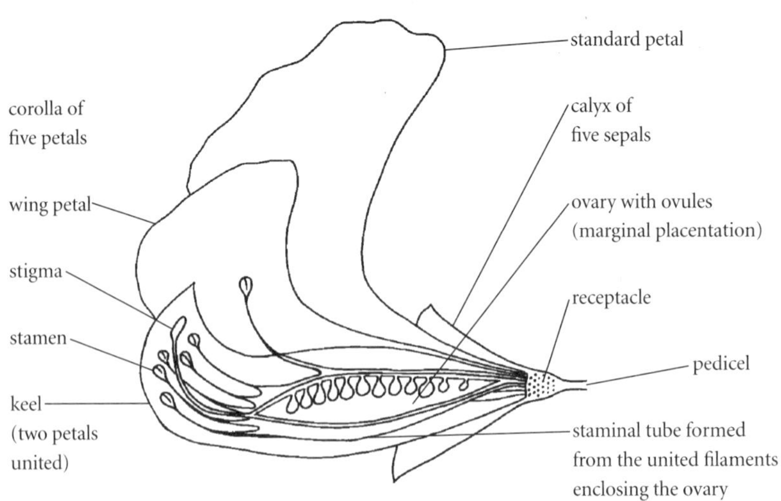

ZYGOMORPHIC (IRREGULAR) DICOTYLEDONOUS FLOWER: FABACEAE (LEGUMINOSAE)

REGULAR AND IRREGULAR FLOWERS: SAMPLE SECTIONS

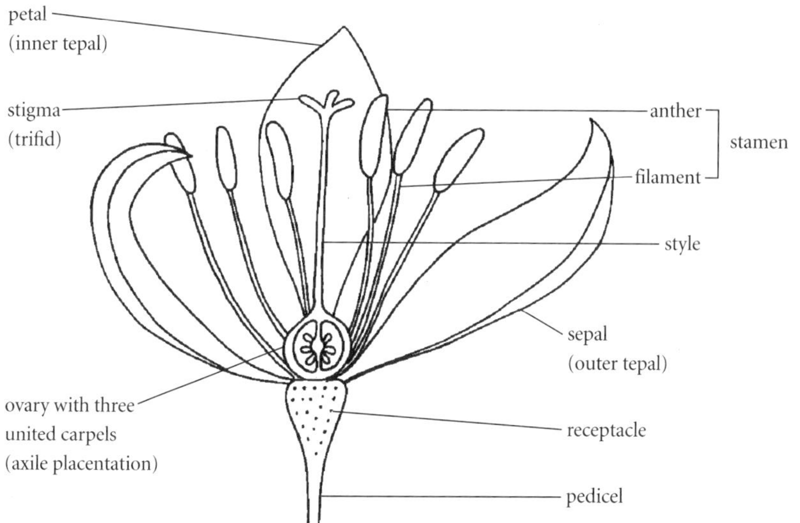

ACTINOMORPHIC MONOCOTYLEDONOUS FLOWER:
LILIACEAE

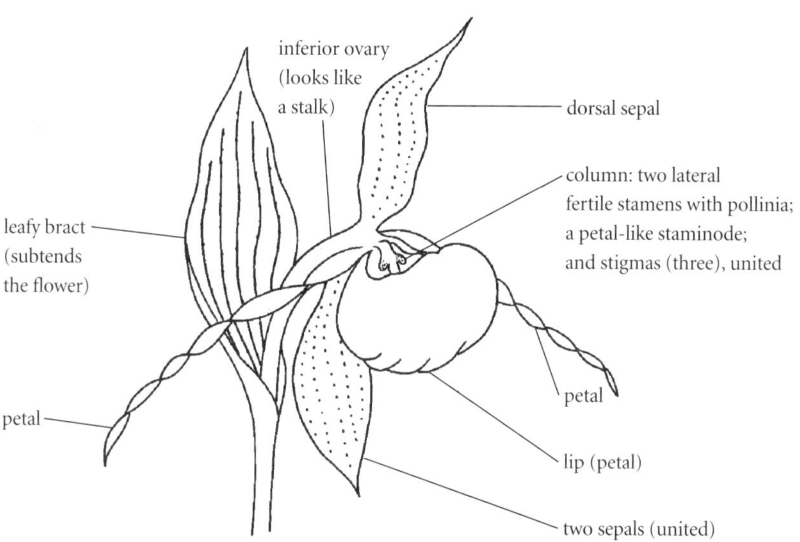

ZYGOMORPHIC MONOCOTYLEDONOUS FLOWER:
ORCHIDACEAE

REGULAR AND IRREGULAR FLOWERS: SAMPLE SECTIONS

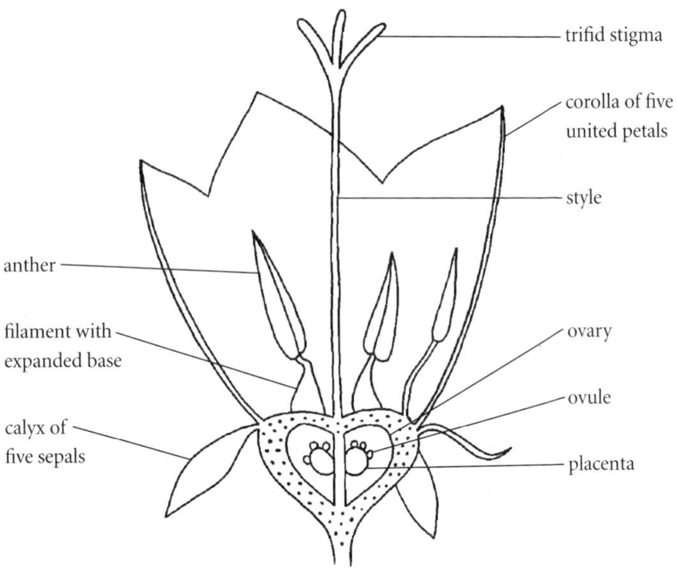

ACTINOMORPHIC EPIGYNOUS SYMPETALOUS FLOWER: CAMPANULA (HAREBELL)

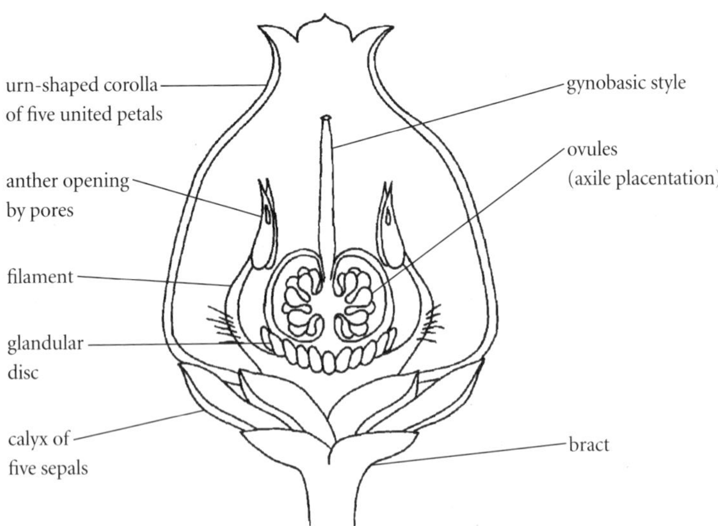

ACTINOMORPHIC SYMPETALOUS FLOWER: ERICACEAE

REGULAR AND IRREGULAR FLOWERS: SAMPLE SECTIONS

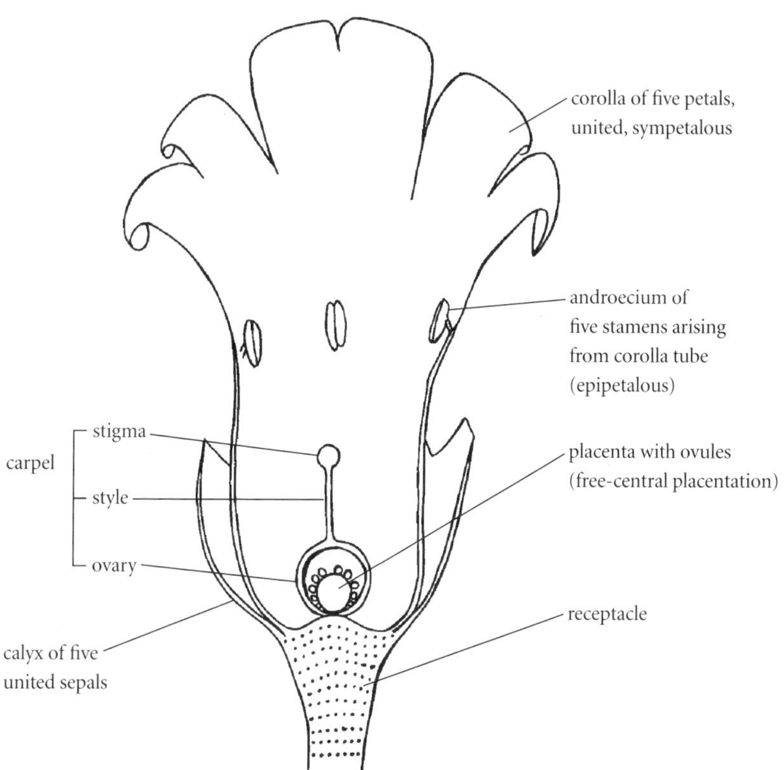

ACTINOMORPHIC SYMPETALOUS FLOWER:
PRIMULACEAE

REGULAR AND IRREGULAR FLOWERS: SAMPLE SECTIONS

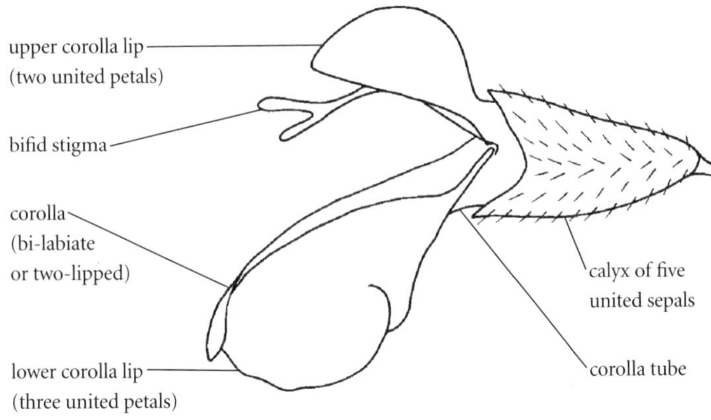

ZYGOMORPHIC SYMPETALOUS FLOWERS:
SALVIA (SAGE)

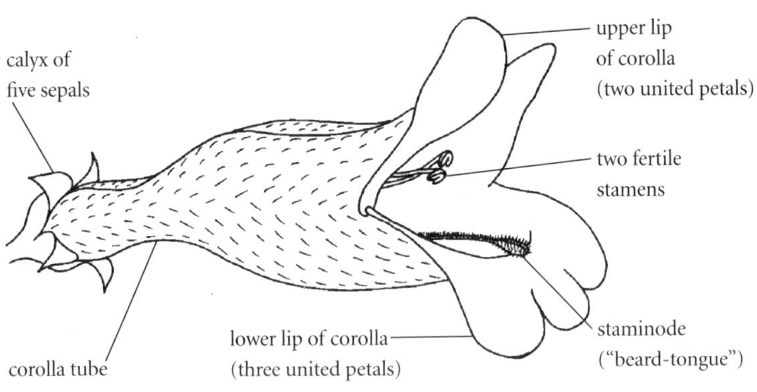

ZYGOMORPHIC SYMPETALOUS FLOWERS:
PENSTEMON (BEARD-TONGUE)

MESOCARP
fleshy part of a drupe

MIDRIB
central vein of leaf

MONADELPHOUS *(stamens)*
united by the filaments to form a tube

MONOECIOUS
separate male and female flowers on the same plant

MONTANE
the eco-region between the foothills and lower subalpine zones, characterized by Douglas fir

MYCORRHIZAL
having an association with a symbiotic fungus

NECTARY
a gland secreting nectar

NODE
place on stem where leaf arises

NUT
a hard, dry indehiscent (non-opening), one-seeded fruit, with a thicker wall than an achene

NUTLET
a small nut, very thick-walled

OB-
broadest at the top, e.g., oblanceolate

OBTUSE
blunt or rounded at the end

OCREAE
a tubular sheathing leaf-stipule, as in dock (*Rumex* sp.)

OIL-DUCTS
tiny black markings on fruits of Apiaceae (Umbelliferae)

OVARY
lower part of carpel; contains the ovules and turns into the fruit

OVATE
oval

OVOID
egg-shaped

OVULE
part which ripens into a seed

OVUM
an egg; a female gamete [pl. *ova*]

PALEA
upper (inner) of two bracts enclosing a flower, in Poaceae (Gramineae)

PALMATE *(leaf)*
having lobes or leaflets radiating from the same point near the tip of the petiole

PANICLE
compound raceme

PAPILLATE, PAPILLOSE
covered with short, rounded projections (papillae)

PAPPUS
tuft of hairs, sometimes scales or bristles, borne on top of an inferior achene, in Asteraceae (Compositae), e.g., thistle

PARASITE
a plant that grows on and derives nourishment from another organism, e.g., *Orobanche* (cancer-root), which grows on the roots of other plants

PEDICEL
individual flower-stalk

PEDUNCLE
main flower-stalk

PELTATE *(leaf)*
with stalk in centre; underneath

PENDULOUS
hanging down

PERENNIAL
a plant that lives for several years

PERIANTH LEAVES
sepals and petals collectively (perianth means "around the flower")

DANDELION FAMILY FEATURES (ASTERACEAE / COMPOSITAE)

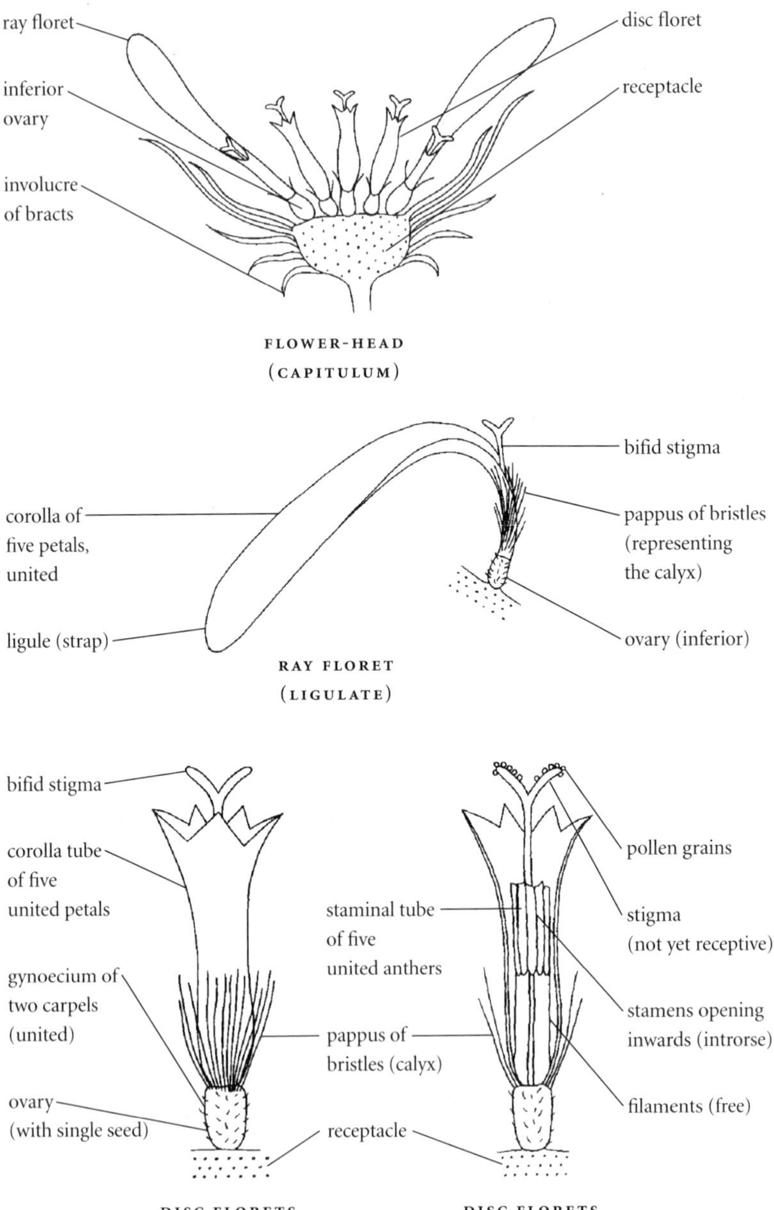

POLLINATION: As the style and stigma grow up through the staminal tube, the stigma gets covered with pollen. The stigma is not receptive at this point but becomes receptive later. Pollen is transferred from the non-receptive stigmas to the receptive ones by insects walking across the flower-head, thus ensuring cross-pollination.

PERIGYNIUM
vase-shaped membrane enclosing the ovary in *Carex* spp.

PERIGYNOUS
flower parts borne around the ovary, on the edge of a hypanthium

PETAL
part of a corolla

PETIOLE
leaf-stalk

PINNA
a primary division of a pinnate or compound pinnate leaf; also, part of the frond of a fern [pl. *pinnae*]

PINNATE *(leaf)*
a compound leaf, with leaflets arranged on two sides of a rachis (a common axis)

PISTIL
a name sometimes used for the carpel

PISTILLATE *(carpellary flower)*
with carpels but no stamens

PLACENTA
part of an ovary that bears the ovules

PLACENTATION *(marginal, axial, free-central)*
arrangement of ovules on the placenta, inside the ovary

PLUMULE
part of the plant embryo

POLLEN
minute grains, produced in the anther of a stamen, that produce the male gametes

POLLINIUM
a mass of coherent pollen, developed in the orchid family, Orchidaceae

POLYPETALOUS
separate petals

POME
a fleshy fruit, derived from a compound, inferior ovary, with a papery endocarp and several seeds, e.g., an apple

PORE
any small aperture; may be found in anthers, epidermis or poricidal capsules

PORICIDAL *(capsule)*
with pores, e.g., poppy fruit

PROCUMBENT
prostrate or trailing, but not rooting at the nodes

PROGLACIAL
in front of a glacier

PROSTRATE
lying flat on the ground

PROTHALLUS
the mature gametophyte in the life-cycles of club mosses, horsetails and ferns; *see Life Cycles, pp. XL-XLIII*

PSEUDOCARP
a "false" fruit—the receptacle forms part of the fruit, e.g., rose-hip

PYXIS
a circumscissile capsule, with a line of dehiscence around the centre, the upper part forming a lid, e.g., plantain (*Plantago* sp.)

RACEME
a "spike" of flowers on which all flowers have stalks (pedicels); see also *panicle*

RACHILLA
axis of a spikelet, e.g., Poaceae (Gramineae)

RACHIS
a central axis

RADICLE
part of the embryo, in the seed, that forms the root

REPRODUCTIVE AND VEGETATIVE PARTS OF RUSH, SEDGE AND GRASS

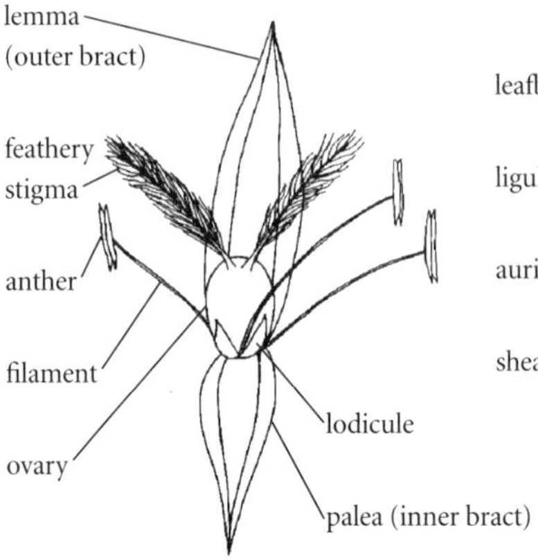

SINGLE FLOWER (FLORET) OF GRASS

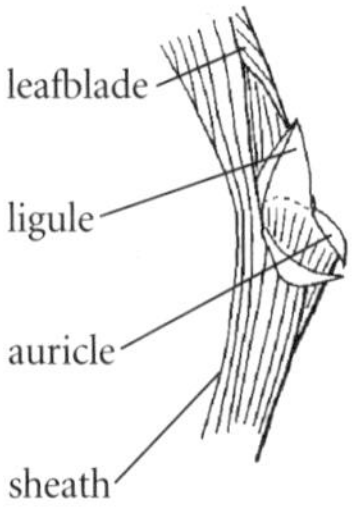

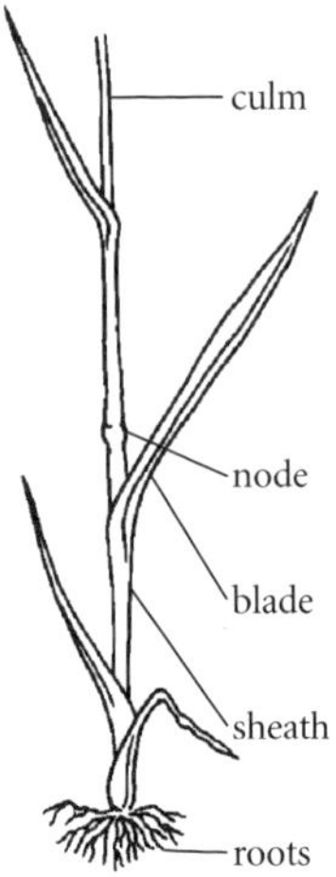

STEM OF GRASS

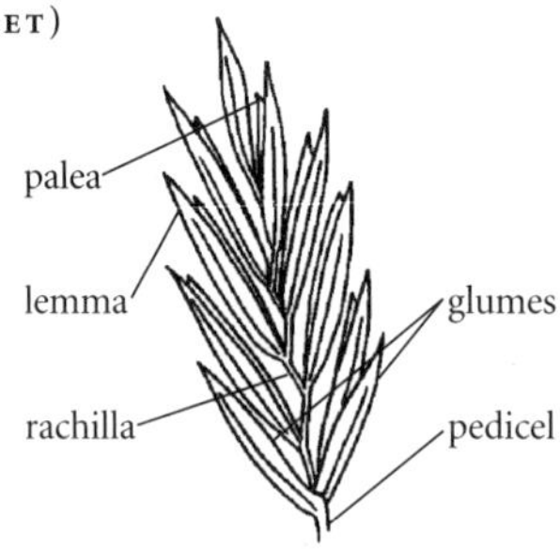

SPIKELET OF GRASS

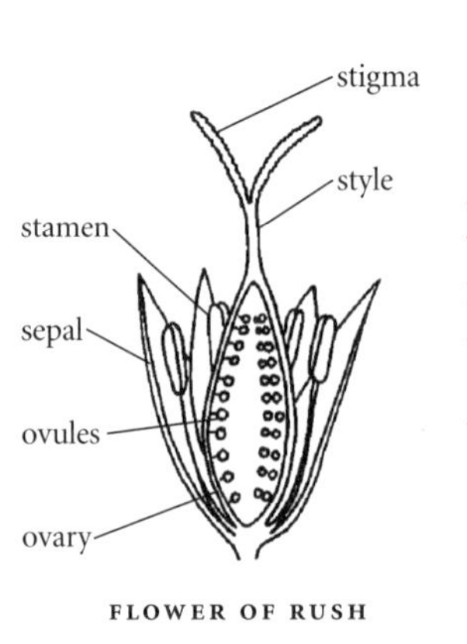

FLOWER OF RUSH

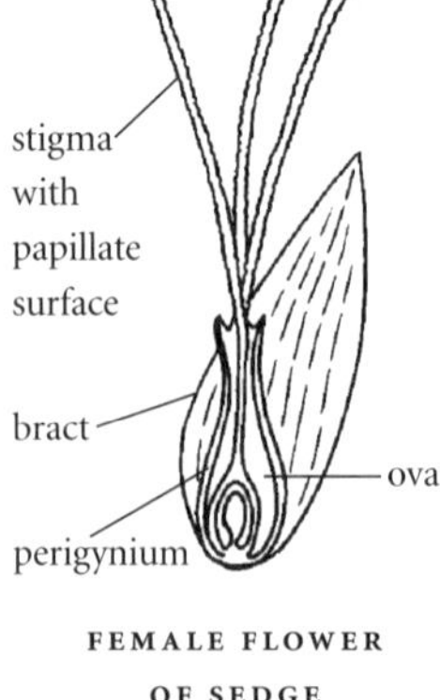

FEMALE FLOWER OF SEDGE

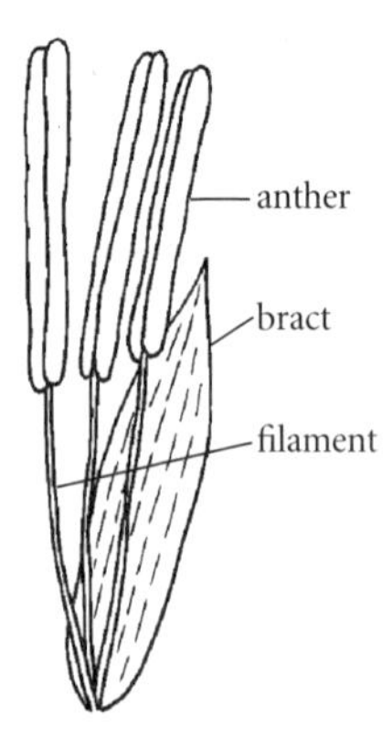

MALE FLOWER OF SEDGE

RAY
the stalk of an umbellet in a compound umbel

RAY FLORET
part of the outer ring of florets in a flower-head (capitulum) of Asteraceae (Compositae)

RECEPTACLE
the swollen top of the flower stalk; sometimes called a *hypanthium*

REFLEXED
bent backwards

REFUGIA
unglaciated areas where plants and animals could live during a glaciation

REGULAR
radially symmetrical

RENIFORM
kidney-shaped

RHIZOME
underground stem

ROSTELLUM
the beak of the style, found in the orchid family, Orchidaceae

RUGOSE
wrinkled

RUNCINATE *(leaves)*
sharply incised, with the segments pointing backwards

RUNNER
a prostrate stem, rooting at the nodes; a stolon

SACCATE
in the shape of a sac or pouch

SAGITTATE
arrow-shaped

SALVERFORM
a flower with a slender corolla-tube abruptly expanded into a flat "limb," e.g., *Phlox* sp.

SAMARA
a winged achene

SCABROUS
rough-surfaced, due to short hairs, e.g., leaves of borage family, Boraginaceae

SCAPE
a leafless flowering stalk

SCHIZOCARP
a fruit of several carpels that splits into separate one-seeded fruits, called mericarps, when ripe, e.g., *Geranium* sp.

SCORPIOID
an inflorescence that gradually uncoils from the tip as the flowers open, e.g., borage family, Boraginaceae

SECUND *(inflorescence)*
uni-laterate, with the flowers all or chiefly on one side of the flowering axis

SEED
fertilized ovule

SEPAL
part of a calyx, usually green and leaf-like

SEPTATE
divided by partitions, or septa

SEPTICIDAL *(capsule)*
splitting open along the internal partitions of septa separating the locules

SEPTUM
a membraneous portion of the fruit of the mustard family, Brassicaceae (Cruciferae)

SERRATE
toothed

SESSILE
without a stalk

SHEATH
part of a leaf-base that clasps the stem

SHRUB
a woody plant with several stems, not as tall as a tree

DRY FRUITS

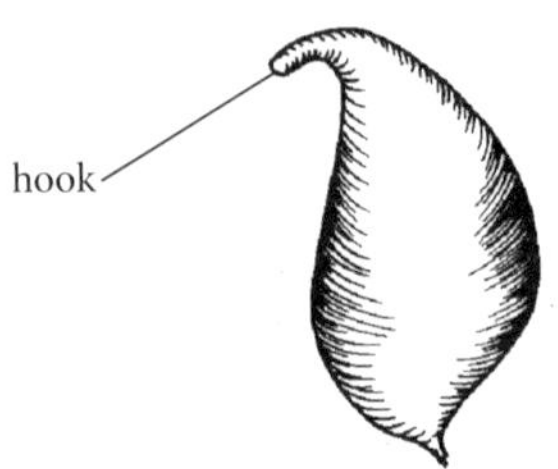

SIMPLE ACHENE
(E.G., BUTTERCUP)

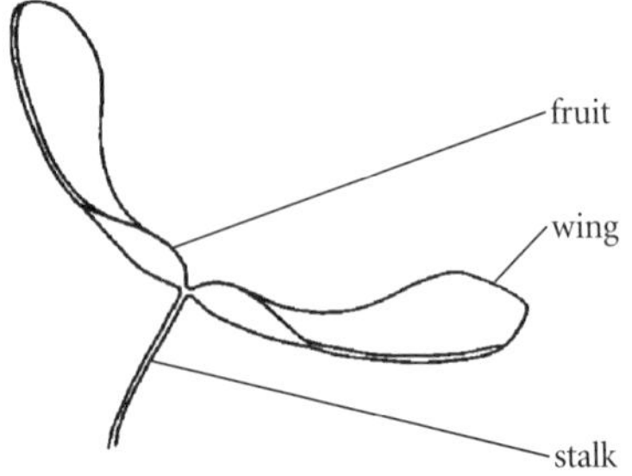

WINGED ACHENE:
A DOUBLE SAMARA
(E.G., MAPLE)

PLUMED ACHENE (SUPERIOR)
(E.G., ANEMONE)

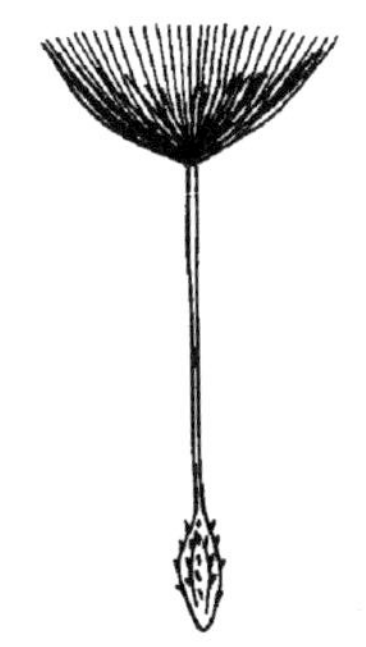

PLUMED ACHENE (INFERIOR)
(E.G., DANDELION)

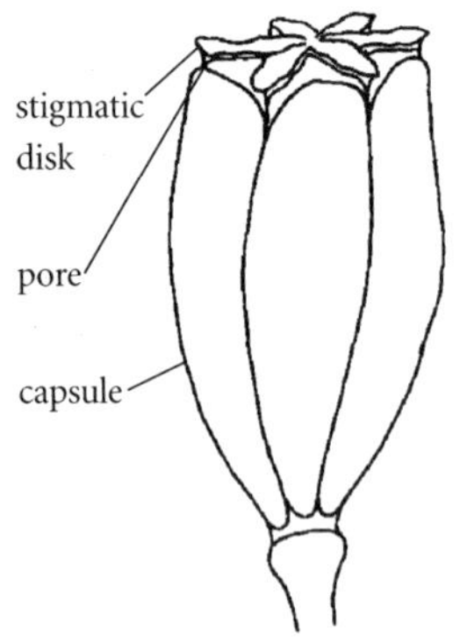

PORICIDAL CAPSULE
(E.G., POPPY)

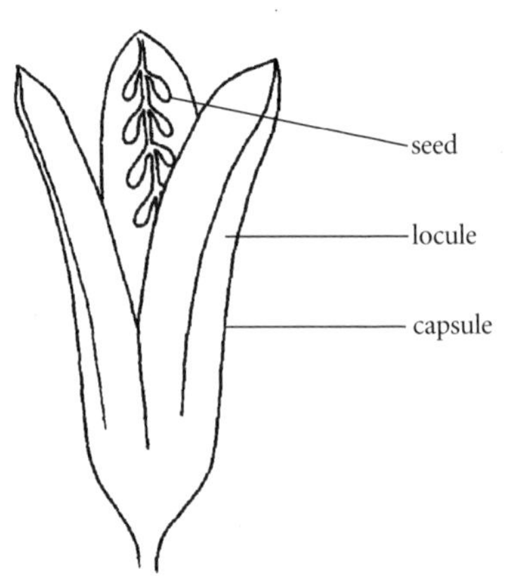

LOCULICIDAL CAPSULE
(E.G., IRIS)

DRY FRUITS

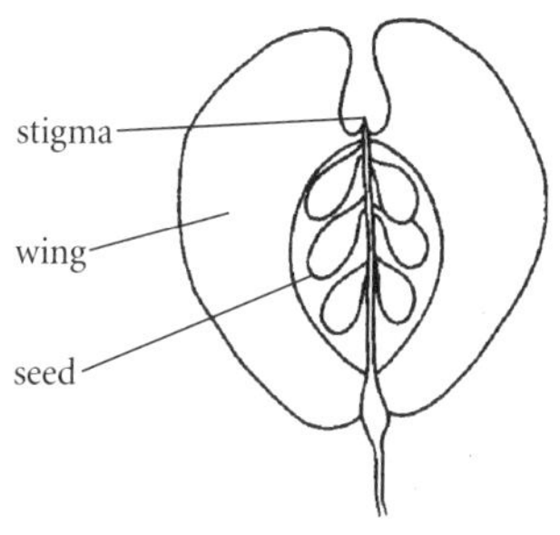

SILICLE
(E.G., STINKWEED)

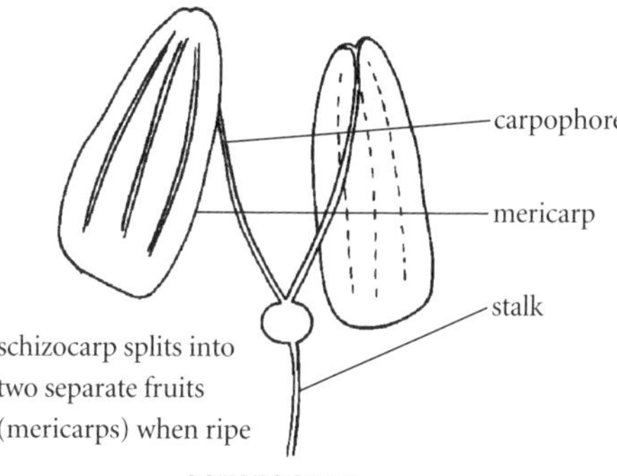

SCHIZOCARP
(E.G., APIACEAE)

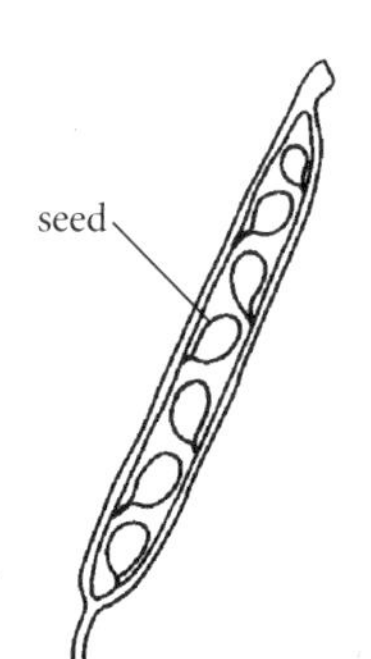

SILIQUE BEFORE DEHISCENCE
(E.G., CRESS)

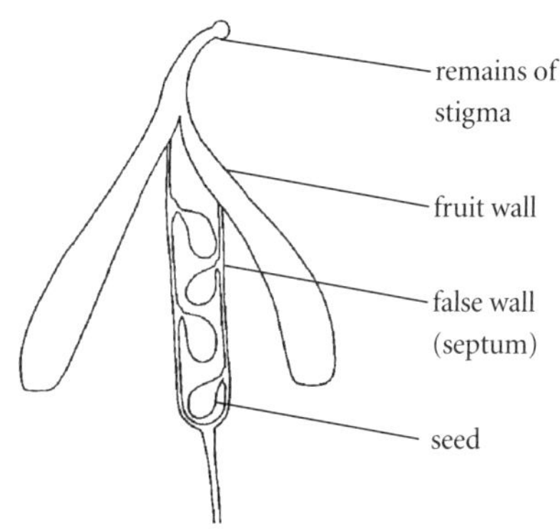

SILIQUE AFTER DEHISCENCE
(E.G., CRESS)

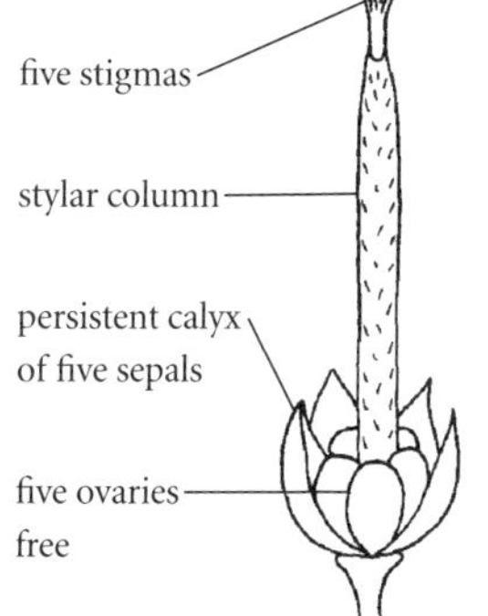

SCHIZOCARP
BEFORE DEHISCENCE
(E.G., GERANIUM)

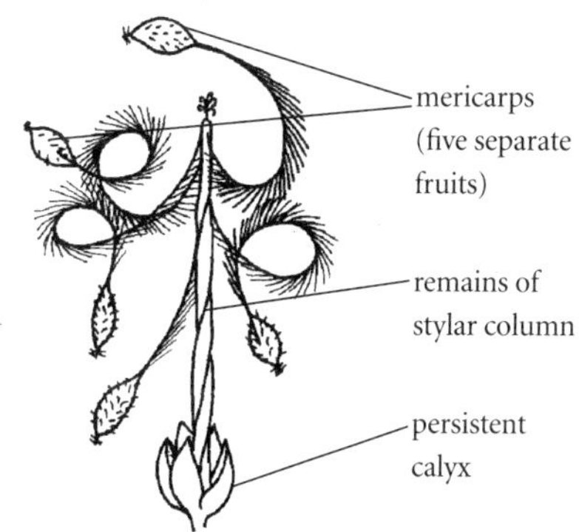

SCHIZOCARP
AFTER DEHISCENCE
(E.G., GERANIUM)

SILICLE
a wide, pod-like fruit of the mustard family, Brassicaceae (Cruciferae)

SILIQUE
a long, narrow, pod-like fruit of the mustard family, Brassicaceae (Cruciferae)

SORUS
cluster of fern sporangia [pl. *sori*]

SPADIX
a spike with a fleshy axis, e.g., arum family, Araceae

SPATHE
a large leaf-like, green or coloured bract enclosing a flower-spike as in arum family, Araceae

SPERM
a male gamete

SPIKE
an elongated flower arrangement where the flowers have no stalks

SPIKELET
a secondary spike: the characteristic unit of the flower-cluster in Poaceae (Gramineae)

SPORANGIUM
spore-sac, in which spores are produced [pl. *sporangia*]

SPORE
a one-celled reproductive structure, not a gamete or a zygote, produced on sporophylls, e.g., a fern spore

SPOROPHYTE
the plant generation which produces spores. In angiosperms, the megaspore develops into the embryo sac and the microspores develop into the pollen grains, *see Life Cycles, pp.* XL–XLIII

SPUR
a hollow tubular projection of a sepal or petal, usually with nectaries, as in the orchid family, Orchidaceae

STALK
any lengthened support of a organ

STAMEN
part of flower which produces pollen; the male reproductive part of the flower. The stamen usually grows in two parts: the anthers, which are sac-like and contain pollen, and the filament (stalk), which is often absent.

STAMINATE FLOWER
male flower with stamens

STAMINAL TUBE
anthers of stamens joined together to form a tube, in Asteraceae (Compositae)

STAMINODE
a sterile stamen

STANDARD PETAL
large upper petal in a flower of Fabaceae (Leguminosae)

STELLATE
star-shaped

STIGMA
upper part of a carpel which receives the pollen; swollen tip of the style

STIGMATIC DISC
several stigmas joined together, as in poppy family, Papaveraceae

STIPE
leaf-stalk of a fern

STIPULE
leafy appendage at the base of a leaf

STOLON
a creeping stem on the surface of the ground; produces new plants at the nodes

STOLONIFEROUS
bearing stolons

STRIGILLOSE
diminutive of strigose

FLESHY FRUITS

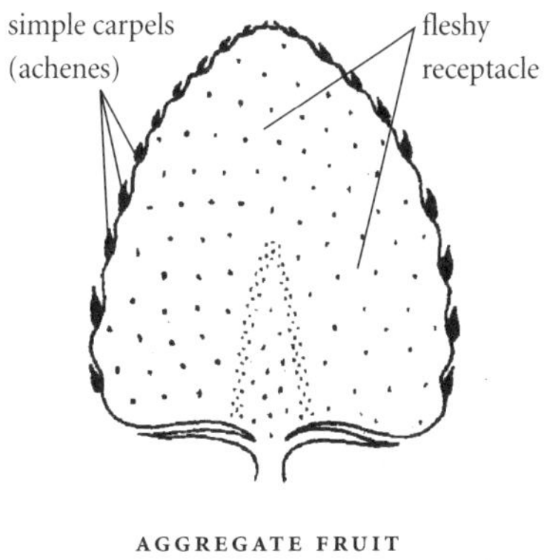

AGGREGATE FRUIT
(E.G., STRAWBERRY)

PSEUDOCARP (A "FALSE FRUIT")
(E.G., ROSE-HIP)

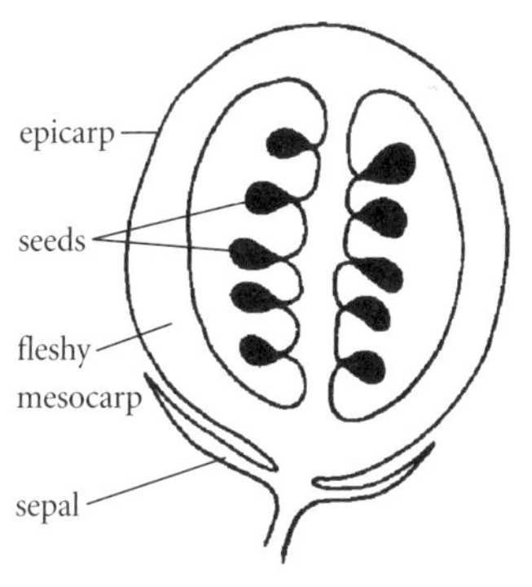

BERRY
(E.G., GOOSEBERRY, TOMATO)

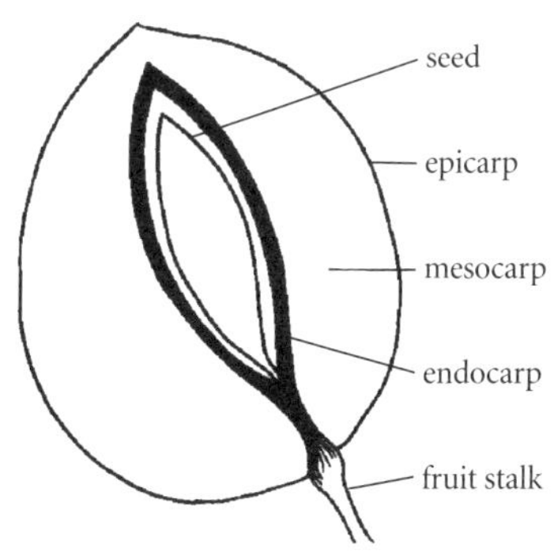

DRUPE (A "STONE FRUIT")
(E.G., PLUM, CHERRY, PEACH)

STRIGOSE
covered with straight appressed hairs all pointing in the same direction

STROBILUS
a teminal cone bearing sporangia, e.g., horsetails (Equisetaceae)

STYLAR COLUMN
several styles (five) joined together, in Geraniaceae

STYLE
part of a carpel between the ovary and stigma through which the pollen-tubes carry pollen to the ovary

SUCKER
a vegetative shoot from a rhizome (underground stem)

SUPERIOR *(ovary)*
ovary on receptacle or hypanthium, above the rest of the flower

SYMPETALOUS
with united petals; gamopetalous

SYNCARPOUS
with united carpels

TAPROOT
the primary root, from the radicle of the embryo in the seed

TEPALS
sepals and petals that are very much alike (e.g., lily family, Liliaceae); see also *perianth leaves*

TERETE
circular in cross-section

TERMINAL
at the tip

TERNATE *(leaf)*
divided into three parts

THALLUS
a plant body not differentiated into root, stem and leaves, e.g., common duckweed (*Lemna minor*)

THRUST SHEETS
slices or blocks of sediments broken off and slid over one another along low angle planes of movement

THYRSE
a tall, mixed panicle, made up of a series of small cymes in a racemose arrangement

TRIFID STIGMA
three-forked

TOMENTOSE
densely hairy, with matted woolly hairs

TUBER
a swollen underground stem bearing buds or "eyes," e.g., potato (*Solanum tuberosum*)

TUBULAR
a tube-shaped flower

TURION
a scaly, often thick and fleshy shoot produced from a bud on an underground rootstock, e.g., *Epilobium* sp.

UMBEL
a flower-cluster with all the flower stalks arising from the same point

UMBELLET
a small umbel

UNISEXUAL
either male or female

UTRICLE
a bladder found on insectivorous plants, e.g., *Utricularia* sp.

VALVE
one of the portions of the wall of a capsule into which it separates when ripe, e.g., the fruit of *Iris* sp.; a valvular capsule

VEIN *(leaf)*
a thread of woody tissue in a leaf or petal

VELU
a membrane that acts as a door to the utricle

VERTICILLATE
arranged in whorls

VILLOSE
covered with long fine hairs, but not matted

VISCID
sticky

VIVIPAROUS
sprouting or germinating on the parent plant, e.g., the bulbils found on alpine bistort (*Polygonum viviparum*)

WEED
any plant growing in a place where it is not wanted

WHORL
a circle of three or more leaves, branches or pedicels (flower-stalks), arising from the same node

WING PETAL
part of the flower of Fabaceae (Leguminosae)

ZYGOMORPHIC *(flower)*
irregular; bi-laterally symmetrical

References

CORE TEXTS

Clark, Lewis J. 1973. *Wild Flowers of British Columbia.* Gray's Publishing, Ltd., Sidney, British Columbia.

Interesting descriptions of many of the flowering plants of British Columbia and Alberta, fully illustrated with 573 excellent colour plates.

Hitchcock, C.C. et al. 1955-1969. *Vascular Plants of the Pacific North West.* 5 volumes. University of Washington Press, Seattle, Washington.

An immensely detailed description of relevant plants, well illustrated with line drawings.

Hultén, Eric. 1968. *Flora of Alaska and Neighbouring Territories.* Stanford University Press, Stanford, California.

Clear descriptions of all the northern species of North America, well illustrated with line drawings.

Kuijt, Job. 1982. *A Flora of Waterton Lakes National Park.* University of Alberta Press, Edmonton, Alberta.

Interesting descriptions of all the plants found in the Park, which lies in the most southerly part of Alberta and features many unusual species. All species are illustrated with clear line-drawings by the author and 8 colour plates.

Moss, E.H. 1959. *Flora of Alberta.* Second edition 1983, revised by J.G. Packer. University of Toronto Press.

The second edition by Dr. Packer, including distribution maps, is an indispensable, authoritative text.

Porsild, A.E. and W.J. Cody. 1980. *Vascular Plants of the Continental Northwest Territories, Canada.* Canadian Museum of Nature, Ottawa.

A detailed account of the northern plants of Canada, well-illustrated by line-drawings.

Porter, C.L. 1959. *Taxonomy of Flowering Plants.* W.H. Freeman and Co., San Francisco and London, England.

A well-illustrated text, with details of flower-structure, including floral diagrams.

Radford, Albert E. 1986. *Fundamentals of Plant Systematics.* Harper & Row, New York.

Inflorescences, flowers and fruits described in detail with clear illustrations.

BEDROCK GEOLOGY

Gordy, P.L., F.R. Frey, & N.C. Ollerenshaw. 1975. Road log - Calgary - Turner Valley - Jumping Pound - Seebe. *Structural Geology of the Foothills between Savanna Creek and Panther River, S.W. Alberta, Canada.* H.J. Evers & J.E. Thorpe, eds. Canadian Society of Petroleum Geologists Guidebook, Exploration Update, 1975.

Harris, S.A. 1985. Evidence for the nature of early Holocene climate and palaeogeography, High Plains, Alberta, Canada. *Arctic & Alpine Research* 17: 49-67.

Harris, S.A. 1990. The geology and climate of Plateau Mountain. *Pica* 10(4): 3-12. Calgary Field Naturalists' Society, Calgary, Alberta.

Harrison, J.E. 1976. *Dated organic material below Mazama tephra, Elk Valley, British Columbia.* Geological Survey of Canada Paper 76-1C: 169-170.

Luckman, B.H. & G.D. Osborn. 1979. Holocene glacial fluctuations in the middle Canadian Rocky Mountains. *Quaternary Research* 11: 52-77.

MacDonald, G.M. et al. 1987. Comparative radiocarbon dating of terrestrial plant macrofossils and aquatic moss from the "ice-free corridor" of Western Canada. *Geology* 15: 837-840.

McCrossan, R.G. et al. 1964. *Geological History of Western Canada.* Alberta Society of Petroleum Geologists, Calgary.

Miller, L.A. 1991. *A paleogeographic reconstruction of glacial lakes Elk and Wigwam, Southeastern British Columbia.* M.Sc. thesis, Department of Geography, University of Calgary.

Storer, J.E. 1978. Tertiary sands and gravels in Saskatchewan and Alberta: Correlation of mammalian faunas. *Western and Arctic Canadian Biostratigraphy.* C.R. Stelck & B.D.E. Chatterton, eds. Geological Association of Canada, Special Paper #18.

CLIMATE OF THE KANANASKIS VALLEY

Hare, F.K. & M.K. Thomas. 1974. *Climate of Canada.* Wiley, Toronto, Canada.

MacHattie, L.B. 1968. Kananaskis Valley winds in summer. *Journal of Applied Meteorology* 7: 348-352.

MacHattie, L.B. 1970. Kananaskis Valley temperatures in summer. *Journal of Applied Meteorology* 9: 574-582.

VEGETATION ZONES

Fryer, G.I. & E.A. Johnson. 1988. Reconstructing the behaviour of a fire in a mountain valley. *Journal of Applied Ecology* 25: 1063-1072.

Glover, R. (Ed.) 1962. *David Thompson's Narrative.* Publications of the Champlain Society, Toronto.

Johnson, E.A. & G.I. Fryer. 1987. Historical vegetation change in the Kananaskis Valley, Canadian Rockies. *Canadian Journal of Botany* 65: 853-858.

Johnson, E.A. & G.I. Fryer. 1989. Population dynamics and age class distributions in lodgepole pine-Engelmann spruce forests. *Ecology* 70: 1335-1345.

Johnson, E.A. & C. Larsen. 1991. Climatically-induced change in fire frequency from 1586 to 1984 in the southern Canadian Rockies. *Ecology* 72: 194-201.

Johnson, E.A. & K. Miyanishi. 1991. Fire and population dynamics of lodgepole pine and Engelmann spruce forests in the southern Canadian Rockies. *The Ecology of Pine and Spruce Forests.* N. Nakagoshi & F.B. Golley, eds. SPB Academic Publ. bv., The Hague.

McDonald, G.M. 1982. Late quaternary paleoenvironments of the Morley Flats and Kananaskis Valley of southwestern Alberta. *Canadian Journal of Earth Sciences* 19: 23-35.

PLANTS OF SPECIAL INTEREST

Darwin, Charles. 1875. *Insectivorous Plants.* D. Appleton & Co., New York.

Lloyd, F.E. 1942. *The Carnivorous Plants.* Chronica Botanica Publ. Waltham, Mass.

Schnell, Donald E. 1976. *Carnivorous Plants of the United States and Canada.* John F. Blair, Publisher, Winston Salem, North Carolina.

Schwartz, Randall. 1974. *Carnivorous Plants.* Avon Books, New York.

OTHER BOTANICAL INVESTIGATIONS

Beke, Gerald. 1969. *Soils of three experimental watersheds.* Ph.D. thesis, Dept. of Soil Science, University of Alberta, Edmonton.

Bird, C.D. and A.H. Marsh. 1973. Phytogeography and Ecology of the Lichen family Parmeliaceae in S.W. Alberta. *Canadian Journal of Botany* 51(1): 261-288.

Bird, C.D. 1990. Alpine plants on Plateau Mountain. *Pica* 10 (3). Calgary Field Naturalists' Society, Calgary, Alberta.

Bryant, J.P. 1968. *Alpine vegetation and frost activity in an alpine fellfield on the summit of Plateau Mountain, Alberta.* M. Sc. thesis, Dept. of Biology, University of Calgary.

Bryant, J.P. and E. Scheinberg. 1968. Vegetation and frost activity in an alpine fellfield on the summit of Plateau Mountain, Alberta. *Canadian Journal of Botany.*

Corbin, John. 1991. Check-list of plants of the Kananaskis Valley. *Pica* 11(2). Calgary Field Naturalists' Society, Calgary, Alberta.

Crossley, D.I. 1951. *The Soils on the Karos Forest Experiment Station in the Subalpine Forest Region in Alberta.* Silvicultural Research Note 100. Canada Dept. Resources and Development, Forestry Branch.

Duffy, P.J.B. and R.E. England. 1967. *A forest land classification for the Kananaskis Research Forest, Alberta, Canada.* Dept. of Forest and Rural Development, Forestry Branch.

Emery, Neil. 1994. *Morphological and physiological phenotype plasticity in Alpine and Prairie ecotypes of* Stellaria longipes. Ph.D. thesis, Dept. of Biological Sciences, University of Calgary.

Harris, S. 1991. Vascular Plant-list of the Plateau Mountain area. *Pica* 11(3). Calgary Field Naturalists' Society, Calgary, Alberta.

Kirby, C.L. and R.T. Ogilvie. 1969. *The Forests of Marmot Creek Watershed Research Basin (in Kananaskis Country).* Ministry of Fisheries and Forestry, Ottawa, Publication No. 1259.

Kondla, N. 1978. *An overview vegetation survey of Kananaskis Provincial Park.* Alberta Parks, Edmonton.

MacGregor, C. 1984. *Kananaskis Country ecological land classification and evaluation.* Alberta Forestry, Lands and Wildlife, Edmonton.

Ogilvie, R.T. 1963. *Ecology of Forests in the Rocky Mountains of Alberta.* Canadian Department of Forestry, Research Branch Report No. 63-A-12.

Ogilvie, R.T. 1971. *Ecology and Taxonomy of Alpine, Subalpine and Foothills Vegetation of Alberta.*

Ogilvie, R.T. 1978. *The alpine and subalpine environment in the Rocky Mountains of Alberta.* Proceedings of the Workshop on alpine and subalpine environments.

H.A. Luttmending and J.A. Shields, eds. British Columbia Ministry of the Environment.

Trottier, Garry C. 1972. *Ecology of the alpine vegetation of Highwood Pass (Kananaskis), Alberta.* M.Sc. thesis, Dept. of Biology, University of Calgary, Alberta.

Wallis, C. and C. Wershler. 1972. *Ecological Survey of Bow Valley Provincial Park.* Alberta Parks Division, Edmonton.

Wallis, C. and C. Wershler. 1972. *Ecological Survey of Bragg Creek Provincial Park.* Alberta Parks Division, Edmonton.

Wallis, C. and C. Wershler. 1981. *Natural History Inventory and Assessment in the Many Springs Area, Bow Valley Provincial Park.* Prepared by Cottonwoods Consultants, Ltd., Calgary, for Alberta Energy and Natural Resources, Fish and Wildlife Division, Calgary, Alberta.

Williams, J.A. 1988. *Vegetation Inventory, Bow Valley Provincial Park, Alberta.* Recreation and Parks, Kananaskis Country, Operations Branch, Canmore, Alberta.

Williams, J.A. 1989. *Many Springs monitoring program 1988.* Alberta Recreation and Parks, Kananaskis Country, Operations Branch, Canmore, Alberta.

Williams, J.A. 1990. *Vegetation Inventory of the Opal, Kananaskis and Misty Ranges in Peter Lougheed Provincial Park.* Recreation and Parks, Operations Branch, Kananaskis Country, Canmore, Alberta.

Williams, J.A. 1990. *Annotated Vegetation Species Check-list.*

Williams, J.A. 1992. *Wetlands classification system, habitat inventory and description of wetlands in Peter Lougheed Provincial Park.* Operations Branch, Alberta Recreation and Parks, Kananaskis Country, Canmore, Alberta.

Williams, J.A. 1992. *Vegetation Classification using Landsat T.M. and SPOT HRV Imagery in Mountainous Terrain, Kananaskis Country, Southwestern Alberta.* M.Sc. thesis, Committee on Resources and the Environment, University of Calgary, Alberta.

Williams, J.A. 1993. *Phytoplankton/Algal sampling of selected wetlands in Peter Lougheed Provincial Park, Kananaskis West District.* Department of Environmental Protection, Government of Alberta.

Williams, J.A. 1994. *An assessment of recreational impact on vegetation of the Visitor Services Meadow, Peter Lougheed Provincial Park.* Prepared for Environmental Programs, P.L.P.P. Ranger Service, Kananaskis West District, Kananaskis, Alberta.

SUGGESTIONS FOR FURTHER READING

Anderson, A.W. 1966. *How We Got Our Flowers.* Dover Publications Inc., New York.

Argus, G.W. and D.J. White. 1978. *The Rare Vascular Plants of Alberta.* Syllogeus No. 17. Canadian Museum of Nature, Ottawa.

Bailey, L.H. 1933, 1963. *How Plants Get Their Names.* Dover Publications Inc., New York.

Bird, C.D. 1990. Alpine Plants on Plateau Mountain. *Pica* 10(3). Calgary Field Naturalists' Society, Calgary, Alberta.

Brown, Annora. 1954. *Old Man's Garden.* Second edition, 1970. Gray's Publishing Ltd., Sidney, British Columbia.

Bush, Dana. 1990. *Compact Guide to the Wildflowers of the Rockies.* Lone Pine Publishing, Edmonton, Alberta.

Cai, Q. and C.C. Chinnappa. 1987. Giemsa C-banded Karyotypes of Seven North American Species of *Allium. American Journal of Botany* 74: 1087-1092.

Corbin, J. 1991. Vascular Plants of the Kananaskis Valley. *Pica* 11(2): 16-25. Calgary Field Naturalists' Society, Calgary, Alberta.

Cormack, R.G.H. 1977. *Wild Flowers of Alberta.* Second edition. Hurtig Publishers, Edmonton, Alberta.

Currah, R. et al. 1983. *Prairie Wildflowers.* University of Alberta Faculty of Extension, Edmonton, Alberta.

Daffern, Gillean. 1985. *Kananaskis Country Trail Guide.* Second edition. Rocky Mountain Books, Calgary, Alberta.

Davies, P.H. and J. Cullen. 1979. *The Identification of Plant Families.* Second edition. Cambridge University Press, Cambridge, England.

Droppo, Olga. 1988. *Field Guide to Alberta Berries.* Calgary Field Naturalists' Society, Calgary, Alberta.

Droppo, O. 1990. Additions to the alpine plants of Plateau Mountain. *Pica* 10(3): 12-13. Calgary Field Naturalists' Society, Calgary, Alberta.

Ealey, David, ed. 1990. *Alberta Vegetation Species List and Species Group Checklists.* Alberta Energy/Forestry, Lands and Wildlife, Edmonton, Alberta.

Emery, R.J.N. and C.C. Chinnappa. 1992. Natural hybridization between *Stellaria longipes* and *Stellaria borealis* (Caryophyllaceae). *Canadian Journal of Botany* 70: 1717-1723.

Ferguson, Mary, and Richard Saunders. 1976. *Canadian Wildflowers.* Van Nostrand Reinhold, Ltd., Toronto.

Flood, W.E. 1961. *Scientific Words, Their Structure and Meaning.* The Scientific Book Guild, London, England.

Ford, Gillian. 1984. *Plant Names Explained.* Publication No. 16. Friends of the Devonian Botanic Garden, Devon, Alberta.

Gadd, Ben. 1986. *Handbook of the Canadian Rockies.* Corax Press, Jasper, Alberta.

Gillett, John M. 1963. *The Gentians of Canada, Alaska, and Greenland.* Research Branch, Canada Department of Agriculture, Ottawa.

Glover, R. (Ed.) 1962. *David Thompson's Narrative.* Publications of the Champlain Society, Toronto.

Hallworth, Beryl. 1978-1982. *Annotated Checklist of the Plants in Norman Sanson's Collection from the Herbarium of the Banff Park Museum.* Volumes 1-9. Biological Sciences Department, University of Calgary, Calgary, Alberta.

Hallworth, Beryl. 1988. *New Canadian Encyclopedia.* Second edition. (Bearberry, p.189, Volume 1. Geranium, p.894, Volume 2. Yarrow, p.2350, Volume 4.) Hurtig Publishers, Edmonton, Alberta.

Harrington, H.P. 1977. *How to Identify Grasses and Grass-like Plants.* Swallow Press, Chicago, U.S.A.

Harris, S.A. 1990. The Geology and Climate of Plateau Mountain. *Pica* 10(4): 3-12. Calgary Field Naturalists' Society, Calgary, Alberta.

Harris, S.A. 1991. Vascular Plant Checklist—the Plateau Mountain Area. *Pica* 11(3). Calgary Field Naturalists' Society, Calgary, Alberta.

Hellson, J.C. and M. Gadd. 1974. *Ethnobotany of the Blackfoot Indians.* Canadian Ethnology Service, Paper No. 19. National Museum of Man, Ottawa.

Hosie, R.C. 1979. *Native Trees of Canada.* Eighth edition. Fitzhenry Whiteside, Ltd., Don Mills, Ontario.

Kananaskis Country Environmental Education Library. 1986. *Vegetation. Climate.* Government of Alberta Publication, Canmore.

Kuijt, Job. 1972. *Common Coulee Plants of Southern Alberta.* University of Lethbridge Production Services, Lethbridge, Alberta.

Lauriault, Jean. 1989, 1992. *Identification Guide to the Trees of Canada.* Canadian Museum of Nature, Ottawa.

Looman, J. 1982. *Prairie Grasses.* Agriculture Canada, Ottawa.

Luer, C.A. 1975. *The Native Orchids of the United States and Canada.* New York Botanical Garden, New York.

Macdonald, Ian D. 1994. *Camp Adventure: Vegetation, Plants and Animals.* The Boys and Girls Clubs of Calgary, Alberta.

Macdonald, S.E. and C.C. Chinnappa. 1989. Population Differentiation for Phenotypic Plasticity in the *Stellaria longipes* Complex. *American Journal of Botany* 76: 1627-1637.

MacKinnon, A., J. Pojar and R. Coupé. 1992. *Plants of Northern British Columbia.* Lone Pine Publishing, Edmonton, Alberta.

Oltmann, C. Ruth. 1976. *The Valley of Rumours—the Kananaskis.* Ribbon Creek Publishing Company, Seebe, Alberta.

Packer, J.G. and C.E. Bradley. 1984. *A Check-list of the Rare Vascular Plants in Alberta.* Natural History Occasional Paper No. 5. Provincial Museum of Alberta, Edmonton, Alberta.

Pinel, H. 1985. *Plant species list for Bow Valley Provincial Park, Exshaw, Alberta.*

Pinel, H. and O. Droppo. 1976. *Plant species list for Bow Valley Provincial Park, Exshaw, Alberta.*

Porsild, C.L. 1974. *Rocky Mountain Wild Flowers.* Natural History Series, No. 2. Canadian Museum of Nature, Ottawa.

Rowe, J.S. 1972. *Forest Regions of Canada.* Publication No. 1300. Canadian Forestry Service, Ottawa.

Scotter, George W. and Hälle Flygare. *Wildflowers of the Canadian Rockies.* Hurtig Publishers, Edmonton, Alberta.

Seaborn, J.R. 1966. *The Bryophyte Ecology of Marl Bogs in Alberta.* M.Sc. Thesis, Department of Biology, University of Calgary, Alberta.

Smith, Bonnie. 1992. *Status Report on Kananaskis Whitlow Grass,* Draba kananaskis *Mulligan.* Committee on the Status of Endangered Wildlife in Canada (COSEWIC), Ottawa.

Stearman, W.A. and Gerry Wheeler. 1983. *Weeds.* Alberta Environmental Centre and Alberta Agriculture, Edmonton, Alberta.

Stearn, W.T. 1966. *Botanical Latin.* Hafner Publishing Company, New York.

Trelawny, John G. 1983. *Wildflowers of the Yukon and Northwestern Canada including adjacent Alaska.* Sono Nis Press, Victoria, British Columbia.

Turner, Nancy and Adam F. Szczawinski. 1978. *Wild Coffee and Tea Substitutes of Canada.* Canadian Museum of Nature, Ottawa.

Turner, Nancy and Adam F. Szczawinski. 1980. *Wild Green Vegetables of Canada.* Canadian Museum of Nature, Ottawa.

Wallis, C.A. and Cleve Wershler. 1972. *An Ecological Survey of Bow Valley Provincial Park.* Volumes 1 and 2. Provincial Parks Planning Department of Lands & Forests, Government of Alberta, Edmonton.

Wallis, C.A. and Cleve Wershler. 1981. *Natural History Inventory and Assessment in the Many Springs area—Bow Valley Provincial Park: Final Report.* Prepared by Cottonwood Consultants, Ltd. for Alberta Fish and Wildlife Division, Alberta Energy & Natural Resources, Edmonton.

White, Helen A., ed. 1974. *The Alaska-Yukon Wild Flowers Guide.* Alaska Northwest Publishing Company, Anchorage, Alaska.

Wilkinson, Kathleen. 1990. *Trees and Shrubs of Alberta.* Lone Pine Publishing, Edmonton, Alberta.

Williams, Joan. 1988. *Vegetation Inventory of Bow Valley Provincial Park.* Alberta Recreation and Parks, Edmonton.

Williams, Joan. 1990. New record for *Orobanche uniflora* L., Small Broomrape (One-flowered Broomrape). *Alberta Naturalist* 20(3). Federation of Alberta Naturalists, Edmonton.

Williams, Joan. 1990. *Vegetation Inventory of the East Ranges in Peter Lougheed Provincial Park, including Lichens, Mosses and Liverworts.* Alberta Recreation and Parks, Edmonton.

Zhang, Xing-hai and C.C. Chinnappa. 1994. Molecular Cloning of a cDNA Encoding Cytochrome c of *Stellaria longipes* (Caryophyllaceae) and the Evolutionary Implications. *Molecular Biology Evolution* 11: 365-375.

Index

** Denotes genera or species not found in Kananaskis Country, but which are mentioned in the text.*

A

Abies lasiocarpa, 14
Acer glabrum, 169
Aceraceae, 169
Achillea millefolium, 248
Actaea rubra, 108, 109
adder's-tongue family, 7
Agoseris
  aurantiaca, 249
  glauca, 249
Agropyron
  dasystachyum, 34
  repens, 34
  spicatum, 34
Agrostis scabra, 35
alder
  green, 86
  river, 86
alexanders, heart-leaved, 184
alfalfa, 159
Alismataceae, 30–31
Allium
  cernuum, 56
  schoenoprasum, 56
Alnus
  crispa, 86
  tenuifolia, 86
Alopecurus aequalis, 35
alumroot, sticky, 133
amaranth family, 99–100
amaranth, prostrate, 99
Amaranthaceae, 99–100
Amaranthus
  graecizans, 99
  retroflexus, 99–100
Amelanchier alnifolia, 142
Anaphalis margaritacea, 250
androsace, sweet-flowered, 198
Androsace
  chamaejasme, 198
  septentrionalis, 198
anemone
  cut-leaved, 108, 109–110
  small wood, 110
Anemone
  multifida, 108, 109–110
  **nemorosa*, 110
  parviflora, 110
  patens, 110–111
angiosperms, 23–276
Antennaria
  lanata, 251
  pulcherrima, 251
  rosea, 251
Apiaceae (Umbelliferae), 179–184
Apocynaceae, 203–204
Apocynum androsaemifolium, 204
Aquilegia
  brevistyla, 111
  flavescens, 112
Arabis
  divaricarpa, 122
  glabra, 123
  lyallii, 123
Aralia nudicaulis, 178–179
Araliaceae, 178–179
*arbutus, 191
Arceuthobium americanum, 90
Arctostaphylos
  rubra, 190
  uva-ursi, 191
Arenaria
  capillaris, 102
  laterifolia, 104
  obtusiloba, 104
  rossii, 104
arnica
  alpine, 252
  heart-leaved, 253
  spear-leaved, 253
Arnica
  angustifolia, 252
  cordifolia, 253
  lonchophylla, 253

arrow-grass family, 29–30
arrow-grass
 seaside, 30
 slender, 30
arrowhead, 31
Artemisia
 biennis, 254
 campestris, 254
 frigida, 254, 255
 ludoviciana, 255
Asclepiadaceae, 204–205
Asclepias
 speciosa, 205
 tuberosa, 204
ash, 152
 sitka mountain, 152
 western mountain, 152
aspen, trembling, 82
asphodel
 dwarf false, 61
 sticky false, 61
Asplenium viride, 8
aster
 golden, 264
 hairy golden, 264
 showy, 256
 siberian, 257
 smooth, 256
Aster
 conspicuus, 256
 laevis, 256
 sibiricus, 257
Asteraceae (*Compositae*), 245–276
Astragalus
 alpinus, 155
 americanus, 155
 frigidus, 155
 striatus, 156
Athyrium filix-femina, 8
avens
 large-leaved yellow (*Geum*), 145
 northern white mountain (*Dryas*), 143
 three-flowered (*Geum*), 146
 white mountain (*Dryas*), 144
 yellow (*Geum*), 145
 yellow mountain (*Dryas*), 143
azalea, false, 193

B

balsam poplar, 81, 82
baneberry, red and white, 108, 109
barley, foxtail, 41
bastard toadflax, 89
 northern, 89–90
bean, golden, 161
bearberry
 alpine, 190
 common, 191
beardtongue
 crested, 227
 shrubby, 227
 slender blue, 228
 smooth blue, 227
 yellow, 226
Beckmannia syzigachne, 36
bedstraw
 northern, 235
 small, 236
 sweet-scented, 235, 236
bergamot, wild, 217
Besseya
 cinerea, 220
 wyomingensis, 220
Betula
 glandulosa, 87
 occidentalis, 87
Betulaceae, 85–87
bilberry
 dwarf, 196
 low, 196
bindweed, black, 94
birch family, 85–87
birch
 black, 87
 bog, 87
 dwarf, 87
 water, 87
bishop's-cap, 135
bistort, alpine, 95
black medick, 158
bittercress, 124
bladder fern, fragile, 9
bladderpod
 double, 129
 northern, 128
bladderwort family, 232–233
bladderwort, common, 233

blue buttons, 242–243
blue-eyed mary, 222
blue-eyed grass
 common, 63
 pale, 64
bluebell, 243
bluebur, 213
bluegrass
 alpine, 45
 Kentucky, 45
bluejoint, 37
bluets, 234
bog cranberry, small, 194
bog orchid
 bracted, 70
 northern green, 70
 tall white, 69
borage family, 211–215
Boraginaceae, 211–215
Botrychium
 lunaria, 7
 virginianum, 7
Brassicaceae (Cruciferae), 121–130
braya, leafy, 123
Braya
 humilis, 123
 purpurascens, 123
brome, awnless, 36
Bromus
 inermis, 36
 tectorum, 36
bronze-bells, 60
brooklime, American, 229
broomrape family, 230–231
broomrape
 clustered, 231
 one-flowered, 231
brown-eyed susan, 262
buckbean, 203
buckbean family, 202–203
buckbrush, 239, 240
buckwheat family, 91–96
buckwheat, wild, 94
buffaloberry, Canadian, 173
bulrush
 great, 51
 tufted, 52
bunchberry, 184
bur-reed family, 27
bur-reed
 narrow-leaved, 27
 slender, 27
burning bush, 98
butter-and-eggs, 223
buttercup family (crowfoot), 108–117
buttercup
 heart-leaved, 115
 mountain, 115
 northern, 116
 tall, 114
butterwort, common, 232

C

Caesalpinioideae, 154
Calamagrostis canadensis, 37
Callitrichaceae, 167–168
Callitriche verna, 168
**Calluna*, 189
Calypso bulbosa, 65, 71
camas
 death, 62
 white, 62
Camassia, 62
Campanula
 rotundifolia, 243
 uniflora, 244
Campanulaceae, 243–244
campion
 Menzies', 106
 moss, 105
 Parry's, 106
canary grass, reed, 43
cancer-root, 231
 one-flowered, 231
Caprifoliaceae, 236–240
Capsella bursa-pastoris, 124
Cardamine umbellata, 124
Carex
 aquatilis, 48
 atherodes, 48
 aurea, 48
 filifolia, 49
 rostrata, 49
 scirpoidea, 49
carrot family, 179–184
Caryophyllaceae, 101–108
Cassiope tetragona, 192

Castilleja
- *lutesens*, 221
- *miniata*, 221
- *occidentalis*, 222

catchfly, 106
cattail family, 26
cattail, common, 26
Cerastium
- *arvense*, 103
- *beeringianum*, 103

Chenopodiaceae, 97–98
Chenopodium
- *album*, 97
- *capitatum*, 97

cherry
- choke, 148–149
- pin, 148

chess, downy, 36
chickweed
- alpine mouse-eared, 103
- field mouse-eared, 103
- long-leaved, 107
- long-stalked, 107–108

Chimaphila umbellata, 186
chives, wild, 56
Chrysosplenium tetrandrum, 133
Chrysopsis villosa, 264
cinquefoil
- graceful, 147
- shrubby, 147
- slender, 147
- snow, 147

Cirsium
- *arvense*, 257
- *flodmanii*, 258
- *hookerianum*, 258
- *undulatum*, 258

Claytonia
- *lanceolata*, 101
- *megarhiza*, 101

Clematis
- *occidentalis*, 112
- *verticellaris*, 112

clematis, purple, 112
cliff-brake
- purple, 9
- smooth, 9

clover
- alsike (*Trifolium*), 162
- Dutch (*Trifolium*), 163
- owl (*Orthocarpus*), 223
- purple prairie (*Petalostemon*), 154
- red (*Trifolium*), 162
- white (*Trifolium*), 163
- white sweet (*Melilotus*), 159
- yellow sweet (*Melilotus*), 158

club-moss family, 3–4
club-moss
- alpine, 4
- little, 5
- stiff, 4

Coeloglossum viride, 69, 70
Collinsia parviflora, 222
Collomia linearis, 206
collomia, narrow-leaved, 206
coltsfoot
- arrow-leaved, 267
- palmate-leaved, 266
- vine-leaved, 267

columbine
- blue, 111
- yellow, 112

comandra, pale, 89
Comandra
- *pallida*, 89
- *umbellata*, 89

Compositae (*Asteraceae*), 245–276
conifers, 12–21
coral-root
- pale, 67
- spotted, 66
- striped, 66

Corallorhiza
- *maculata*, 66
- *striata*, 66
- *trifida*, 67

Cornaceae, 184–185
Cornus
- *canadensis*, 184
- *stolonifera*, 185

corydalis
- golden, 120
- pink, 120

Corydalis
- *aurea*, 120
- *sempervirens*, 120

cotton grass
- green-keeled, 50

thin-leaved, 50
couch grass, 34
cranberry
low-bush (*Viburnum*), 240
small bog (*Oxycoccus*), 194
Crassulaceae, 131–132
Crepis
elegans, 259
nana, 259
cress
Lyall's rock (*Arabis*), 123
marsh yellow (*Rorippa*), 129
mountain (*Cardamine*), 124
purple rock (*Arabis*), 122
silver rock (*Smelowskia*), 129
crocus, prairie, 110–111
crowberry, 168, 194
crowberry family, 168
crowfoot family (buttercup), 108–117
Cruciferae (*Brassicaceae*), 121–130
Cupressaceae, 12–13
currant family (gooseberry), 139–140
currant
bristly black, 140
golden, 140
Cyperaceae, 47–52
cypress family, 12–13
cypress, summer, 98
Cypripedium
calceolus, 67
passerinum, 68
Cystopteris fragilis, 9
Cytisus scoparius, 230

D

daisy, tufted, 261
dandelion family, 245–276
dandelion
common, 274
mountain, 273
northern, 273
orange false (*Agoseris*), 249
red-seeded, 274
yellow false (*Agoseris*), 249
Danthonia californica, 37
Delphinium
bicolor, 113
glaucum, 113
nuttallianum, 113
Deschampsia caespitosa, 37
Descurainia
pinata, 125
richardsonii, 125
dewberry, 151
dicots, 73–276
Dipsacaceae, 242–243
Disporum trachycarpum, 56
dock
golden, 96
western, 96
Dodecatheon
conjugens, 199
pulchellum, 199
dogbane family, 203–204
dogbane, spreading, 204
dogwood family, 184–185
dogwood
dwarf, 184
red osier, 185
draba, golden, 126
Draba
aurea, 126
incerta, 126
kananaskis, 126
oligosperma, 127
Dracocephalum parviflorum, 216–217
dragonhead, American, 216–217
Drosera
anglica, 130–131
linearis, 131
rotundifolia, 131
Droseraceae, 130–131
dryad (mountain avens)
white, 143, 144
yellow, 143
Dryas
drummondii, 143
integrifolia, 143
octopetala, 144
duckweed family, 52–53
duckweed
common, 53
ivy-leaved, 53

E

Elaeagnaceae, 172–173

Elaeagnus commutata, 172–173
elder, red-fruited, 239
elderberry, 239
Eleocharis palustris, 50
elephant's head, 225
Elymus innovatus, 38
Empetraceae, 168
Empetrum nigrum, 168, 194
Epilobium
    alpinium, 175
    angustifolium, 174
    clavatum, 175
    latifolium, 175
    palustre, 175, 176
Equisetaceae, 5–6
Equisetum
    arvense, 6
    hyemale, 6
    scirpoides, 6
    sylvaticum, 6
**Erica*, 189
Ericaceae, 189–197
Erigeron
    aureus, 260, 263
    caespitosus, 261
    peregrinus, 261
Eriogonum
    flavum, 92
    ovalifolium, 92
    umbellatum, 93
Eriophorum viridi-carinatum, 50
Erysimum
    asperum, 127
    cheiranthoides, 127
    inconspicuum, 127
Erythronium grandiflorum, 57
Euphorbia
    esula, 167
    serpyllifolia, 167
Euphorbiaceae, 166–167
evening primrose family, 173–176
evening primrose, yellow, 176
everlasting
    pearly (*Anaphalis*), 250
    showy, 251
    woolly, 251

F

Fabaceae (*Leguminosae*), 154–163
fairy candelabra, northern, 198
fairy-bells, 56
false asphodel
    dwarf, 61
    sticky, 61
false dandelion
    orange, 249
    yellow, 249
false melic, 45
false mitrewort, 137
felwort, 201
fern family, 7–11
fern
    fragile bladder, 9
    lady, 8
    northern holly, 10
ferns and fern allies, 3–11
fescue
    red, 39
    rough, 39
    sheep, 38
Festuca
    ovina, 38
    rubra, 39
    scabrella, 39
figwort family, 219–230
fir
    Douglas, 21
    subalpine, 14
fireweed, common, 174, 176
flax family, 165
flax, wild blue, 165
fleabane
    aster, 261
    golden, 260, 263
    tall purple, 261
    tufted, 261
forget-me-not, alpine, 215
foxtail, water, 35
Fragaria virginiana, 144
Fraxinus, 152
Fumariaceae, 119–120
fumitory family, 119–120

G

gaillardia, 262
Gaillardia aristata, 262
Galium
  boreale, 235
  trifidum, 236
  triflorum, 236
gentian family, 200–202
gentian
  four-parted, 202
  moss, 201
  spurred, 202
  *yellow, 200
Gentiana prostrata, 201
Gentianaceae, 200–202
Gentianella
  amarella, 201
  propinqua, 202
Geocaulon lividum, 89–90
Geraniaceae, 164
geranium family, 164
geranium
  sticky purple, 164
  wild white, 164
Geranium
  richardsonii, 164
  viscosissimum, 164
Geum
  allepicum, 145
  macrophyllum, 145
  triflorum, 146
ginseng family, 178–179
glacier lily, 57
globe-flower, 117
Glyceria
  grandis, 40
  striata, 40
goat's-beard, yellow, 276
golden bean, 161
goldenrod
  alpine, 272
  late, 271
  low, 271
  Missouri, 271
  mountain, 272
  tall, 271
Goodyera
  oblongifolia, 68
  repens, 68
gooseberry family (currant), 139–140
gooseberry, northern, 140
goosefoot family, 97–98
goosefoot
  spear-leaved, 98
  white, 97
Gramineae (*Poaceae*), 32–46
grapefern, Virginia, 7
grass family, 32–46
grass
  bluebunch wheat, 34
  California oat, 37
  Columbia needle, 46
  common blue-eyed
    (*Sisyrhinchium*), 63
  common tall manna, 40
  couch, 34
  few-seeded whitlow, 127
  fowl manna, 40
  golden whitlow, 126
  green-keeled cotton
    (*Eriophorum*), 50
  Hooker's oat, 40
  June, 42
  Kananaskis whitlow, 126
  Kentucky blue, 45
  marsh reed, 35
  mountain hair, 46
  mountain timothy, 44
  pale blue-eyed
    (*Sisyrhinchium*), 64
  quack, 34
  reed canary, 43
  rough hair, 35
  slough, 36
  thin-leaved cotton
    (*Eriophorum*), 50
  tickle, 35
  timothy, 35, 44
  tufted hair, 37
  white-grained mountain rice, 42
  whitlow, 126
grass-of-Parnassus family, 138–139
grass-of-Parnassus
  fringed, 138
  marsh, 139
  northern, 139
Grindelia squarrosa, 263
gromwell

western false, 215
woolly, 214
Grossulariaceae, 139–140
groundsel
balsam, 269
black-tipped, 269
prairie, 268
grouseberry, 197
gumweed, 263
gymnosperms, 12–21

H

Habenaria
dilatata, 69
hyperborea, 70
viridis, 69, 70
Hackelia
floribunda, 212
jessicae, 212
Halenia deflexa, 202
Haloragaceae, 177
Haplopappus lyallii, 263
harebell, 243
alpine, 244
harebell family, 243–244
hawks-beard, 259
dwarf, 259
hawkweed, narrow-leaved, 265
heal-all, 218
heath family, 189–197
heather
purple, 194
red, 194
white mountain, 192
yellow, 195
hedge nettle, marsh, 218
hedysarum
alpine, 156
Mackenzie's, 157
northern, 157
yellow, 157
Hedysarum
alpinum, 156
boreale, 157
mackenzii, 157
sulphurescens, 157
Helianthus nuttallii, 264
Helictotrichon hookeri, 40
hemp, Indian, 204
Heracleum lanatum, 180
Heterotheca villosa, 264
Heuchera cylindrica, 133
Hieracium
canadense, 265
umbellatum, 265
Hierochloe odorata, 41
Hippuridaceae, 177–178
Hippuris vulgaris, 178
holly fern, northern, 10
honeysuckle family, 236–240
honeysuckle
bracted, 238
twining, 236, 238
Hordeum jubatum, 41
horsetail family, 5–6
horsetail
common, 6
field, 6
woodland, 6
huckleberry, 196
Hydrophyllaceae, 209–211

I

Indian hemp, 204
Indian pipe, 189
Indian pipe family, 188–189
inflated oxytrope, 160
Iridaceae, 63–64
iris family, 63–64
**Iris missouriensis*, 63
iron plant, Lyall's, 263

J

jack-go-to-bed-at-noon
(*Tragopogon*), 276
Jacob's-ladder, showy, 208
jasmine, rock, 198
Juncaceae, 53–55
Juncaginaceae, 29–30
Juncus
alpinus, 54
alpinoarticulatus, 54
balticus, 54
June grass, 42
June-berry, 142

juniper
- common, 13
- creeping, 13
- ground, 13
- Rocky Mountain, 13

Juniperus
- *communis,* 13
- *horizontalis,* 13
- *scopulorum,* 13

K

Kalmia microphylla, 192
Kalm's lobelia, 244
kinnikinnick
- (*Arctostaphylos*), 191
- (*Cornus*), 185

kittentails, 220
Knautia arvensis, 242–243
knotweed
- common, 94
- yard, 94

kobresia, Bellard's, 51
Kobresia myosuroides, 51
Kochia scoparia, 98
Koeleria macrantha, 42

L

Labiatae (*Lamiaceae*), 216–218
Labrador tea, common, 189, 193
ladies'-tresses, hooded, 72
lady fern, 8
lady's-slipper
- northern, 68
- yellow, 67

lamb's-quarters, 97
Lamiaceae (*Labiatae*), 216–218
Lappula
- *echinata,* 213
- *squarrosa,* 213

larch
- Lyall's, 15
- subalpine, 15
- western, 16

Larix
- *lyallii,* 15
- *occidentalis,* 16

larkspur
- low, 113
- Nuttall's, 113
- tall, 113

Lathyrus
- *ochroleucus,* 157
- *venosus,* 157

laurel
- mountain, 192
- swamp, 192

leafy braya, 123
Ledum groenlandicum, 189, 193
Leguminosae (*Fabaceae*), 154–163
Lemna
- *minor,* 53
- *trisulca,* 53

Lemnaceae, 52-53
Lentibulariaceae, 232–233
Lepidium
- *densiflorum,* 128
- *ramosissimum,* 128

Leptarrhena pyrolifolia, 134
Lesquerella arctica, 128
Liliaceae, 55–62
Lilium philadelphicum, 57
lily family, 55–62
lily
- glacier, 57
- western wood, 57

Linaceae, 165
Linaria vulgaris, 89, 223
Linnaea borealis, 236, 237
Linum lewisii, 165
Listera borealis, 71
Lithophragma
- *bulbifera,* 134
- *glabrum,* 134

Lithospermum
- *incisum,* 213
- *ruderale,* 214

little club-moss family, 4–5
little red elephant, 225
Lobeliaceae, 244
Lobelia kalmii, 244
lobelia family, 244
lobelia
- brook, 244
- Kalm's, 244

locoweed
early yellow, 161
showy, 161
Lomatium
dissectum, 181
triternatum, 181
Lomatogonium, 200
Lonicera
dioica, 236, 238
involucrata, 238
Loranthaceae (Viscaceae), 90
lousewort
bracted, 224
flame-coloured, 225
lucerne, 159
lungwort, tall, 214
lupine, silky perennial, 158
Lupinus sericeus, 158
Luzula
parviflora, 54
spicata, 55
Lycopodiaceae, 3–4
Lycopodium
alpinum, 4
annotinum, 4

M

madder family, 234–236
mallow family, 169–170
mallow, scarlet, 170
Malvaceae, 169-170
manna grass
common tall, 40
fowl, 40
maple family, 169
maple, mountain, 169
mare's-tail family, 177–178
mare's-tail, common, 178
Matricaria matricarioides, 265
meadow rue
veiny, 117
western, 108, 117
meadowsweet
pink, 153
white, 153
Medicago
lupulina, 158
sativa, 159
medick, black, 158
melic, false, 45
Melilotus
alba, 159
officinalis, 160
Mentha arvensis, 217
Menyanthaceae, 202–203
Menyanthes trifoliata, 203
Menziesia ferruginea, 193
mertensia, tall, 214
Mertensia paniculata, 214
milfoil, 248
milk vetch
alpine, 155
American, 155
ascending purple, 156
milkweed family, 204–205
milkweed, showy, 205
milkwort family, 165–166
Mimosoideae, 154
mint family, 216–218
mint
horse, 217
wild, 217
Minuartia
elegans, 104
obtusiloba, 104
mist maiden, 210–211
mistletoe family, 90
mistletoe, 90
dwarf, 90
Mitella nuda, 135
mitrewort, 135
false, 137
Moehringia lateriflora, 104
Monarda fistulosa, 217
Moneses uniflora, 186
monocots, 23–72
Monolepis nuttalliana, 98
Monotropaceae, 188–189
Monotropa uniflora, 189
moonwort, 7
mooseberry, 240
mountain ash
sitka, 152
western, 152
mountain avens (*Dryas*)
northern white, 143
white, 144

yellow, 143
mountain heather, white, 192
Musineon divaricatum, 182
musineon, leafy, 182
mustard family, 121–130
mustard
gray tansy (*Descurainia*), 125
green tansy, 125
tower (*Arabis*), 123
wormseed (*Erysimum*), 127
Myosotis alpestris, 215
Myriophyllum exalbescens, 177

N

needle grass, Columbia, 46
nettle family, 87–88
nettle
common, 88
marsh hedge, 218
stinging, 88
nodding onion, 56

O

oat grass
California, 37
Hooker's, 40
Oenothera biennis, 176
old man's whiskers, 146
Oleaceae, 152
oleaster family, 172–173
Onagraceae, 173–176
onion, nodding, 56
Onosmodium molle, 215
Ophioglossaceae, 7
orchid family, 64–72
orchid
bracted bog, 70
northern green bog, 70
round-leaved, 71
sparrow's egg, 68
tall white bog, 69
Orchidaceae, 64–72
Orchis rotundifolia, 71
Orobanchaceae, 230–231
Orobanche
fasciculata, 231
uniflora, 231
Orthilia secunda, 187
Orthocarpus luteus, 223
Oryzopsis
asperifolia, 42
exigua, 43
Osmorhiza depauperata, 182
owl-clover, 223
Oxycoccus microcarpus, 194
Oxyria digyna, 93
oxytrope, inflated, 160
Oxytropis
podocarpa, 160
sericea, 161
splendens, 161

P

paintbrush
alpine yellow, 222
common red, 221
stiff yellow, 221
Papaver
kluanensis, 118
nudicaule, 119
**somniferum*, 118
Papaveraceae, 118–119
Papilionoideae, 154
Parnassia
fimbriata, 138
montanensis, 139
palustris, 139
Parnassiaceae, 138–139
parsley
mountain wild, 181
prairie, 181, 182
western wild, 181
parsnip
cow (*Heracleum*), 180
meadow (*Zizia*), 184
pasque-flower, 110–111
pea family, 154–163
pea vine, 157
pearlwort, mountain, 105
Pedicularis
bracteosa, 224
flammea, 225
groenlandica, 225
Pellaea
atropurpurea, 9

glabella, 9
pennycress, 130
Penstemon
confertus, 226
eriantherus, 227
fruiticosus, 227
nitidus, 227
procerus, 228
peppergrass, common, 128
pepperwort, branched, 128
Perideridia gairdneri, 183
Petalostemon, 154
Petasites
frigidus, 266
palmatus, 266
sagittatus, 267
vitifolius, 267
phacelia, mountain, 210
Phacelia
franklinii, 209
sericea, 210
Phalaris arundinacea, 43
Phleum
alpinum, 44
commutatum, 44
pratense, 35, 44
phlox family, 205–208
phlox
blue, 206
moss, 207
Phlox
alyssifolia, 206
hoodii, 207
Phyllodoce
empetriformis, 194
glanduliflora, 195
Physaria didymocarpa, 129
Picea
engelmannii, 17
glauca, 17, 187
mariana, 18
pigweed, 97
red-root, 99–100
Pinaceae, 14–21, 90
pine family, 14–21, 90
pine
Bank's, 90
limber, 20
lodgepole, 20, 90
scrub, 19
white-bark, 19
pineappleweed, 265
Pinguicula vulgaris, 232
pink family, 101–108
Pinus
albicaulis, 19
banksiana, 90
contorta, 20, 90, 186, 187
flexilis, 20
pipsissewa, 186
Plantaginaceae, 233–234
Plantago major, 232
plantain family, 233–234
plantain
common (*Plantago*), 234
lesser rattle-snake (*Goodyera*), 68
creeping rattle-snake (*Goodyera*), 68
Platanthera
dilatata, 69
hyperborea, 70
Poa
alpina, 45
pratensis, 45
Poaceae (*Gramineae*), 32–46
Polemoniaceae, 205–208
Polemonium
pulcherrimum, 208
viscosum, 208
Polygala senega, 166
Polygalaceae, 165–166
Polygonaceae, 91–96
Polygonum
amphibium, 95
arenastrum, 94
aviculare, 94
convolvulus, 94
viviparum, 95
Polypodiaceae, 7–11
Polystichum lonchitis, 10
pondweed family, 28–29
pondweed
alpine, 28
clasping-leaf, 29
thread-leaved, 28
poplar, balsam, 81, 82
poppy family, 118–119

poppy
alpine, 118
Iceland, 119
Kluane, 118
*opium, 118
Populus
balsamifera, 82
tremuloides, 82
Portulacaceae, 100–101
Potamogeton
alpinus, 28
filiformis, 28
richardsonii, 29
Potamogetonaceae, 28–29
Potentilla
fruiticosa, 147
gracilis, 147
hookeriana, 147
nivea, 147
prairie crocus, 110–111
prairie smoke, 146
primrose family, 197–200
primrose
dwarf Canadian, 200
fairy, 200
yellow evening, 176
Primula mistassinica, 200
Primulaceae, 197–200
prince's pine, 186
Prunella vulgaris, 218
Prunus
pensylvanica, 148
virginiana, 148–149
Pseudotsuga menziesii, 21
Pteridophyta, 3–11
puccoon, 214
incised, 213
narrow-leaved, 213
purslane family, 100–101
pussytoes
pink, 251
rosy, 251
showy, 251
woolly, 251
Pyrola
asarifolia, 187
chlorantha, 188
secunda, 187
Pyrolaceae, 185–188

Q

quack grass, 34

R

ragwort, 269
brook, 270
giant, 270
triangle-leaved, 270
Ranunculaceae, 108–117
Ranunculus
acris, 114
cardiophyllus, 115
eschscholtzii, 115
flammula, 116
pedatifidus, 116
reptans, 116
raspberry
dwarf, 150
running, 151
wild red, 151
reed grass, marsh, 37
red-root pigweed, 99–100
Rhinanthus
crista-galli, 228
minor, 228
Rhododendron albiflorum, 189, 195
rhododendron, white-flowered, 189, 195
Ribes
aureum, 140
lacustre, 140
oxyacanthoides, 140
rice grass, 43
white-grained mountain, 42
river beauty, 175
rock cress
Lyall's, 123
purple, 122
silver (*Smelowskia*), 129
rocket
prarie, 127
small-flowered, 127
rockstar, 134
Romanzoffia sitchensis, 210–211
Romanzoffia
cliff, 210–211
sitka, 210–211

Rorippa
- *islandica*, 129
- *palustris*, 129

Rosa
- *acicularis*, 149
- *woodsii*, 150

Rosaceae, 141–153
rose family, 141–153
rose
- common wild, 150
- prickly, 149
- Woods', 150

roseroot, 132
Rubiaceae, 234–236
Rubus
- *arcticus*, 150
- *idaeus*, 151
- *pubescens*, 151

Rumex
- *acetosa*, 96
- *maritimus*, 96
- *occidentalis*, 96

rush family, 53–55
rush
- alpine (*Juncus*), 54
- common scouring (*Equisetum*), 6
- creeping spike (*Eleocharis*), 50
- dwarf scouring (*Equisetum*), 6
- small-flowered wood (*Luzula*), 54
- spiked wood (*Luzula*), 55
- wire (*Juncus*), 54

rye, hairy wild, 38

S

sage, pasture, 255
sagewort
- biennial, 254
- pasture, 255

Sagina saginoides, 105
Sagittaria cuneata, 31
Salicaceae, 81–84
Salix
- *bebbiana*, 83
- *glauca*, 84
- *reticulata*, 84
- *vestita*, 84

Sambucus racemosa, 239
sandalwood family, 88–90
sandberry, 191
sandwort
- arctic (*Minuartia*), 104
- blunt-leaved (*Moehringia*), 104
- linear-leaved (*Arenaria*), 102
- purple alpine (*Minuartia*), 104

Sanicula marilandica, 183
Santalaceae, 88–90
sarsaparilla, wild, 178–179
Sarothamnus scoparius, 230
saskatoon, 142
Saussurea
- *densa*, 267
- *nuda*, 267

saw-wort, dwarf, 267
Saxifraga
- *bronchialis*, 136
- *caespitosa*, 136
- *lyallii*, 136–137
- *oppositifolia*, 137

Saxifragaceae, 132–137
saxifrage family, 132–137
saxifrage
- common, 136
- golden (*Chrysosplenium*), 133
- leather-leaved (*Leptarrhena*), 134
- Lyall's, 136–137
- purple, 137
- red-stemmed, 136–137
- spotted, 136
- tufted, 136

Schizachne purpurascens, 45
Scirpus
- *acutus*, 51
- *caespitosus*, 52

scorpion weed
- Franklin's, 209
- silky, 210

scouring rush
- common, 6
- dwarf, 6

Scrophulariaceae, 219–230
sedge family, 47–52
sedge
- awned, 48
- beaked, 49
- bog (*Kobresia*), 51

golden, 48
rush-like, 49
single-spiked, 49
thread-leaved, 49
water, 48
Sedum
lanceolatum, 131
rosea, 132
selaginella, prairie, 5
Selaginella densa, 5
Selaginellaceae, 4–5
self-heal, 218
seneca snakeroot, 166
Senecio
canus, 268
lugens, 269
pauperculus, 269
triangularis, 270
service-berry, 142
Shepherdia canadensis, 173
shepherd's purse, 124
shooting star
mountain, 199
saline, 199
sibbaldia, 151
Sibbaldia procumbens, 151
Silene
acaulis, 105
menziesii, 106
parryi, 106
silver-berry, 172–73
silver-plant, 92
single delight, 186
Sisyrinchium
montanum, 63
septentrionale, 64
skunkweed, 208
sky pilot, 208
slough grass, 36
smartweed, water, 95
Smelowskia calycina, 129
Smilacina
racemosa, 58
stellata, 59
trifolia, 59
Smilax, 179
snakeroot (*Sanicula*), 183
seneca (*Polygala*), 166
snowberry, 239
Solidago
decumbens, 272
gigantea, 271
missouriensis, 271
multiradiata, 272
spathulata, 272
Solomon's-seal
false, 58
star-flowered, 59
three-leaved, 59
Sonchus arvensis, 272–273
Sorbus
scopulina, 152
sitchensis, 152
sorrel
green (*Rumex*), 96
mountain (*Oxyria*), 93
sow thistle, perennial, 272–273
Sparganiaceae, 27
Sparganium
angustifolium, 27
minimum, 27
spearwort, creeping, 116
speedwell
alpine, 229
thyme-leaved, 230
Spermatophyta, 12–21
Sphaeralcea coccinea, 170
spiraea, 153
pink, 153
Spiraea
betulifolia, 153
densiflora, 153
lucida, 153
Spiranthes romanzoffiana, 72
spleenwort, green, 8
spring beauty
alpine, 101
western, 101
spruce
black, 18
bog, 18
Engelmann, 17
white, 17
spurge family, 166–167
spurge, 167
leafy, 167
squawroot, 183
Stachys palustris, 218

Stellaria
borealis, 107
longifolia, 107
longipes, 107–108
Stenanthium occidentale, 60
stickseed
Jessica's, 212
large-flowered, 212
stinkweed, 130
Stipa columbiana, 46
stitchwort, northern, 107
stonecrop family, 131–132
stonecrop
common, 131
lance-leaved, 131
strawberry
blite, 97
*tree, 191
wild, 144
Streptopus amplexifolius, 60
sugar scoop, 137
summer cypress, 98
sundew family, 130–131
sundew
long-leaved, 130–131
oblong-leaved, 130–131
sunflower, common tall, 264
sweet cicely, spreading, 182
sweet clover
white, 159
yellow, 160
sweet grass, 41
Symphoricarpos
albus, 239
occidentalis, 240

T

tamarack, western, 16
Taraxacum
ceratophorum, 273
laevigatum, 274
officinale, 274
teasel family, 242–243
Thalictrum
occidentale, 108, 117
venulosum, 117
Thermopsis rhombifolia, 161
thistle
Canada, 257
creeping, 257
Flodman's, 258
Hooker's, 258
perennial sow (*Sonchus*), 272–273
wavy-leaved, 258
white, 258
Thlaspi arvense, 130
Tiarella unifoliata, 137
tickle grass, 35
timothy grass, 35, 44
mountain, 44
toadflax (*Linaria*), 89, 223
bastard (*Comandra*), 89
northern bastard (*Geocaulon*), 89–90
Tofieldia
glutinosa, 61
pusilla, 61
Tolmachevia integrifolia, 132
townsendia
low, 275
Parry's, 275
Townsendia
exscapa, 275
parryi, 275
Tragopogon dubius, 276
Trifolium
hybridum, 162
pratense, 162
repens, 163
Triglochin
maritima, 30
palustris, 30
Trisetum spicatum, 46
trisetum, spike, 46
Trollius albiflorus, 117
twayblade, northern, 71
twinflower, 236, 237
twinberry, black, 238
twisted stalk, clasping-leaved, 60
Typha latifolia, 26
Typhaceae, 26

U

Umbelliferae (*Apiaceae*), 179–184

umbrella-plant
subalpine, 93
yellow, 92
Urtica
dioica, 88
gracilis, 88
lyallii, 88
Urticaceae, 87–88
Utricularia vulgaris, 233

V

Vaccinium
caespitosum, 196
membranaceum, 196
myrtillus, 196
scoparium, 197
Vahlodea atropurpurea, 46
valerian family, 241–242
valerian
mountain, 242
northern, 241
sitka, 242
Valerianaceae, 241–242
Valeriana
dioica, 241
septentrionalis, 241
sitchensis, 242
Venus'-slipper, 65, 71
Veronica
alpina, 229
americana, 229
serpyllifolia, 230
vetch
purple, 157
tufted, 163
wild, 163
vetchling, cream-coloured, 157
Viburnum edule, 240
Vicia
americana, 163
cracca, 163
sparsifolia, 163
Violaceae, 170–172
Viola
adunca, 171
canadensis, 171
nephrophylla, 172
violet family, 170–172
violet
bog, 172
dog-tooth (*Erythronium*), 57
early blue, 171
western Canada, 171
Virginia grapefern, 7
Viscaceae (Loranthaceae), 90
**Viscum album*, 90

W

wapato, 31
water-milfoil family, 177
water-milfoil, spiked, 177
water-plantain family, 30–31
water smartweed, 95
water starwort family, 167–168
water starwort, vernal, 168
waterleaf family, 209–211
wheat grass
bluebunch, 34
northern, 34
whiteman's-foot, 234
whitlow grass, 126
few-seeded, 127
golden, 126
Kananaskis, 126
wild rye, hairy, 38
willow family, 81–84
willow
beaked, 83
Bebb's, 83
dwarf, 84
rock, 84
smooth, 84
wolf (*Elaeagnus*), 172–173
willowherb, 175
great, 174
marsh, 175, 176
wind-flower, 110–111
wintergreen family, 185–188
wintergreen
common pink, 187
greenish-flowered, 188
one-flowered, 186
one-sided, 187
wolf willow, 172–173

wolfberry, 240
wood rush
  small-flowered, 54
  spiked, 55
woodsia
  mountain, 11
  Oregon, 10
Woodsia
  oregana, 10
  scopulina, 11
wormwood, 254
  cut-leaf, 254
  plains, 254

Y

yampa, 183
yarrow, common, 248
yellow rattle, 228
youngia, 259

Z

Zigadenus
  elegans, 62
  venenosus, 62
Zizia aptera, 184

About the Authors

BERYL HALLWORTH is a botanist, teacher, curator and naturalist associated with the Herbarium of the Department of Biological Sciences at the University of Calgary since 1967. She is the author and co-author of several papers and book chapters on plants and natural history.

C.C. CHINNAPPA is Professor of Botany and Curator of the Herbarium at the University of Calgary. His main research interests are the evolutionary strategies of polyploid plant species. He has published more than one hundred papers on biosystematics, cytogenetics, palynology and phenotypic plasticity of plants.

KANANASKIS FIELD STATIONS is an institute of the University of Calgary and consists of two facilities: the Barrier Lake Field Station (built in 1967), located in the Kananaskis Valley, and the R.B. Miller Station (founded in 1950), located in the Sheep River Wildlife Sanctuary. The mandate of the Stations is to provide research and educational facilities for the University of Calgary, other universities, government agencies, private research institutes, schools and other educational groups. These programs are multidisciplinary, with an emphasis on global change, biological diversity and sustainable ecological systems. The Stations also assists the development, implementation and evaluation of research and education programs with user groups. Kananaskis Field Stations was a principal supporter of the publication of this book.